Sound Streams

Sound Streams

A Cultural History of
Radio-Internet Convergence

A N D R E W   J .   B O T T O M L E Y

UNIVERSITY OF MICHIGAN PRESS

ANN ARBOR

For questions or permissions, please contact um.press.perms@umich.edu

Published in the United States of America by
the University of Michigan Press
Manufactured in the United States of America
Printed on acid-free paper

First published June 2020

A CIP catalog record for this book is available from the British Library.

Library of Congress Cataloging-in-Publication Data

Names: Bottomley, Andrew, author.
Title: Sound streams : a cultural history of radio-internet convergence /
    Andrew J. Bottomley.
Description: Ann Arbor : University of Michigan Press, 2020. | Includes
    bibliographical references and index.
Identifiers: LCCN 2020004125 (print) | LCCN 2020004126 (ebook) |
    ISBN 9780472074495 (hardcover : alk. paper) | ISBN 9780472054497
    (paper : alk. paper) | ISBN 9780472126774 (ebook)
Subjects: LCSH: Internet radio broadcasting—History. | Radio programs—
    History.
Classification: LCC TK5105.887 .B68 2020 (print) | LCC TK5105.887
    (ebook) | DDC 384.54/402854678—dc23
LC record available at https://lccn.loc.gov/2020004125
LC ebook record available at https://lccn.loc.gov/2020004126

Portions of Chapter 2 and Chapter 6 were previously published in the following
journal articles:

Bottomley, Andrew J. "Giant Pools of Content: Theorizing Aggregation in
Online Media Distribution." from *JCMS: Journal of Cinema and Media Studies*,
Volume 59, Number 1. Fall 2019, pp. 149-156. Copyright © 2019 by the
University of Texas Press. All rights reserved.

Bottomley, Andrew J. (2015) "Podcasting, *Welcome to Night Vale*, and the Revival
of Radio Drama," *Journal of Radio and Audio Media*, 22:2, 179-189, copyright
© The National Association of Broadcasters, www.nab.org, reprinted by
permission of Taylor & Francis Ltd, http://www.tandfonline.com on behalf of
The National Association of Broadcasters.

For Diana

Contents

Digital materials related to this title can be found on the Fulcrum platform via the following citable URL: https://doi.org/10.3998/mpub.9978838

Acknowledgments

Scholarly endeavors, even single-authored monographs like this one, are always the work of many hands. *Sound Streams* was crafted over the course of more than a half-decade, and it is overwhelming to consider the long list of individuals and institutions who have provided me with assistance and support along the way.

This project began life as a doctoral dissertation at the University of Wisconsin–Madison, benefiting immeasurably from the input of my committee members, Jonathan Gray, Jeremy Wade Morris, Ronald Radano, and Greg Downey. Michele Hilmes, my dissertation advisor, deserves every word of praise I can summon. I would not have undertaken the study of radio history if not for her, and she has been a generous mentor and staunch supporter of me and this project at every step. At UW–Madison, financial support was received from the University of Wisconsin Graduate School and Department of Communication Arts, and through a Mellon-Wisconsin Summer Fellowship. I must also thank the many wonderful classmates in the Media and Cultural Studies graduate program who provided me with insight, advice, and friendship, most of all Sarah Murray, Alyx Vesey, Kyra Hunting, Myles McNutt, Danny Kimball, Lindsay Hogan, Nick Marx, Josh Jackson, Eleanor Patterson, Caroline Leader, Evan Elkins, Kit Hughes, and Josh Shepperd. Christopher Cwynar's camaraderie and intellectual support were, and continue to be, indispensable.

A good many friends and colleagues have aided the progress of *Sound Streams* from idea to writing and revision to publication. Among the numerous scholars who deserve thanks for their guidance and support: Alex Russo, Neil Verma, Cynthia Meyers, Jason Loviglio, Derek Kompare, Tim

Anderson, Shawn VanCour, Chenjerai Kumanyika, Andrew Leland, Jennifer Hyland Wang, Eric Harvey, Mack Hagood, Brian Fauteux, Kathy Fuller-Seeley, Jonathan Sterne, Jennifer Stoever, Jacob Smith, Richard Berry, Siobhan McHugh, Lauren Bratslavsky, Bill Kirkpatrick, Elana Levine, Michael Z. Newman, William Boddy, Megan Sapnar Ankerson, Mark Williams, Kyle Wrather, Ben Aslinger, Michael Curtin, Jennifer Waits, Noah Arceneaux, Zachary Davis, John Sullivan, Anne MacLennan, Phylis Johnson, Branden Buehler, Isabel Pinedo, and Bill Herman. Amanda Keeler and Kyle Barnett have been particularly enthusiastic, and their knowledge and encouragement are cherished.

Throughout the development of this project, it has been a pleasure to work with the wonderful community of radio scholars in North America, both through the National Recording Preservation Board's Radio Preservation Task Force and the Society for Cinema and Media Studies' Radio Studies Scholarly Interest Group. It is a privilege to now work at SUNY Oneonta, and my thanks go to my current colleagues in the Department of Communication and Media and, especially, to my students, past and present.

Historical research is collaborative work, and I am grateful to Andy Lanset at New York Public Radio and Rebecca Schulte and Kathy Lafferty at the University of Kansas's Spencer Research Library for their help finding archival materials. Jeff Robins and Megan O'Hern at WXYC, Ali Foreman at WREK, and John Dillingham at KJHK also helped locate important documents. I conducted numerous interviews for this project and owe a debt of gratitude to the people who shared their time and memories with me, including Ellen Horne, Paul Jones, Mike Rinzel, Megan Ryan, John Selbie, Michael Shoffner, Laura Walker, and Soren Wheeler. Robert Burcham, Gary Hawke, and Anthony Nguyen also shared information and clarified details. I am especially grateful to Carl Malamud for answering my questions as well as sharing a cache of personal documents related to the Internet Multicasting Service. Tom Hjelm also deserves special thanks for arranging my interviews and participant observation research at New York Public Radio.

At University of Michigan Press, my editor, Mary Francis, has shown steadfast support for this project going all the way back to its dissertation days. My friend Daniel Murphy created *Sound Streams*' cover art. Additional thanks go to Susan Cronin, and to my anonymous reviewers who offered close readings and thoughtful suggestions.

Final thanks go to my family, whose unfailing confidence and support are impossible to quantify or repay: my parents, Judy and Bernie Bottomley; my brother Peter Bottomley; my mother-in-law, Agnes Willis; my grandparents-in-law, Emily and Charles Preston; my sister-in-law Teresa Willis and brother-in-law Hamish Wright; my nieces Maia, Sophia, and Lily; and my extended family of aunts, uncles, and cousins. I never would have gotten any of this done without Diana Willis Bottomley, a constant source of solace and inspiration. Thank you all.

Introduction

R.O.I.: Radio on the Internet

In the 2015 second season of HBO's television comedy series *Silicon Valley*, the founders of the fictional start-up Pied Piper—after being denied funding by all the computer technology industry's major venture capital firms—turn in desperation to a vulgar dot-com-era billionaire, Russ Hanneman (played by Chris Diamantopoulos), whose claim to fame is that he "put radio on the internet" twenty years before. There is no mistaking that the fictional Hanneman is a nod to the very real Mark Cuban, who similarly became an "internet billionaire," as his introduction in the reality television series *Shark Tank* announces, when he sold his web audio-video portal Broadcast.com to Yahoo! for $5.7 billion in 1999. *Silicon Valley* is set in the present day, and in it Hanneman's radio on the internet—or "R.O.I.," as he calls it—is held up as a joke: a feat of dumb luck that signifies the irrationality of the 1990s dot-com bubble, a time when you could simply say "internet" and put an "e-" or a "net-" prefix in your company name and make hundreds of millions of dollars in a stock IPO. Moreover, Hanneman's so-called invention is framed as frivolous compared to Pied Piper's lossless data compression algorithm, which viewers are meant to understand as a truly useful and groundbreaking technological innovation, a holy grail in the modern age of ubiquitous computing, cloud storage, and high-definition streaming media. *Silicon Valley* gives the impression that anyone could have put radio online, yet it is a technology that no one wanted or needed. "We can take this thing called *the radio* and put it on this new thing called *the internet*," the buffoonish Hanneman boasts as he recalls his eureka moment, "and no one was doing it, no one."[1]

The modern internet effectively flattens all media: it connects in one place everything from film and television to music, games, books, newspapers, personal documents and photos, and phone calls. Although taken for granted today, this multimedia bazaar is still a relatively recent development: up until the early 1990s the internet delivered only rudimentary text and graphics. Not until the mid-1990s did technology develop that could transmit audio across computer networks. And as online sound arrived, it did so in the form of a much older medium: radio. Radio on the internet—Hanneman's R.O.I.—played a fundamental yet little-acknowledged role in popularizing the internet and making it what it is today. Likewise, the internet transformed the century-old medium of radio, opening it up to new voices and new practices like podcasting and music streaming.

While *Silicon Valley* gets laughs by making it out to be quaint and simplistic, internet radio really was a significant breakthrough: in terms of its technology, the new streaming media industry that built up around it, and what it represented culturally. With the rapid growth of the World Wide Web into a mainstream mass medium in the late 1990s, streaming internet radio was one of the primary forms of "rich media" content (to use the industry jargon of the day) that helped make the web attractive to both ordinary users and investors. Internet radio was the first major instance of digital *convergence* in the contemporary media era, preceding web television and video, online social networking, and even digital music. It is an exemplary case of media convergence and one that is instructive for media studies analyses of the convergences that are now ubiquitous throughout all media in these first decades of the twenty-first century.

Situating *Sound Streams*

Sound Streams traces the history of radio-internet convergence from 1993 up to the present day. A key premise of this book is to historicize the so-called new media of the internet, streaming audio, and podcasting, showing how their emergence has been more a case of evolution than revolution, and how new media always refashion old media. My interest in writing *Sound Streams* stems from a lack of awareness about the most basic facts of radio's relationship to the internet within the media industries, the popular press, and academia too. Podcasting, in particular, routinely gets referred to as a new "medium" and championed as a disruptive technology that is especially "intimate" and "authentic" and so on. These discourses typically divorce the contemporary media from the past, treating them as though they are purely a product of new technologies and recently developed modern sentiments. With this book, I aim to shift the conversation away from a language of newness and toward a more carefully contextualized analysis of the process of cultural change. While radio on the internet possesses unique characteristics, to be sure, ultimately there is much about online radio and podcasting today that is remarkably old, when the full range of radio's history and forms is taken into account. Over its century-long history, radio has repeatedly encountered the same issues and debates that have surrounded online audio media of late. Mapping these continuities between the new and old is culturally significant because there are important lessons contemporary scholars, producers, and audiences can learn from the past. In this way, my book draws on and contributes to theories of remediation, as well as work in the emerging field of new media history. There is also an important element of documentary history to this book: in most cases, the history I have uncovered is completely unknown to media scholars. *Sound Streams* exposes that unknown history while correcting many of the myths and misinformed dominant narratives that widely circulate about online audio.

There are two central arguments threaded throughout this book. First is that the various iterations of internet radio, from streaming audio to podcasting, are all *radio*. That is, these are new radio *practices* rather than each being a separate new medium. This conceptual emphasis on practice—loosely defined as fields of action populated by routine activities and discourses—enables an appreciation of internet radio's new empirical and symbolic dimensions without sacrificing a focus on distinguishing features of radio that have remained historically consistent.[2] Key to this argument are theories of convergence and remediation. Drawing on the work of Raymond Williams, Paddy Scannell, and David Hendy, among others, I identify the concept of *sociability* as a primary marker of radio-internet convergence. Rather than a particular set of technologies or institutional structures or textual conventions, radio at its core represents a sonically mediated set of cultural relations that produce a social space that structures our everyday lives—metaphorically a space that psychologically fulfills our basic human need for companionship and community and that scores the rhythms of our daily activities, while also a physically mediated space that can function as a public forum for social interaction. Ultimately, I define radio as any nonmusic sound medium that is purposefully crafted *to be heard* by an audience.

Over the past twenty-five years, media convergence has pulled radio apart—altering what it sounds like, who produces it and how it is distributed, where and how it is listened to, how the audience interacts with radio institutions, and so on—while it has also put it back together. What is more, rather than push radio into obsolescence, many of radio's main social and cultural functions (i.e., why we listen, when and where we listen), as well as its programming and aesthetic conventions, have proliferated online. Radio or radio-like services can today be found through practically any media device or platform: radio is on our computers and smartphones; it flows through our smart home devices; there are podcasts and radio channels through all the major digital media distribution platforms like iTunes and Spotify; and so on. And the logics of radio permeate our networked digital media culture, from the centrality of participation and audience cocreation—hallmarks of radio dating back to its origins in amateur wireless communities and prominent in mass media broadcasting formats like call-in talk radio—to the notions of listening and voice. In particular, a plurality of voices has become a crucial form of engagement in contemporary internet culture, social media in particular. Other radio functions, such as curation and the need for cultural intermediaries like disc jockeys (DJs), have found renewed cultural value despite, or even as an unintended result of, trends toward media personalization. Media convergence has transmogrified radio—sometimes distorting or fragmenting it, sometimes expanding on or strengthening it—but something distinctly *radio* not only persists but thrives.

The second central argument in this book is that radio's unique *affective* dimensions and special relationship to everyday lived experience make the medium especially suitable for our contemporary networked digital media culture. This argument extends the prior notions of radio's sociability and the propagation of radio logics throughout internet culture. Radio programming is sociable and attentive to listeners' everyday lives, and moreover, there is a distinctive conversationality and reflexivity to radio talk. In particular, there is a propensity in radio for *sharing* and emotional disclosure. Radio talk is a form of what Erving Goffman calls "fresh talk," by which he means formal (institutionalized, public-directed) speech that is spontaneous and sensitive to its immediate context and the identity of the particular audience.[3] Moreover, broadcast radio has a history of inviting listener interaction via call-in shows, music requests, and the like, and the *aurality* of radio carries a whole subtext of affect and meaning (tone, stress, volume, speed, pitch, accent, dialect, and other emotional elements of speech and the ways in which the words are spoken). Historically, this form of loose, inclusive, and often deeply personal and visceral discourse is atypical of other mass media like television or print journalism. Yet these characteristics have come to dominate the contemporary mass media landscape through modes of "storytelling" that emphasize affect, sentiment, and subjective expression rooted in everyday lived experience.[4] For instance, memoir-esque introspection and personal narrative have become driving forces in modern journalism (what some call the "Ira Glass effect").[5] John Hartley describes bottom-up, socially networked "self-representation" as the driving force in modern journalism, while Rosalind Coward has observed that we are now a "confessional society" in which the "personal voice" has become increasingly prevalent in mass media.[6] These are radio logics that are now everywhere present in modern media culture.

The Web 2.0 convergence of media production and consumption also entails a social convergence of private and public life that is accompanied by a redefinition of the political. Christian Fuchs writes that "the emergence of 'social media' is embedded into the trend that boundaries between the dualities of modernity have become somewhat liquid and blurred: we find situations where the distinctions between play and labour, leisure time and work time, consumption and production, private and public life, the home and the office have become more porous."[7] This blurring of private and public life, and the ensuing emphasis on personal stories and expressions of affect, resembles radio over its long history. Media convergence has enabled radio's logics of connectivity and sharing—sharing of information and opinions but also emotion—to become more widespread and intensified, as the creation and distribution of personal expressions for public consideration is now everywhere commonplace in social media, news, and popular entertainment.

The overarching goal of *Sound Streams* is to construct a history of radio-internet convergence that strikes a balance between continuity and change. The chief research questions driving this book are these: How has radio's convergence with the internet transformed radio technologically, industrially, aesthetically, and most importantly, culturally and discursively? And in return, how has radio shaped the internet and its development? Contra to bullish and ahistorical "digital revolution" or "disruptive innovation" narratives that divorce the internet and digital media from the past, I carefully position internet radio within the nearly century-long history of radio broadcasting. Simultaneously, I separate out the specific characteristics, affordances, and constraints of internet radio that are unprecedented, and the unique challenges wrought by radio's new digital dimensions.

This is an enormous story, and the scope of the book is purposefully broad. *Sound Streams* is not a chronicle of everything that has happened in internet radio and podcasting over the past quarter-century. Rather, I analyze a cross-section of that history through a set of representative case studies. These case studies are designed to guide and structure the analysis, illustrating overall trends in radio's technology, industry structure, production culture, programming and aesthetics, and audiences and social function. This also is not a global history; the focus is specifically on internet radio's emergence and diffusion within the US context. The reasons for that choice are multiple, including that the bulk of existing scholarship on internet radio and podcasting has emerged primarily from Europe, especially the United Kingdom. Moreover, while the internet is a global medium, internet radio originated in the United States, and for most of its history it was predominantly US citizens and communities driving the major innovations in technology, industry, and programming. Thus, while there are many interesting and important stories to tell about internet radio in global contexts, in this instance the history itself calls for a US-centric focus.

How to Read *Sound Streams*

At its core, *Sound Streams* is a work of academic scholarship. It is my hope, however, that this book will appeal to a wide range of readers, including undergraduate media studies students, radio and podcast professionals, and dedicated radio and podcast listeners. In order to make the text as accessible as possible for these multiple audiences, the presentation and structure of *Sound Streams* deviate slightly from that of a standard academic book. For starters, it is divided into a larger number of shorter chapters. These chapters are topically focused, as well as roughly ordered chronologically. Although the book is best read in progression and in its entirety, you may choose to read the chapters selectively, depending on your interests. For example, if you are primarily concerned with podcasting, then you will find chapters 3, 6, and 7 most relevant to your tastes. Music radio and online music-streaming services are confined mainly to chapter 5. To guide these choices, a summary of the chapter subjects is provided in the "Chapter Overview" section at the end of this introduction.

Generally, I have sought to minimize lengthy discussions of theory and method, and to isolate these subjects when they are needed. Undoubtedly, a rigorous methodology is essential for any academic endeavor, especially historical research and research that applies a mixed methods approach—as is the case with *Sound Streams*. I realize, though, that many nonacademic readers do not share mine and my colleagues' curiosity for this metadiscourse, and thus the particulars of my interdisciplinary methodological approach appear in an appendix at the end of the book. The main theoretical concepts and arguments that frame my analysis of radio and the internet are set out primarily in the remaining sections of this introduction. Importantly, it is here that I will define key terms, such as *radio* and other keywords that will be invoked regularly: convergence, remediation, practices, sociability, liveness, flow, temporality, streaming, intentionality, witnessing, presencing, conversationality, sharedness, intimacy. Nonacademic readers may choose to skip the remainder of this chapter—and if you encounter any of these concepts later in the book and they cause confusion or appear to lack context, know that you may return here to the introduction for elaboration.

Locating Media Convergence

Sound Streams is a study of so-called new media and digital media, most definitely, but it is foremost a study of the relationships between old and new media. It aims to highlight the many continuities between existing and emerging media. The concept of *convergence* is almost anachronistic in the fields of media and communication studies today. When Henry Jenkins first wrote of "convergence culture" some fifteen years ago, it became a go-to theory to describe the process by which the boundaries were blurring between "old" and "new" media forms, "grassroots" and "corporate" organizations, and "producers" and "consumers" of media.[8] Overused as a buzzword, convergence turned so pervasive that it is nowadays ambiguous and taken for granted as an analytical concept. It is a deceptively simple concept: the idea that in our networked digital media culture there are now relations among and between the multitudinous forms and sites of media. The technology, content, production, and reception of any one medium is no longer separate; it now spans across various media.[9] Thus, the convergence media environment is marked by characteristics of hybridity, flexibility, complexity, and "multi-" this and "cross-" that. But there is a certain level of abstraction to currently in-vogue terms and concepts like "multiplatform," and the full extent of what exactly is converging routinely gets glossed over.

Turning back to the *Silicon Valley* opening example, the Russ Hanneman character's description of "putting" one medium onto another—radio on the internet—also conjures up images of *recombination*. It privileges the media object (i.e., the technological devices of "the radio" and "the internet") while implying a debased form of innovation that, following Todd Gitlin's canonical critique of US television's recombinant culture, invents new things by repurposing previously successful formulas in an entirely predictable, lowest-common-denominator fashion.[10] However, there is much more to media convergence than a slapdash melding together of preexisting media technologies. James Hay and Nick Couldry have identified four key iterations of media convergence prevalent in media studies: an economic synergy among media companies and industries; a multiplication of "platforms" for media content; a "technological hybridity" merging the uses of separate media into one another; and a "new media aesthetic" mixing previously distinct styles and forms, including the intermingling of fiction and nonfiction forms.[11] Jenkins summarizes convergence similarly, positioning it as an ongoing process that takes place at "various intersections of media technologies, industries, content, and audiences."[12] He divides convergence into five distinct but interrelated processes: technological convergence; economic convergence; social convergence; cultural convergence; and global convergence. Many studies of convergence tend to focus on only one of these processes, however, while merely paying lip service to the others. For instance, focusing only on the horizontal integration of the media industries (economic convergence) or analyzing transmedia content that combines multiple communication channels (cultural convergence). The result is that convergence is routinely broken into fragments; the process is rarely studied holistically, as it demands to be.

While the concept of convergence has fallen somewhat out of favor in media studies, it remains useful as a guiding principle for analyzing supposedly "disruptive innovations" that prompt large-scale shifts in media and culture.[13] A principal objective of this book is to revitalize the concept of media convergence, examining the convergence of radio and the internet across a range of processes—technological, economic, social, cultural—rather than attending to them in isolation. It is also a story of historical convergence, of the histories of the medium of radio and the medium of the internet intertwining. While it is tempting to describe internet radio as a type of *new media* (seeing as most anything that is digital or has ties to the internet gets labeled "new media"), more accurately it is an amalgam of two older media: radio and the internet. Indeed, the internet was not exactly "new" when it first converged with radio in 1993; it was already more than twenty years old.

In our society's collective rush to celebrate digital media and discursively frame technological change as cultural progress, the similarities between various media often get ignored, along with many of the important lessons older media might teach us. The simple fact is that old media rarely disappear upon the introduction of new media. Rather, the new(er) media incorporate the older media, while the older media are refashioned to compete and coexist with the new media. This is the basic premise of *remediation*. Jay David Bolter and Richard Grusin challenge linear narratives of media history, in which new media technologies are supposedly invented in isolation and replace existing media in a tidy, progressive line of succession. In actuality, new media always negotiate with other existing media in their formation. Bolter and Grusin's theory of remediation explains how all emerging media forms, from Renaissance painting to the internet, present themselves "as refashioned and improved versions of other media."[14] That is, "What is new about new media comes from the particular ways in which they refashion older media and the ways in which older media refashion themselves to answer the challenges of new media."[15] Not only are new media forms and practices always constructed, they are constructed in direct relation to old media. Moreover, this remediation is a two-way process. "Convergence is remediation under another name, and the remediation is mutual," they write, explaining that "the Internet refashions television even as television refashions the Internet."[16] I invoke this concept of remediation throughout *Sound Streams* to reveal the historical linkages between radio's past and present.

Much new media theory and scholarship, in addition to much popular discourse about the internet, has championed the arrival of Web 2.0 and networked digital media for decentralizing media production and consumption, mobilizing media engagement, and putting increased choice and control in the hands of audiences.[17] These shifts allow individuals to define and personalize their own media experience as well as directly contribute to the mass media they consume, in effect turning all audience members into media producers. Mark Deuze suggests that "the convergence of the cultures of production and consumption of media" is the single process of convergence that has had the largest impact on both the media industries and ordinary people's everyday lives.[18]

In the past quarter-century, digital media have enabled the development of a wealth of new technologies for radio transmission and reception, new forms of programming, and new economic models for production—from podcasting and mobile phone apps to content aggregators and personalized music-streaming services. Yet, despite the prevalence of progress narratives that separate internet "new media" from broadcast "old media," in actuality these various media coexist in a fairly symbiotic relationship. The internet has by no means annihilated radio. It is my contention that it has extended radio's reach and given the medium new vitality. Still, the convergence culture has undoubtedly transformed radio in a host of ways that I will explore throughout this book. Suffice it to say, radio persists as a central site of American culture and daily life, even if more and more of its listeners are "tuning in" via their web browsers or smartphone apps rather than their car stereos and tabletop radio sets. Moreover, it is notable how many of internet radio's new formats and services discursively construct themselves as broadcast radio through language and imagery: TuneIn, Radiotopia, Rdio, and so on. It seems that the idea of radio is as important today as ever before, even if its technology, institutions, programming, and cultural meaning are shifting.

What Is Radio, Anyway?

Radio has been an integral part of American life for 120 years. Beginning with Guglielmo Marconi's wireless telegraph in 1899, radio as a person-to-person or point-to-point communication system flourished in the 1900s and 1910s, pushed forward by an assortment of business concerns and the military, as well as ordinary citizens (typically referred to as amateurs or hobbyists).[19] In the 1920s, entrepreneurial individuals developed the practice of using radio as a means of bringing culture and entertainment to a mass audience—a concept that was soon exploited by a few emergent commercial interests, launching the network era of radio broadcasting.[20] The amateur radio operators continued to build and operate two-way radios, however, growing a recreational "ham radio" culture that was especially vibrant in the mid-twentieth century.[21] Following the introduction of television in the 1940s and the end of what is referred to as the Golden Age of Radio, commercial radio turned into a platform for recorded music, rock and roll in particular, and format-driven radio stations prevailed.[22] Car radios and portable transistor radios became widespread during the 1950s, turning radio into a mobile medium that could penetrate all areas of public and private space.[23] Citizens band radio (or CB radio), another system for two-way interpersonal communication, was introduced in the 1940s, and it was especially popular among long-distance truck drivers in the 1970s.[24] The expansion of FM radio facilitated the countercultural underground radio of the 1960s and 1970s, including the experimental freeform radio format, as well as a boom in locally produced programming and noncommercial public and college radio.[25] Programming syndication and automation systems also led to the national uniformity of format radio beginning in the 1960s.[26] Satellite radio networks appeared in the 1990s, introducing another infrastructural system for national and even transnational broadcasting. Despite all of these technological, industrial, and cultural shifts—and there are certainly other notable developments not mentioned here—we as a society have collectively understood each of these different devices and activities as one thing: *radio*.

Yet a good many people today conflate *radio* with *broadcasting*, or the "one to many" linear transmission and reception of a continuous channel of audio content via radio waves. This is what has come to be known as "terrestrial radio" in recent decades—a true neologism of the internet age. Nevertheless, broadcasting is merely one of multiple ways to use radio. I refer to these different ways of using or engaging with radio as *practices*.[27] Moreover, broadcasting in this sense rarely refers to technology or distribution alone. It is an arrangement or disposition: a shorthand for a particular configuration of radio production, consumption, regulation, identity, and representation. In the United States today, broadcasting refers to the one-way transmission of formally structured programs created by an industry of recognized institutions and received by a mass audience, all officially sanctioned and governed in the United States by the regulatory body of the Federal Communications Commission (FCC). That practice is itself a historical and social construct. Hence, broadcasts that deviate from these norms get labeled things like "pirate radio." Furthermore, these different practices of radio overlap and coexist. For instance, broadcasting did not replace ham radio in a tidy line of progression; ham radio continues to prosper to this day, even though broadcasting is culturally dominant. At any given time, multiple different radio practices may exist, all operating within the larger field of "radio." Yet what connects these varied radio practices? What is the central characteristic, or set of characteristics, that defines the medium?

Over the years, media and communication scholars, radio practitioners, regulators, and cultural critics have all grappled with this question, "What is radio?" The answers to this question largely come down to what each analyst considers to be the measure of the medium.[28] Radio is most typically distinguished by either *broadcasting* (i.e., linear production and reception) or *on demand* (e.g., podcasting's downloadable audio files or the click-to-play streaming of automated music services like Spotify). The mode of transmission, then, is the key determinant in such analyses, radio being designated as broadcasting alone. This is why a radio scholar like Richard Berry does not see podcasting as radio, even though he freely admits that most audiences perceive podcasting and radio as identical at the textual level.[29] The bulk of radio studies scholarship is also focused exclusively on mass media broadcasting, in particular the activities of the national networks (e.g., NBC, CBS, NPR; BBC in the UK)—a situation that only strengthens the association of radio and broadcasting.

Radio can also be distinguished by the technology or apparatus itself (e.g., wireless technology). Or it can be identified by the industrial forma- tion: radio is that which is produced by radio institutions—which is an- other way podcasts have been separated from radio, since until recently most podcasts emerged from sources beyond the major broadcast net- works. Others sidestep distribution and technology altogether, assessing radio purely by its textual properties and the elements of genre, form, and aesthetics.[30] There are many different measures of the medium.

The very question of "What is radio?" seems to suggest an object— something material and concrete with particular (usually technological) functions. In line with radio scholars like Paddy Scannell and Chris Priest- man, I believe the better question to ask is "What is radio *for*?"[31] This restatement of the question shifts the focus to the social, cultural, and po- litical functions of radio, making the medium a thing that humans act upon instead of a fixed object.[32] Scannell implores us also to ask *who* radio is for—seeing as media like radio are always shaped by assumptions about the audience for whom they are made.[33] First and foremost, I prefer to see ra- dio as a set of cultural relations that is identifiable by a composite of textual codes and representational strategies, cultural and industrial structures, and production and listening practices. As such, radio as a medium is broad and flexible, and able to accommodate new practices like on-demand streaming and podcasting. Here I am drawing principally on cultural studies with its focus on the culture of everyday life, as well as phenomenological theories of media that emphasize the ways we experience media like radio and the meaning these media have in our daily lives.[34]

With this inclusive definition of radio, I want to unsettle the notion that broadcasting, podcasting, and other practices are separate and distinct me- dia. Particularly in discourses surrounding so-called new media or digital media, there is a tendency to enthusiastically refer to the latest invention, platform, or format as a new "medium." Much of this discourse is driven by the tech industry itself, which has a vested interest in differentiating its products from those of its competitors, old or "legacy" media included. There is an increasing push among tech companies, the nascent podcasting industry, and even some entrenched radio producers to simply call all radio and podcast content "audio." For some, this audio rhetoric reads as a sincere effort to come up with an all-encompassing term that puts on equal footing a potpourri of content: traditional radio programming, podcasts, digital field recordings of public events, user-generated sound recordings, and so on. For others, it reads as a calculated marketing attempt to distance contemporary radio and podcasting from the image of broadcast radio as old-fashioned, unhip, even obsolete. This audio rhetoric takes advantage of the popular notion of new media as progressive and disruptive: "radio" is the past, antiquated and exhausted; "audio" is the future, dynamic and full of possibilities. Whatever the case, *audio* is an insufficient term. It is far too ambiguous; it can literally mean any recorded or reproduced sound—it makes no distinction between music, speech, environmental sound. What we call these different media matters. It is more than mere semantics: embedded in these terms are histories, politics, and cultural relations that risk being made invisible.

It is flawed logic to base any notion of a medium around technology alone. A *medium* must first and foremost be understood as a social configuration. As Jonathan Sterne reminds us, while the technology is certainly important, a medium consists of a whole set of social and cultural interconnections and "recurring relations among people, practices, institutions, and machines (rather than simply machines in and of themselves)."[35] He astutely points out that if we wish to talk about a medium like radio connecting a nation or the world, then it requires social formations and structures—it is the social interconnection and articulation that transform a mere technology into a medium.[36] Similarly, Brian Winston and other theorists of the social shaping of technology challenge technologically deterministic conceptions of "revolution," arguing instead that technology evolves slowly and is socially shaped, rather than emerging rapidly, naturally, or as the result of a predetermined logic or single determinant.[37] Likewise, in the media ecology intellectual tradition, a medium is viewed as an environment, in which there is a complex interplay between people, technology, and the physical world.[38] In each of these theories and approaches, a medium is a dynamic process that can—and almost always does—consist of multiple machines or devices, formats, platforms, and so on.

Any definition of radio, therefore, must emphasize the recurring practices and relations of the medium along with the network of technologies and institutions in which they take place.[39] Radio as a medium cannot be identified by technology or industry or text alone; rather, it is an assemblage of technologies and institutional structures, government policies and regulations, production practices and aesthetics (formal and stylistic elements), audience formations and reception practices, and the broader sociohistorical context of the moment in which the medium is being studied. As a result, I choose to work inclusively, adopting a very broad notion of radio that encapsulates a range of practices including broadcasting, internet simulcasting, podcasting, music streaming, audiobooks, and even CB and other two-way radio systems. Each of these practices, while distinct in ways that are significant, shares with the others core characteristics and functions that make them more similar than dissimilar. I applaud efforts to identify production techniques, textual signs and codes, and so on that might distinguish podcasting from broadcasting; however, my concern is that from a historical and analytical perspective, separating podcasting from radio ultimately loses much more than it gains.

This expansion of radio to include new practices like podcasting is closely aligned with Michele Hilmes's notion of *soundwork*. This is the term Hilmes uses "to designate creative/constructed aural texts that employ the basic sonic elements of speech, music, and noise."[40] She distinguishes soundwork from music, noting that speech is usually the dominant element of soundwork, with music and noise secondary. Hilmes readily acknowledges that the boundaries between radio/soundwork and the "music" field can be blurry. I further assert that music consists of compositions (songs or albums) that are intended to be heard in isolation or as part of a choreographed performance (e.g., concert, theater, soundtrack). Radio makes heavy use of prerecorded music and live musical performances initially created for other purposes. When presented on radio, though, that music is recontextualized and invested with new meaning, typically through the addition of speech: assembled by a DJ into a playlist, used as background, framed as part of a special "live" broadcast.[41] Radio is very much a music medium—radio programming is overflowing with music content. However, there is a crucial distinction to be made between music *presentation* and music itself.

Ultimately, I define *radio* as any sound medium that is purposefully crafted *to be heard* by an audience—even if it is only an audience of a few people.[42] The "purposefully crafted *to be heard*" provision implies intent on the behalf of the producer, as well as an acknowledgment that the text is designed to make use of the special properties of sound.[43] The "by an audience" provision means that it must be *public*. It need not be broadcast, although it must be distributed across a communications network, such as the internet, for it to at least potentially reach an audience. The reason for this stipulation is that sociability is the primary characteristic of radio.

Sociability and Radio in Everyday Life

Despite the significant technological and industrial shifts that radio-internet convergence has wrought, radio as a medium has stayed relatively consistent across its century-long history, if and when it is viewed principally as a set of cultural relations instead of as a technology, an industry, or an aesthetic formation alone. And it is the characteristic of *sociability*—rooted in the element of *liveness*—that creates the social space that makes radio *radio*, in the last instance.[44] Here the "social" in sociability is meant in a double sense: it refers to our basic human need for companionship and community, as well as an informal social gathering. Liveness, too, is not conceived of simply as a technology, production strategy, or textual feature, but rather as an ideology and discursive formation that shapes both how radio is presented and how it is perceived by its audience. Thus, sociability remains integral to our understanding of something we call "radio"—or some sense of "radioness," no matter when, where, or how it is encountered.

Perhaps the most widely accepted of radio's distinctive qualities—apart from its being a nonvisual, "blind" medium—is that it exists primarily in *time* rather than space. According to Andrew Crisell, radio is "a 'present-tense' medium," by which he means that "it seems to be an account of what *is* happening rather than a record of what *has* happened."[45] Liveness has long been central to the ontology of broadcasting.[46] Broadcasting's one constant feature is its capacity to allow audiences to hear what is being transmitted, when it is transmitted, as it is transmitted. Paddy Scannell calls liveness the "most basic character of the medium of broadcasting."[47] Here liveness is a question of temporality. Unlike film or print publishing, broadcasting is immediate, permitting audiences to encounter its content "here," "now," "as it happens," "in the moment." The effect of liveness is more than just one of immediacy: it is also one of *presence*, of absent listeners feeling a sense of "being there" in the moment. For Mary Anne Doane, a broadcast medium's "greatest technological prowess is its ability to be there—both on the scene and in your living room."[48] Audiences often feel as though they have witnessed an event because they heard it broadcast live, despite the fact they were not physically present for it. This phenomenon is possible because, as John Durham Peters describes, *witnessing* requires presence in time more than space—listening to something live on the radio "serves as an assurance of access to truth and authenticity . . . because events only happen in the present."[49] Thus, witnessing and liveness are both rooted predominantly in the nature of time (temporality rather than spatiality), which enables broadcast radio to disembody and reembody sound while allowing audiences to feel a sense of presence, even intimacy and interaction.

Perhaps unsurprisingly, much of the cultural analysis of liveness focuses on media events, as these are the occasions that most clearly possess the element of immediacy. The subtitle of Daniel Dayan and Elihu Katz's canonical book *Media Events* is *The Live Broadcasting of History*. Dayan and Katz describe media events as happenings that offer a "sense of occasion."[50] Scannell writes, "The liveness of broadcast coverage is the key to its impact since it offers the real sense of access to an event in its moment-by-moment unfolding. This *presencing*, this re-presenting of a present occasion to an absent audience, can powerfully produce the effect of being-there, of being involved (caught up) in the here-and-now of the occasion."[51] Thus, embedded in the concept of liveness are notions of witnessing, presence, and also *eventfulness*.

This liveness is a performance—an effect—more than a reality. Radio producers are trained to narrate audio stories, and also to get interview subjects to recount events, in the *historical present tense*, since it expresses action and creates an effect of immediacy. It makes it seem as though the past is happening again. Through programming strategies and textual practices, producers insist on broadcasting's *immediacy* and *spontaneity*. They also use direct address that creates a colloquial, almost conversational relationship of "copresent intimacy" between the hosts and audience.[52] They attract and sustain audiences through a rhetoric of liveness that engages listeners in a "spontaneous, informal, unscripted 'here and now.'"[53] This rhetoric is carried over from broadcasting to podcasting—the chattiness and direct address a primary source of podcasting's perceived "intimacy."

Although radio was predominantly broadcast live during its first few decades, for more than a half-century now the bulk of radio content has actually been prerecorded and only presented as "live." Even a live news or sports broadcast contains what Jane Feuer calls a "mosaic" of canned and live production elements, including prerecorded bumpers, voiceover commentary, interviews and remote reports, instant replays, advertisements, and clip reels.[54] Through a mixture of audience assumptions and institutional discourses and practices, broadcasters are able to overcome this fragmentation and its inherent contradictions to maintain an illusion of directness and presence. This, for Feuer, is broadcasting's "ideology of liveness." She argues that the ambiguity of the term "live" is exploited beyond just the live/recorded binary to the point that "the live" is equated with "the real."[55] Audiences are led to perceive all broadcasting as fundamentally live. Thus, a concept of liveness cannot simply be collapsed into the idea of an "as it happens" real-time broadcast of a live event.

Radio's temporality is fundamentally not a question of content but rather a question of *transmission*. Broadcasting is always live in the sense that it is immediate—it is received in the same moment as it is transmitted.[56] It is the broadcasting signal or feed itself that is endlessly copresent. Liveness does not require simultaneity because liveness exists in the "now" of output and the experience of audience reception. This relates to Raymond Williams's theory of *flow*, which describes the defining characteristic of broadcasting as *sequence*.[57] For Williams, the distinguishing characteristic of flow refers not to the programming itself but to the technology of transmission combined with the institutionalized programming practices and, most importantly, the audience's experience of consuming live broadcast media as one long, uninterrupted text. Flow, while technologically and economically determined, ultimately describes the social function of broadcasting within the home and within people's everyday lives.

Liveness, then, is best represented by broadcasting's perpetual presence and the everyday experience of casually tuning in to a diverse, miscellaneous mix of often very banal programming. The very notion of a "broadcast clock"—a template used to program radio stations and networks—and the programming practice of "dayparting" indicate that there is a routinization to broadcasting. The structure of much of broadcast radio's daily schedule has been built around the social activities and daily routines of the audience (e.g., "drive time" programs that coincide with rush hour commutes).[58] Even many on-demand podcasts apply these same logics: there are daily news podcasts, including ones discursively positioning themselves for morning consumption, like NPR's *Up First*; other podcasts are designed for nighttime listening, such as *Sleep with Me*; producers also set podcasts' runtime to fit within the length of an average commute. Podcatcher apps, such as Apple Podcasts and Overcast, replicate the predictability of broadcast transmission by ensuring that new episodes of your subscribed podcast series download automatically. Moreover, the sheer volume of podcasts and the fact that there is quite certainly a podcast waiting to be heard on any niche topic you can conjure aligns the practice of podcasting with the continuous, never-ending flow and sheer ordinariness that closely connects broadcasting to "the real."[59] It is precisely the regularity and typicality of radio—broadcasting *and* podcasting—that defines liveness.

Certainly, modern media consumption habits diverge from those of the broadcast era, since much radio listening is *on demand* and *time-shifted* rather than live. The internet affords on-demand "anytime, anywhere" media practices.[60] Although *streaming* is a term closely associated with online media, the internet is in many ways a *de-streaming* technology, when viewed through the lens of the experience of time for the maker and listener of radio. Time in broadcasting is linear, while time online is nonlinear, displaced, and repetitive.[61] And yet broadcast-esque sociability and liveness are being recreated jointly by media producers and audiences. Derek Kompare argues that streaming services today replicate broadcasting's continuous succession of content, constantly prompting us to listen to or watch similar programs when the initial chosen program is finished playing, plus bombarding us with ads and queues and other ephemeral pieces of content that connect all the discrete objects into a seemingly endless stream.[62] These new digital flows suggest a desire, on behalf of both media institutions and audiences, to reproduce broadcasting's flow model of attention and constant engagement. The practice of binging on media originates from the same "impulse to go on watching" that was so central to Williams's idea of flow and the broadcasting experience.[63]

Liveness still has the qualities of presence and immediacy, only they take on very different meanings in this context. Scannell offers a phenomenological approach for studying radio, one that analyzes the "dailiness" of broadcasting. Broadcasting's medium is *time*, and it produces a "daily service," in that it is always always on.[64] Broadcasting's content is mostly preplanned, yet audiences encounter it in the now, on their personal time. In this way, liveness serves as a vital connection between the medium and the world of the individual audience member. "The liveness of the world," Scannell writes, "returns through the liveness of radio and television— their most fundamental common characteristic."[65] Moreover, there is a *conversationality* to broadcasting that, in its resemblance to the living reality of everyday existence, produces its "reality effect."[66] It is sociable and attentive to our everyday lives. Even though it is a simulated construction, an affective bond is developed between the listener and the radio personalities or institution.

Radio's liveness is defined through the experience of the listener. It is less a matter of the content literally being produced in real time than it is of the radio signal being endlessly copresent. This is as true of podcasting as it is of broadcasting: programs are constantly flowing onto podcast platforms and into our feeds. The audiences know they can tune in whenever they want and hear some radio, reasonably expecting to find something that will suit their tastes and the rhythm of their daily life. That radio is typically very casual and spontaneous in character. And there is an element of communicative exchange in this experience. Since radio is a "blind" medium consisting of nothing but *sound* (words, music, noise, silence), it communicates fundamentally through speech. It is also a very particular form of speech that, compared to other media, closely resembles ordinary conversation.[67] Television and film tend to use speech to direct the audience's attention to images and text on the screen. Print newspapers, magazines, and books are, like radio, also limited to communicating primarily through words—though they can also use images. Unlike radio, there is a structured formality to much print media. And print lacks the dimension of *voice* (accent, tone, inflection, and other expressive elements). Writing to be read (literary) differs considerably from writing to be heard (oral). On the radio, producers must describe everything, conveying context audibly; print communication exists in space rather than in time, meaning audiences can jump backward and forward—options not traditionally available to radio listeners. The presentation style on radio is linear and direct, which often translates to being more informal and colloquial. In effect, speech on the radio is designed to sound like a (very well-planned and polished) conversation between friends.[68] It is no coincidence that contemporary radio and podcasting are routinely described through the language of "storytelling," as they share many of the same forms and structures as oral storytelling.

This conversational style is a source of the "intimacy" and affective closeness that is frequently attached to broadcast radio and podcasting. Especially when placed alongside other hypermediated experiences like contemporary television, radio seems closer to interpersonal modes of communication. This conclusion might seem to privilege talk radio formats, but, as Andrew Crisell points out, the primary code of radio is verbal context.[69] Even pop songs played on the radio take on unique meanings through the way they are presented, identified, and explained by the DJ. The meaning being conveyed with even a simple Top 10 playlist is part of the sociability that defines radio.

This brings us back to *what* and *who* radio is "for." There is still an eventfulness central to the conception of liveness, even if it is now an event-fulness rooted in typicality rather than atypicality. The radio event simply becomes an experience that cannot be easily reproduced or revisited—it is something unique that engages the audience. More and more, eventfulness is achieved through *sharedness*. Social participation has always been central to media events. In these situations, the liveness of broadcasting constructs a sense of *copresence* or belonging among a community. But that sense of belonging does not exist only in mass media events—it is also a crucial part of everyday media consumption. Nick Couldry has written of liveness as being a sense of "continuous connectedness," extending the concept for the online media environment to include "social co-presence."[70] Copresence here refers to how broadcasting's liveness is extended across media, such as when audiences chat online about a podcast episode via social networking sites like Twitter. This is similar to "second screen" use in contemporary television, wherein fan communities and television networks use social media and hashtags to enhance interactivity around broadcast television programs.[71] Thus, everyday social communication is being continuously mediated, and it is taking on event-like qualities: immediate, intense, par-ticipatory. Put another way, the stuff of the everyday is achieving an even greater sense of immediacy in networked digital media culture due to the ease and speed of social media. The stuff of the everyday, to evoke Lance Sieveking, is the stuff of radio.[72]

Despite rhetoric about "digital revolution," there is plenty of evidence to indicate that the cultural function of radio has not changed all that much. We still turn to radio to help us make sense of the world, as we always have. Listening to the radio is about reaching out to others and, in the process, relinquishing some control. Discursively there is an element of conversation in radio listening that lends the medium its sociability. Chris Priestman refers to it as "radio as conversation."[73] The radio experience is not defined entirely on the listener's terms; it is a process of exchange. Graeme Turner argues that the online consumption of broadcast media replicates liveness in that it is also about "sharing an experience that is un-predictable and continually immanent."[74] Again, liveness is produced when we meet the radio broadcast or podcast *in* time, and that is true whether we encounter the content live over the air or online days or months after its initial release. As Scannell describes, radio's "unobtrusively unfolding sequence of events" gives substance and structure to our everyday lives.[75] It is precisely the natural flow and rhythms of daily media like broadcast-ing and podcasting that make them appear alive to us and give them their meaning in our everyday existence. The sociability of radio—its power to construct a sense of copresence and to score the rhythms of our everyday lives—is still the same as it ever was.

Chapter Overview

This book is framed as a cultural history of internet radio, examining the cultural, industrial, and aesthetic consequences of radio's convergence with the internet and related networked digital media. Despite origins going back more than twenty-five years, internet radio's development is still very much underway—making this study equal parts a history of the past, the present, and the near future. Indeed, a recurring theme throughout this book is that a medium is always transforming, never fixed—this is increasingly salient in today's mode of "perpetual beta," in which digital technology companies like Google regard their products as always being tentative works in progress.[76] The chapters in this book loosely follow a chronological progression from the more distant past (1990s) to the recent past (2000s) to the contemporary moment (2010s), concluding with a few future speculations.

Chapter 1, "Soundtracking the Information Superhighway: The Origins of Internet Radio and Streaming Audio," maps out internet radio's initial period of emergence between 1993 and 1995. It captures the story of technological but also cultural and industrial convergences—of the moments when the previously distinct media of "the radio" and "the internet" first came together. I present two case studies of internet radio's early adopters, focusing especially on discourse analysis of these key historical actors' motives for pursuing their innovations. The first case study is of the Internet Multicasting Service, which was the first internet "cyberstation," founded in 1993 in Washington, D.C. The IMS operated as a nonprofit and was driven by public service principles and utopian views of virtual communities and of the internet as a democratizing force. Its developers were computer engineers who cared little for radio as a medium and instead saw it as a means to an end: programming was designed to lure audiences online and demonstrate to them the potential of the internet medium, while also influencing telecommunications policy governing the internet. The second case study is of a group of student-run college radio stations— WXYC at the University of North Carolina at Chapel Hill, WREK at the Georgia Institute of Technology, and KJHK at the University of Kansas in Lawrence—that in 1994 were the first radio stations to put their broadcast signals online for live 24-7 simulcast streaming. Although operating independently of one another, each station's move into streaming audio was pushed by young computer engineers motivated by a hacker ethic and a desire to tinker with the technological apparatus and push the boundaries of possibility in media technology, much like the amateur hobbyists in the 1910s–1920s prenetwork broadcast radio era. None of internet radio's key actors were experienced radio producers or broadcasting institutions: it was outsiders—unburdened by the past and eager to gamble on new innovations—who demonstrated that radio on the internet had a future.

Chapter 2, "Radio Dot-Com: Internet Radio Goes Mainstream," continues that early history of internet radio, tracing its rise in popularity from 1996–2000 and, especially, internet radio's crucial role in the growth of the nascent commercial World Wide Web and the origins of the modern-day multimedia internet. The chapter begins with case studies of RealAudio, the first standardized streaming audio software format, and AudioNet (later Broadcast.com), the largest internet radio web portal of the 1990s. These two companies represent the emergence of a commercial market for internet radio, as well as a shift toward aggregation as the primary distribution model in the online media business. Pseudo.com, a for-profit internet-only radio network centered entirely around its own original content, comprises a third case study: a well-funded start-up and the cultural life-center of New York's Silicon Alley, Pseudo and its outlandish programming embody the exuberance of 1990s youth culture and the dot-com era. Rooted in online chat software, Pseudo illustrates an early move toward interactive programming and the participatory culture of Web 2.0. Again, internet radio's innovators were internet rather than radio professionals. Yet when it came to content, they often lacked vision, replicating the style and conventions of traditional radio broadcasting or repurposing content from terrestrial radio. Thus, during the 1990s, the internet opened up new opportunities for radio while it also reproduced many of broadcasting's established structures and cultures.

Chapter 3, "Everybody Speaks: Audioblogging and the Birth of Podcasting," is the first in a pair of chapters exploring talk and news radio online. It investigates the roots of podcasting in the embryonic Web 2.0 culture, specifically as an audible version of blogging called audioblogging. I identify this period (2000–2005) as the first wave of podcasting, which was then followed by a second wave from 2005 to 2010 and an ongoing third wave from 2010 to the present. Rather than simply regard audioblogging as a stepping-stone to podcasting (the latter of which did not actually take its name or present form until 2004), I argue that audioblogging represents a failed attempt to develop a truly divergent type of internet audio media, one in which masses of ordinary individuals would post short, personal audio messages online. The practice of podcasting as it developed following its integration into the mainstream media landscape in 2005 has largely adopted the basic structures and forms of broadcasting, such as the episodic series. Nevertheless, the informal, self-reflexive, often mundane conversational style of audioblogging—which was actually a convergence of not only radio and the internet but also blog/diary writing and the telephone—carried over into many areas of podcasting, not to mention social media platforms like Twitter (which itself started as an audioblogging service).

Chapter 4, "On the Line and Online: Talk Radio Meets the Internet," takes a very different approach to talk and news radio online. It explores the impact of networked digital media, including social networking sites and collaborative software tools, on broadcast radio production, and the ways in which the internet helped producers make traditional talk radio more accessible and inclusive of the public during the 2000s and 2010s. A case study of the New York City weekday public radio program *The Brian Lehrer Show*, produced by NPR member station WNYC, outlines these shifts in broadcast radio production culture. I exhibit how call-in talk radio was itself already an interactive format, and, in the hands of an ambitious production team, it was combined with online social networking tools to create an even more robust and effective public forum. Through the context of crowdsourcing and citizen journalism, the internet has opened up new channels and platforms for the audience to speak, and for public service-oriented talk radio producers like Lehrer to listen to them and elevate their voices.

Chapter 5, "Hang the DJ?: Music Radio and Sound Curation in the Algorithmic Age," turns attention to music radio in the 2000s and 2010s. Unlike the other areas of internet radio covered in this book, which were mostly propagated by amateurs, small start-ups, and independent and public media groups, online music has almost exclusively been the domain of highly centralized and commercialized services, such as Pandora, Spotify, Rdio, Rhapsody, and iTunes Radio / Apple Music. Eschewing the issues of copyright regulation and music industry economics that have dominated discussions of online music radio, I turn instead to questions of labor and curation. Through a case study of Pandora Internet Radio and its Music Genome Project, I reveal the vast amounts of invisible labor that go into computer algorithms, databases, and other digital technologies. While the cultural work of DJs has largely been offloaded to computers—or at least it appears that way to listeners who rarely hear human voices on the stream—there are still workers very actively involved behind the scenes. There are multiple radio practices and logics present in the way platforms like Pandora build their music libraries and filter content. What is more, I point to recent shifts, including Apple's investment in Beats 1, that suggest a push-back against the hyperpersonalization and nichification of online music listening, leading to a growing desire for intermediaries and experiences replicating radio.

Chapter 6, "Touch at a Distance: The Remediation of Radio Drama in Modern Fiction and Nonfiction Audio Storytelling," is the first in a pair of chapters to explore the various modes of audio storytelling thriving during the third wave of podcasting. Audio storytelling is defined as narrative-driven fiction and nonfiction programs that focus on individuals and their personal experiences. I identify six modes of audio storytelling: live storytelling; personal documentary; historical narrative; the nonnarrated story; the feature-documentary; and audio drama. An emphasis is placed on the areas of aesthetics and narrative form, and how at the production and textual levels contemporary podcasts are reviving—and expanding—older genres, techniques, and styles of radio broadcasting that, in some cases, have been silent for decades. The first of two case studies in this chapter centers on the audio drama podcast *Welcome to Night Vale*, arguing that podcasting has opened up spaces for it and a new wave of other creative audio productions to flourish. These are spaces that have been mostly closed off in US broadcasting since the end of radio's Golden Age in the mid-twentieth century. Next, attention turns to the aesthetic innovations of the feature-documentary broadcast-turned-podcast *Radiolab*. I draw on the concept of remediation again to highlight *Radiolab*'s linkages to the sounds of radio's past, including long traditions of radio documentary, radio art, and radio drama. Connecting back to radio's central characteristic of sociability, I emphasize the series' conversational narrative style and the close attention it pays to individuals and personal experience.

Finally, Chapter 7, "Make Them Feel: Nonnarrated Audio Storytelling and Affective Engagement," extends that central focus on affect and people's lived experiences through an examination of the nonnarrated mode of audio storytelling. The podcast *Love + Radio* is offered as a case study of a (mostly) nonnarrated podcast that uses personal, conversational narratives to capture the contemporary structure of feeling. Grounded in a long tradition of US public radio production that includes *StoryCorps* and documentary soundwork featured on NPR programs like *All Things Considered*, this approach to audio storytelling focuses on the voices and lives of ordinary people, and functions politically by listening to others and expanding the range of public discourse. Although not without their shortcomings, these podcasts give value to the voices of underrepresented persons, revitalizing radio as a tool for participatory democracy. It is argued that the rising popularity of these long-form, affect-driven audio stories hints at an expanding cultural desire to slow down and listen attentively and empathetically to our fellow citizens during a time of growing incivility and social discord.

The book's conclusion looks toward the future by learning from the past. It begins with an evocation of utopian discourses of radio that have long accompanied the medium, specifically Bertolt Brecht's 1930s vision for a two-way radio system. Since its earliest days, political activists and cultural theorists have optimistically looked to radio as a truly democratic participatory media culture. Most recently, the internet has emerged as the democratic media platform de rigueur. There is an uncanny resemblance in the discourses surrounding radio and the internet during their moments of newness, before either medium's identity and protocols became firmly established. Each has been viewed as an agent of democratization and participatory community, giving rise to the media-empowered citizen. Contrary to the popular belief that broadcast radio is homogeneous and overly institutionalized and commercialized, however, I argue that the radio landscape in the United States, historically and still today, is expansive and more inclusive of the audience than any other mass communications medium. The distinctive conversationality of radio talk and the widespread focus on quotidian activities and personal expression in radio's content are symbolic of radio as the quintessential "people's medium." Radio has a special capacity for facilitating community and human connectedness and for bringing the listener a sense of community participation. This sociability transcends a particular technology or platform. It is also a model for the principal components of the internet and digital culture: radio's convergence with the internet has significantly transformed radio, but radio has also impacted the internet, especially through the form of participatory social media. Yet, in a final short case study, I contemplate the recent emergence of podcast networks and the trends of professionalization and consolidation within the podcasting industry, observing that the supposedly "disruptive" practice of podcasting is in actuality replicating well-established institutional structures and economic models from broadcasting. This is a mixed blessing, as it is fostering a burgeoning field of innovative radio content while also potentially crowding newcomers and alternative voices out of the space.

CHAPTER ONE

Soundtracking the Information Superhighway

The Origins of Internet Radio and Streaming Audio

Ten seconds of light pop synthesizer fanfare leads up to the announcement, "Internet Talk Radio: asynchronous times deserve asynchronous radio." The forty-seven-minute program that follows features a protracted interview between the host, a cerebral yet affable man in his thirties, and Marshall T. Rose, a computer network engineer and member of the Internet Engineering Task Force (IETF). Their sobering conversation about SNMP internet protocols, OSI layers, and other issues of internet architecture is punctuated by short segments: "The Incidental Tourist" with amusing reviews of restaurants in Prague, a "Book Byte" commentary on a newly released computer system administrators guidebook, and "Name That Acronym," a humorous bit of computer engineering trivia. At the end of the program, the host signs off with another tagline, "Internet Talk Radio: same day service in a nanosecond world." While the production quality is bare-bones and the audio is tinny and echoing, the program is not unlike an episode of National Public Radio's *All Things Considered* or *Fresh Air* with its talk radio mix of interviews, commentary, and special features and its erudite yet relaxed style—that is, apart from the highly specialized subject matter about computer networking technology and the instructional interjections about what listeners were permitted to do with the show's "files" and how to send the producers "electronic mail" correspondence.

The date is March 31, 1993. The host is Carl Malamud. And the program is *Geek of the Week*. Although it makes no bold proclamations or ostentatious displays of its landmark status, this is the first-ever internet radio broadcast.[1]

This chapter tells the story of how the previously distinct media of "the radio" and "the internet" first converged, and the handful of the people and institutions who helped enliven the internet with sound, creating new industries and new understandings of media culture in the process. This is as much a story about the origins of the modern-day multimedia internet as it is about radio. After all, the original World Wide Web standards were not designed for transmitting audio or moving images.[2] The web was originally conceived as a means for exchanging text and information, and it was not obvious in the early 1990s that the technology could or should be used for distributing radio or other audiovisual media. Radio was the trailblazer that led the way for the modern streaming media era. So who first decided to bring radio to the internet, and why? What genres and structures did early internet radio programming take? Was it merely terrestrial radio transposed onto the internet, or was its style, form, and content distinctly new in any notable ways? These are a few of the questions explored in this chapter, as the initial development of internet radio from 1993 to 1995 is historicized.

Two case studies form the core of this chapter. The first is the Internet Multicasting Service, noted in the introduction as the first internet "cyberstation," founded in 1993 in Washington, D.C., by technologist and open-source advocate Carl Malamud. And the second is a group of student-run college radio stations—WXYC at the University of North Carolina at Chapel Hill, WREK at the Georgia Institute of Technology, and KJHK at the University of Kansas in Lawrence—that in late 1994 were the first radio stations to put their broadcast signals online for live, 24-7 simulcast streaming. In both cases, internet radio's innovators were outsiders to traditional radio, unburdened by the past and eager to experiment with new technologies. They were attracted to online radio as a platform for pushing the boundaries of possibility in the emerging technology and culture of the internet. Yet, when it came to content, they mostly lacked a vision for radio as an artistic or storytelling medium, and replicated the style and conventions of traditional radio broadcasting or repurposed content from terrestrial radio. Moreover, theirs was a predominantly white male production culture, and their audience was similarly white, highly educated, and upper-middle class, resulting in a media environment that lacked cultural diversity and inclusion. Thus, during the mid-1990s, the internet opened up new opportunities for radio while it also reproduced many of broadcasting's established power structures and taste cultures.

The Internet Gets a Voice: Internet Multicasting Service

The *Geek of the Week* series, the Internet Talk Radio "channel," and the Internet Multicasting Service (IMS) "cyberstation" that hosted them were the brainchild of Carl Malamud, a thirty-three-year-old technology writer, open-standards advocate, and computer industry consultant. When Malamud launched his project from an office space above the Young Chow Chinese restaurant on Washington, D.C.'s Capitol Hill, it was little more than an experiment.[3] Indeed, Malamud himself described it as an idea that "started as a hobby and kind of went nuts."[4] Although the internet itself was over twenty years old in the spring of 1993, it was still a long way from being a mainstream medium. Barely 2 percent of Americans had access to the internet, and the majority of that niche population worked within scientific research institutions, universities, or large corporations involved in the computer or telecommunications industries.[5] Tim Berners-Lee's World Wide Web system had been in operation for two years, though it had yet to catch on in a sizable way. In fact, it was not until April 30, 1993—a month after *Geek of the Week*'s premiere—that CERN placed Berners-Lee's web software in the public domain, making it available for anyone to use for free.[6] It was only in 1993 and 1994 that the first widely used graphical web browsers (Mosaic and Netscape Navigator) and search engines and web directories (ALIWEB, Yahoo!, et al.) were introduced, giving users convenient access to the web. Between 1992 and 1995, the US government gradually opened up its National Science Foundation Network (NSFNET) backbone service for commercial use; previously it had been restricted to military, government, scientific, and educational institutions. At that same time, multiple new network service providers emerged to provide internet access to commercial customers. All of these events worked together to popularize the internet among the general public during the mid-1990s. However, the IMS cyberstation was ahead of this curve, launching at a time when the internet user base still consisted primarily of technologists. Terrestrial radio stations and networks were not even online at this time, yet alone considering using the computer network to transmit programming to listeners. For the most part, the people on the internet were the people building the internet.

There was no "live" broadcast of IMS's programming at first. Episodes were prerecorded and made available to listeners as audio file downloads in multiple formats, including .au (developed by Sun Microsystems and common on NeXT systems) and .gsm (Global System for Mobile Audio, originally designed for mobile phones). The audio quality of these formats was fairly low, since files were compressed to keep their size down, usually using 8-bit resolution. This made fidelity roughly equivalent to AM radio or sound over a landline telephone (for comparison, CD quality sound—then the standard for "high quality" audio—was 16-bit resolution).[7] This was before the MP3 format was released—that format, introduced in 1995, would go on to become the de facto standard for digital audio coding in the late 1990s. It was also before the RealAudio format (discussed in chapter 2), which debuted in 1995 and made streaming audio practical for most users. (Workstations with high-speed connections could actually stream the .gsm files—meaning play the audio in "real time" at the same time as it was downloading—since the files were encoded small enough. This option, however, would not have been universally available in 1993–95.)[8] A typical half-hour episode of *Geek of the Week* took up 15 megabytes of hard disk space—a sizable amount of storage capacity, considering that most personal computer hard drives at the time were only 400 megabytes. This would require nearly two hours to download over a standard modem connection. As a result, most people would have listened to shows like *Geek of the Week* on their own time, in what today would be called an on-demand or time-shifted manner. It was essentially proto-podcasting: digital, internet-distributed, downloadable, mobile, and time-shifted audio content. Indeed, Malamud's declaration of "asynchronous radio," or what he alternately called "random access radio," referred to the fact that the audience was in control of their personal listening experience.[9] The weekly episodes of *Geek of the Week*, at least for the first few months of its run, were also divided up into multiple smaller files, in order to both ease the download time and enable listeners to pick and choose which segments to listen to and what order to listen to them in. The notion of user control that is closely associated with podcasting today was very much present here at the birth of internet radio in 1993.

Distribution of IMS's programming was more involved than simply posting the radio program files to a website. Again, the web was a little-utilized system in March 1993, meaning there was no universal point of access for the internet.[10] Rather, the files were initially uploaded to UUNET servers in Virginia—the central hub of the Alternet network—and then copied to regional networks in some 30 countries where local users could access them.[11] Therefore, most people would have been downloading the files from one of a few dozen anonymous file transfer protocol (FTP) sites dispersed across the global internet. These were not websites but rather server-based systems used to exchange data between computers on a net-work, located at "dotted decimal" IP addresses and usually only featuring a textual interface. The "multicasting" in the Internet Multicasting Service's name referred to internet protocol (IP) multicasting, an experimental net-work protocol that allowed for real-time streaming of data from one or a few sources to many receivers on the internet. The alternative was to uni-cast, transmitting the same data over and over to each individual recipient separately—a much less efficient and more resource-intensive process.[12] Although the transmission of voice across computer networks had been tested as early as 1974, it was not until the IP multicasting standards—first developed by Steve Deering at Stanford University in 1986—that multimedia content could be effectively carried over the internet. The in-frastructure for multicasting was built up by the cooperatively run IETF during the early 1990s with a virtual network called the IP Multicast Back-bone, or MBone for short. At the March 1992 IETF meeting in San Di-ego, California, Deering (then a researcher at Xerox's Palo Alto Research Center, aka Xerox PARC) and associates multicasted audio of multiple sessions over the internet to twenty sites spanning three continents. The open-source audio conferencing software VAT (Visual Audio Tool), writ-ten by Van Jacobson and Steve McCanne, was used to transmit and receive the broadcast. This "audiocast," as they called it, was the first marked use of livestreaming media on the internet.[13]

Malamud witnessed the IETF demonstration and cites it as his inspira-tion for venturing into internet radio. He used their software tools and also helped out Deering and others with IETF broadcasts and additional mul-ticasting events from 1993 on.[14] IMS did not use live multicasting for all its programming, though, and ultimately multicasting failed to achieve wide-spread adoption during the 1990s.[15] Regardless, the concept of network multimedia was still experimental when Malamud came to it in 1993. The initial use of the technology had also been limited to audio and video con-ferencing in an institutional, mostly business-to-business setting. These were point-to-point communication applications rather than broadcasting. Malamud would be the first to bring the technology to "prime time," as he put it, and display its practical implications beyond the small technical community of the IETF.[16] Even IP multicasting's innovator Steve Deering professed that Malamud was "making the technology do things we never expected. . . . We don't have the imagination to go make these things hap-pen, or the drive. He does."[17]

Malamud's original impetus was to start a magazine about the internet.

It would be a trade publication for the computer networking community that went beyond the usual detailed reference works, mixing reliable technical information with "a few funny stories, a little gossip, some editorials."[18] He explained that there was already enough computer code in the world, yet not enough informed discussion in the media helping to relate the meaning of the technology to the broader public.[19] He found the publishing costs of a magazine to be too expensive, however, and opted to do "something on the network" instead.[20] "I decided it was time to start using the tools that I've been teaching others how to use."[21] As the IETF pilot experiment demonstrated, the internet infrastructure had expanded to where it could support the circulation of multimedia data—not just text but also audio and video. Few US homes had so much as dial-up connections using modems and telephone lines in 1993. Many large institutions, however, were becoming equipped with permanent, high-speed connections via local area networks (LANs) and early broadband technology. Most modern computers also had the storage capacity and hardware (i.e., sound card, speakers) to support audio playback: any Macintosh or NeXT computer and any IBM PC with a CD-ROM drive or Sound Blaster board. Malamud estimated in 1993 that there was a potential audience of "several million" people worldwide who had the connectivity and computing power to access multimedia internet content.[22]

No doubt this was a niche audience consisting primarily of (white male) computer programmers and technology researchers. That is why Malamud purposefully targeted his flagship internet radio program at the technorati.[23] The guest list for *Geek of the Week*—which produced about forty-five episodes over a sixteen-month span in 1993–94—featured a who's who of internet architects, including Tim Berners-Lee, Brewster Kahle, Tim O'Reilly, David Crocker, and Steve Deering. The show's subject matter about internet infrastructure and governance, while made relatively accessible, still was highly specialized and destined only to appeal to listeners who already had a vested interest in computer science and data communications. *Geek of the Week* was soon followed by other programs, including *TechNation*, a talk show about the cultural impact of science and technology, and *Soundprint*, a general interest, nonfiction documentary series featuring eclectic, creative radio pieces from diverse freelance producers. These were both independent public radio series that IMS syndicated rather than produce itself. (*TechNation* was produced by KQED San Francisco and *Soundprint* was distributed by American Public Radio, the predecessor of Public Radio International.) While *Soundprint* was not a tech-centric program, it is notable that the Bill Siemering executive-produced series had recently struggled to secure funding and distribution through the US public radio establishment, in particular the Corporation for Public Broadcasting (CPB), because it lacked a single thematic focus, often featured controversial content, and did not attract the large audiences that public radio mangers were beginning to prioritize.[24] In other words, IMS was not programming blockbuster entertainment content designed to attract a mass audience; this was niche content for a particular subset of intellectual elites. Nonetheless, its programs were well suited to make an impact among the subculture of highly educated techies who constituted a large portion of the online community in 1993.

IMS attracted an audience immediately. A few weeks prior to its launch, the *New York Times* ran a front-page story on the fledgling internet radio operation, declaring that "it signals the first step in a transformation in which national and even global computer networks will fiercely compete with—or even replace—traditional television and radio networks."[25] The article sparked a flood of interest. Within its first few months, Malamud was already claiming one hundred thousand regular listeners in thirty countries.[26] There were reports of regional networks running their own local "radio stations" online by streaming IMS's programming continuously over the network. One network manager apparently uploaded the sound files into his company's voicemail system so workers could call up and listen on their office speakerphones. Other people copied the audio files onto their laptop computers and played them through their car radio—protopodcasting yet again. Some terrestrial radio stations even (re)broadcast *Geek of the Week* over the airwaves. All of this reuse was possible because Malamud embraced copyleft and free software principles born from the 1980s hacker community, or what today would be recognized as *open source* (a term that did not come into widespread use until the late 1990s).[27] The software being used was nonproprietary, allowing universal access. Malamud permitted "unlimited noncommercial copies," as long as no derivative works were made and the sponsorship messages were not removed.[28] (This was, in effect, a Creative Commons No Derivative Works license, albeit a decade before Creative Commons existed.) "Listeners are encouraged to copy the data as widely as possible," Malamud declared in Internet Talk Radio's mission statement.[29] The goal, first and foremost, was to push the internet technology forward.

It was not long before IMS's programming began to grow in quantity and broaden in scope. By January 1994, it was distributing approximately ten hours of programming per week.[30] In addition to *Geek of the Week*, *TechNation*, and *Soundprint*, a number of series were added to the programming lineup, the majority of them independent US public radio productions. The Pacifica (KPFA) series *Communications Revolution* utilized an hour-long format that offered short documentary features about telecommunications policy followed by a panel discussion. Produced by the Woodrow Wilson International Center for Scholars and Radio Smithsonian, the half-hour magazine format program *Dialogue* covered topics from foreign policy to cultural history and modern poetry. *Common Ground*, a half-hour "world affairs" program produced by the public policy think tank The Stanley Foundation, featured documentary-style investigative reporting on issues related to the environment, public health, and human rights in Africa, Russia, and elsewhere. NASA even produced an original series of short five-minute featurettes for IMS called *Space Story*. IMS also distributed a few multiepisode radio features, including *Hell's Bells*, a documentary about the history of the telephone, and *The Ballad of Ned Ludd*, a dramatized "techno-folk opera" about the Luddite movement. IMS collaborated with the Kennedy Center for the Performing Arts, making available audio from a number of lecture series and live theater performances.[31] A partnership with HarperCollins Publishers resulted in a daily *HarperAudio!* segment featuring excerpts from audiobooks, including recordings of Ernest Hemingway, Robert Frost, J. R. R. Tolkien, and T. S. Eliot reading their own works.[32] It was an eclectic slate of programming, not dissimilar from that of an NPR-affiliated news and talk radio station.

Most of the aforementioned programs were part of the Internet Multicasting Service's first channel, Internet Talk Radio, which was known as its "science and technology channel."[33] Although, as Malamud was fond of saying, "channels" was only a metaphor. Since most of the programming was simply uploaded to the internet, there were no broadcast channels or stations to tune in to; the "channels" were really just categories to separate out IMS's different types of programming. It is nevertheless notable how IMS carried over the language of broadcast radio to help make its efforts legible to the audience.

The biggest expansion of IMS's programming came through a partnership with the National Press Club, a professional organization for journalists and public relations practitioners. With headquarters just blocks from the White House, in the mid-1990s the National Press Club was host to regular events with prominent politicians, foreign dignitaries, and business leaders. It held weekly luncheons with a featured speaker; even before

Internet Talk Radio officially launched, Malamud made an arrangement to record and distribute the audio of these National Press Club luncheon speeches online.[34] These speeches became the centerpiece of IMS's second channel, its "public affairs channel" named the Internet Town Hall.[35] Between April and August 1993, Internet Town Hall transmitted recordings of twenty luncheons, including speeches by US attorney general Janet Reno, Senator Bob Dole, and the Dalai Lama. The relationship with the National Press Club grew to the point that IMS was given an office in the club's building. Its production studio remained in Malamud's space above the Young Chow Chinese Restaurant, while the larger server systems were housed on the eleventh floor of the National Press Club. IMS linked up the club with a high-speed internet connection, built the organization's first website, and helped computerize its research library and train its staff and members.[36] In addition, IMS either hosted or provided the technical support for a number of state-sponsored events, including the Global Schoolhouse Project's video conferences between American and foreign elementary school students organized by the National Science Foundation. It even facilitated an internet broadcast of the closing ceremony of the United Nations' Fiftieth Anniversary celebration.[37] More and more, IMS's programming began to shift away from the highly technical and instead moved increasingly toward government and public policy. In fact, *Geek of the Week* quietly ended in July 1994, barely a year and a half after it began.

IMS went to a 24-7 livestreaming format on January 4, 1995.[38] This occurred with the addition of the network's third channel, RT-FM (short for "Radio Technology for Mankind" and a play on the techie epigram "Read the fucking manual"). The channel carried a continuous feed of music—streamed off Malamud's Denon CD jukebox, no less.[39] IMS also added even more syndicated radio content, from the CBC, World Radio Network, Radio France International, and the *Christian Science Monitor*'s Monitor Radio.[40] Malamud had begun to experiment with live multicasting as early as May 1993, when he helped NPR link up its May 21 *Science Friday* installment of *Talk of the Nation* for the first-ever live internet radio simulcast.[41] *Science Friday* host Ira Flatow "asked me to be on the program," Malamud said. "When I was talking to the producers, I asked them if they wanted to be a bit adventurous, so we gave the live hookup a try."[42] Malamud linked NPR to the MBone for the event, and he continued to use the special network for live internet broadcasts in the years to come. Use of the MBone was tightly restricted, requiring cooperation between numerous network administrators. However, IMS maintained privileged access because of Malamud's stature within the computer networking community.[43] Beginning with Larry King on December 15, 1993, the National Press Club luncheons were broadcast live on the internet.[44] IMS also live-streamed audio of congressional proceedings in 1995. Its first broadcast from the floor of the House of Representatives—Newt Gingrich's inaugural speech as Speaker of the House—was heard live by twelve people.[45] A modest audience to say the least, though Malamud's goal was always to pursue a vision of what the technology could do rather than garner ratings. "The fun work is before there's money in it, because it's never been done," he said.[46] Perhaps the single biggest exhibition of multicasting technology was IMS's massive "Radio Technology for *Manana*" booth at the Networld+Interop computer industry trade show in Las Vegas in May 1994. For four days, Malamud's crew, with assistance from the Advanced Research Projects Agency (ARPA), produced a continuous live mix of interviews with the likes of Ralph Nader and Congressman Edward Markey (a prominent member of the House Subcommittee on Communications and Technology), US commerce secretary Ron Brown's speech about the Clinton-Gore administration's telecommunications initiatives, newscasts from Monitor Radio and the CBC, special broadcasts of *TechNation* and *Sunergy* (a corporate talk show produced by Sun Microsystems), and music performances.[47] It also included livestreaming video of the PBS television series *Computer Chronicles*. The multimedia event was designed to "demonstrate the revolutionary technical advances that allow personal computer users around the world to 'tune in' to interactive programs in cyberspace."[48]

The 1995 live audio broadcasts of Congress, which eventually came to be known as the Congressional Memory Project, was part of a larger, highly ambitious project. It was in effect an online audio version of C-SPAN—itself a private nonprofit organization funded by cable and satellite television fees. At the time, however, there was no easy or inexpensive way to access past C-SPAN broadcasts. IMS's goal was to archive all the congressional proceedings and offer them free to the public, as well as make them searchable using specially developed voice-recognition software. Malamud's hope was that this service would improve citizens' access to government activity and increase transparency. Noting that many politicians edit their comments in the *Congressional Record*, he said, "Imagine they wouldn't be able to say, 'No, I never said that,' because it's all recorded."[49] Developed by Deb Roy of the MIT Media Lab, the searchable audio archive that coupled the *Congressional Record* text database to the audio recordings made on the floors of the House and Senate was, in fact, operational. It collected

132 hours of audio during a test period in January–February 1995, and the searchable database was put online in October 1995.[50] With a working prototype in place, the intention was to get the institution of Congress to take over the operation, though Congress failed to follow through.[51] (Malamud tried again in 2007 and 2011 to get Congress to make recordings of all committee proceedings available online, this time in web video format, but failed yet again despite support from Speaker of the House Nancy Pelosi and, later, Speaker John Boehner.)

It is hardly a coincidence that the first internet radio station was operating out of Washington, D.C., as opposed to Silicon Valley or a major media center like New York or Los Angeles. In the 1990s, the D.C. metropolitan area was a central node in the global internet and a hub for high-tech companies, so much so that it was dubbed "The Netplex."[52] The D.C. region had long been a global center for both satellite and long-distance phone telecommunications, with major companies like COMSAT and MCI headquartered nearby. As it still is today, the US federal government was the country's biggest buyer of information technology, and it maintained some of the most advanced data communications networks in the world. The government, of course, played an integral role in developing the first interconnected computer networks in the 1960s (the "network of networks" that is the internet), principally through the Department of Defense's ARPANET, and it continued to operate and control much of the internet backbone up through the mid-1990s.[53] D.C. was home to a number of the major commercial network owners (MCI, Sprint, Metropolitan Fiber Systems) and leased-line providers (UUNET, PSI, Sprint again) that connected corporate and other large group networks to the internet. It was also home to some of the biggest commercial online services (AOL, Genie), and it had previously been home to The Source, which was the first online service, founded in 1979 and purchased by CompuServe in 1989. Over half of the internet traffic between the United States and foreign nations in 1994 was routed through the D.C. area.[54]

Furthermore, the early to mid-1990s was the era of the "information superhighway," a policy initiative touted by the recently elected Clinton-Gore administration seeking to link every US home and business via a high-speed fiber-optic network.[55] This was the birth of the digital convergence era, a time of great enthusiasm for the internet, cable television, and related multimedia technologies. While IMS could have operated from anywhere, a D.C. headquarters had significant advantages in terms of network access and technical support. "We all talk to each other—we have to," Malamud said of the Netplex high-tech community, "because the net is a very complex machine."[56] Furthermore, if Malamud wanted to affect public policy and social change—as he most certainly did—proximity to the nation's capital was a must.

Although IMS was not publicly funded, it was formed as a nonprofit corporation operating in the public good. Malamud explicitly modeled his operations on NPR and PBS.[57] Yet, IMS's financial backers deviated notably from the media businesses, financial services companies, and private charitable foundations that typically support US public broadcasting. Information technology firms were IMS's biggest donors: Sun Microsystems, O'Reilly & Associates, MCI, and Interop Company. IMS benefited considerably from the largesse of computer systems and networking companies. For instance, Metropolitan Fiber Systems (MFS) donated a fiber-optic ethernet line to IMS, which it used to connect with the Alternet network, another company that donated its service as well. Such lines could cost $60,000–$100,000 to set up in 1993, plus thousands more per year to operate.[58] With these state-of-the-art network connections, IMS had "the fastest Internet links in D.C.," claimed Malamud, "so fast that the White House borrowed some of our bandwidth in 1993" for an event on the White House lawn.[59] Commercial tech companies like MFS, Sun, and MCI were willing to donate considerable amounts of money, equipment, and services to Malamud because he was essentially doing research and development out in public view. "It's the kind of out-of-the-box thinking MCI needs right now," MCI vice president Lance Boxer said about Malamud's efforts with IMS. "He's got ideas about how to take this thing called Internet, which was not a profitable business, and make it into something we could sell."[60] Indeed, if Malamud could successfully demonstrate new uses for computers and the internet and attract new consumers to the technology, it would be a boon for the entire industry.

This use of private entrepreneurship for the public good was very much in keeping with the neoliberal policies of the Clinton-Gore administration. The strategy was to shift functions previously delegated to government to the private sector instead (corporations, nonprofit organizations, individual citizens).[61] Whereas the internet, and who used it and for what purposes, was once the domain of the US federal government, its future development was now being left to the free market. For his part, Malamud believed fervently in the liberal ideal of virtual spaces that fostered community and served public interests. He was, after all, a product of the 1970s–1980s hacker community, a culture imbued with creativity, sharing, and a desire to "produce differences in computer, network and communications technologies," as sociologist Tim Jordan puts it.[62] In other words, to test the boundaries of what is possible, in the hopes of moving both the technology and society forward. Even a program like *Geek of the Week*, which on the surface could appear as little more than techie navel-gazing, was created with education and transparency in mind. "Many of these people participate in working groups collectively known as the IETF, or Internet Engineering Task Force," said Malamud. "Now, through the *Geek of the Week* interview, you have a chance to get to know some of the key people who make the Internet work."[63] Nevertheless, there were clearly potential commercial applications to IMS's experiments, and it was these opportunities that MCI and its ilk were investing in.

Internet radio was actually only one among a series of initiatives Malamud engaged in under the Internet Multicasting Service name, all of them seeking to demonstrate new uses for computers and the internet. While the organization started with radio and typically referred to itself as "a non-profit radio station in cyberspace," on occasion Malamud defined IMS more broadly as "an organization that is providing large-scale dissemination of data onto the Internet."[64] Even from the start, then, internet radio was only one piece of a much larger strategy. While IMS had up to a half-dozen employees at its peak and frequently relied on help from outsiders like MFS, it was in no small way Malamud's organization. Essentially whatever Malamud was involved in during the mid-1990s got lumped under the IMS banner. In addition to the Global Schoolhouse Project distance education initiative, Malamud coordinated initiatives like TPC.INT, an "experiment in remote printing" that allowed people to send faxes via email. The service paired the internet and telephone networks, and greatly reduced the cost of sending instant printed documents—an email being a small fraction of the cost of a fax, which relied on then-expensive long-distance telephone services.[65] Even more notably, Malamud spearheaded an effort to put online the Securities and Exchange Commission (SEC) filings database, known as EDGAR (Electronic Data Gathering, Analysis, and Retrieval). With support from the National Science Foundation and the NYU Stern School of Business, IMS made available for free to financiers, investors, and researchers a cache of government information that up to that point had only been accessible through expensive paid services—private companies that were making millions through the sale of public information. Although Congress and the White House backed it, the SEC protested the project, claiming that putting the data archives online would cost upwards of $40 million and that no one wanted it.[66] At the end of the two-year project, with fifty thousand people per day accessing the database, the SEC begrudgingly took it over in October 1995.[67] Malamud and company successfully proved that the internet could be used to inexpensively provide government information to the general public. That is, that technology could open the government to the people. Malamud's post-IMS career has been fueled by these open government goals, most recently with the nonprofit group Public.Resource.Org, which digitizes and archives public domain materials online.

Crucially, Malamud was not a radio professional. Although he was a fan of NPR and had participated in radio during high school at the Interlochen Center for the Arts in Michigan, he never worked professionally in broadcasting prior to launching IMS.[68] His interests were always split between technology and politics. At the young age of seventeen, Malamud managed the campaign for an Illinois State Senate candidate.[69] He himself has never held political office, though he has worked in and around the Washington establishment for decades and considers himself a "free-lance civil servant."[70] Malamud studied economics and public policy at Indiana University but dropped out of graduate school before finishing his doctoral dissertation in order to help build the university's computer and information services systems.[71] While at IU, he was particularly focused on developing computer literacy among students and opening up access to campus computer facilities and databases.[72] During the 1980s, he authored a number of professional reference books on computer networks, as well as wrote regularly for technology industry trade publications. He also consulted for government groups and major computer firms, including the US Joint Chiefs of Staff, the Office of the Secretary for Defense, the Federal Reserve, Sun Microsystems, Hewlett-Packard, and Xerox.[73] Malamud became a particularly vocal advocate for open standards and network interoperability, founding the nonprofit Document Liberation Front, which succeeded in making the technical standards that undergird the internet free to all—a crucial step in the development of simple, open, universal rules that enabled the network to be accessible to anyone who wanted to participate, not only the governments and select technical experts who could afford access to the expensive and tightly controlled standards documents.[74] In the early 1990s, just prior to IMS, Malamud circled the world to write the "technical travelogue" *Exploring the Internet*, a book documenting the development of internet infrastructure at locations across the globe. "He is no different from Bill Paley or Edward

R. Murrow in creating the new cyberspace model," Eric Schmidt, then the chief technology officer of Sun Microsystems and later the CEO of Google, remarked about Malamud in 1995. "He sees himself as an information entrepreneur and a change agent."

A common thread woven throughout these varied career activities was Malamud's optimistic desire to connect people to computers and the internet, so they could then use the technology to connect to information and to each other for the betterment of society. Malamud was an avid believer in Marshall McLuhan's utopian notion of the "global village," the idea that media technology would overcome the challenges of space and time through instantaneous global communication, resulting in "a simultaneous happening" that transcended geographical and cultural boundaries.[75] Malamud adamantly believed that the internet had made the global village a reality, and that it could bring into being a more open and inclusive public sphere, a new field for political action that could enhance civil society.[76] This vision was most clearly evident in IMS's Internet Town Hall channel. "Our goal is to develop a presence on the internet, a professional, informative source of news and information about public affairs and science and technology," Malamud wrote, adding, "We want to be your place in cyberspace, the town crier to the global village." By partnering with the National Press Club, he went on, "the Internet Multicasting Service hopes to be able to provide a crucial link between the traditional media and the internet."[77] The concept of an "electronic town hall," where politicians would use the information superhighway to interact directly with citizens, drew parallels to the town hall-style presidential debates that were fashionable during the 1992 presidential campaign, invoked enthusiastically by Bill Clinton and Ross Perot. According to Malamud, though, theirs was hollow rhetoric, the campaign trail town halls amounting to "large banks of 800 numbers that would provide instantaneous polling . . . some glorified telemarketing operation." Instead, he wanted to build "a town hall on a general infrastructure [that] allows a truly unique form of participatory democracy, a forum where national leaders can engage in a dialogue with the American people."[78]

This idealized public forum was the Internet Town Hall. "We want to see a day where ordinary citizens can walk up to their computer, see the president of the United States, and tell him what they think," Malamud wrote in a March 1993 op-ed announcing Internet Talk Radio. "The Internet is a powerful force for democratic values and for an informed, responsive government."[79] The full vision of civic participation—"a place where citizens can engage in a dialogue with their leaders"—was never entirely fulfilled.[80] Malamud was able to get the content out to the audience, yet enabling the audience to talk back—getting sound to come in from lots of sources without disruption—provided to be technically impossible at the time.[81] (They did actually do it for the *Science Friday* radio segment, however; all the listener calls came in over the internet.)[82] Still, the fundamental idea was bold, and in retrospect evokes many of the principles of Web 2.0 and social media that would emerge a decade later. Call-in programs had, of course, been a staple of talk radio for decades. Yet Malamud insisted that the "talk-back" format permitted by internet technology like email was more democratic, enabling multifarious comments and questions from a much wider audience, not just "the guy who has time to keep hitting the redial button."[83] "Is this the listener you want on the air?" Malamud queried in an October 1993 op-ed in the public media trade newspaper *Current*. "With electronic mail, a single phone line can be used to bring in hundreds of messages."[84] He also suggested that radio stations could "run a discussion group around topics that you feature on programs"—although, curiously, he himself never integrated chat functionality into IMS.[85]

Notably, Malamud envisioned monthly Internet Town Hall events at the National Press Club where a national leader, such as Secretary of Labor Robert Reich, would speak on new media-related topics and then answer questions that came in live from the internet audience via audio-video conferencing software. Instant polls could also be administered online.[86] Perhaps the closest IMS ever came to achieving these goals was Vice President Al Gore's National Press Club luncheon speech of December 21, 1993. During the live internet broadcast, more than two hundred listeners emailed questions and comments.[87] Then, immediately following the speech, Gore was presented with a floppy disk containing the instant feedback from the public.[88] Even if IMS never fully realized the potential for truly interactive, two-way communications, this pluralist ideal informed nearly all of the organization's programming and activities.

This is partly what Malamud meant when he said that "radio" was only a metaphor for what IMS was doing. He was fond of saying things like, "We call this radio, but it is a different kind of radio."[89] He routinely referred to internet radio as a "new medium" since it deviated considerably from the conventional broadcast listening experience. "This is a different beast from radio," he wrote, "one in which sound and video and images and text can all coexist easily."[90] That is, internet radio was not *only* radio (i.e., audio) because it was now integrated into the convergent, multimedia space of the internet. This is a technologically deterministic definition of radio. Nevertheless, with the Congressional Memory Project, for instance, the idea was that listeners could not only search archived proceedings and compare the *Congressional Record* to the actual live proceedings, but they could also reference related materials online (e.g., reports, charts, photos) that were attached to the segment via hyperlinks.[91] This is perhaps an obvious and seemingly simple idea today, as multitasking while listening to radio online is a commonplace experience. However, the idea that people could listen to the radio (at their convenience, no less) while instantaneously pulling up a referenced news story or name-checked white paper was relatively unprecedented in the mid-1990s. I would argue, though, that this is all very much in keeping with traditional radio's sociability, and rather than an entirely new mode of communication it simply extends and enhances the radio listener's already existing sense of immediacy and copresence. Nevertheless, Malamud's interests lay in championing the internet rather than radio, and his driving notion of a "general-purpose Internet infrastructure" implied that it flattened all media and, in effect, combined the best elements of each communications medium.[92]

Yet, as much as Malamud was not a radio professional and he sought to expand the boundaries of what radio could do, IMS's programming remained remarkably faithful to radio conventions of genre, style, and form. Malamud very purposefully modeled the sound and image of his programming on traditional radio, NPR in particular. He said he wanted to capture the "feel" of shows like *All Things Considered*.[93] "We'd love to be *All Things Considered* for the internet," he explained to *TechNation*'s Moira Gunn.[94] As previously described, most of IMS's content consisted of news and talk programming, in particular interview-based programs and news magazines that mixed talks, discussions, arts, and documentary-style reports. Of course, IMS was filling out its schedule by syndicating content from the public radio community, with shows such as *TechNation*, *Soundprint*, and *Communications Revolution*. These may have been relatively fringe programs coming out of smaller public radio outlets, at least when compared with NPR mainstays of the era like *Morning Edition*, *Fresh Air*, *The Diane Rehm Show*, or *Car Talk*. Nevertheless, the production style and subject matter of these programs hardly challenged the listener's expectations of what a radio show was or could be. Indeed, even for original IMS productions like *Geek of the Week*, Malamud intentionally aimed to replicate what he called "the old-time radio feel, complete with corny theme song and Larry King–style interviews."[95] He mimicked this style with the hopes that it would be familiar and appealing to the tech industry and policy wonk audiences he was targeting with IMS—groups that fit squarely within the NPR public radio audience demographic.[96] Even IMS's concentrated focus on science, technology, and public policy issues was not unusual for public radio; IMS just delved a bit more deeply into these niches than was the norm. IMS may have been narrowcasting in a unique way and using the internet itself as a sandbox to explore the possibilities of the technology. At the content level, however, there was little effort to push radio as an art form forward.

On April 1, 1996—nearly three years to the day after the first episode of *Geek of the Week* was posted online—the Internet Multicasting Service ceased its internet radio operations. The shutdown received little notice apart from a *Washington Post* article, which appeared more than a month after the fact. "We finished our work and disbanded," Malamud said cryptically. He referred to himself as an architect, someone who creates and then moves on. "Now there are craftsmen out there who know their tools much better than I do," he added, referring to the fact that numerous internet radio stations had sprung up by mid-1996. The internet, he concluded, "has grown up and gone mainstream, which is good."[97] The full story is a bit more complicated, and as with so many things, it came down to finances. "We ran out of money," Malamud admitted twenty years later. "Couldn't raise any more. The Internet was starting to go big time and there was lots of money for corporate stuff but it was really hard to knock on foundation doors and say, 'Hi, I put big databases on the Internet and I do multimedia streaming, can we have a contribution?'"[98] He was also surely distracted by another colossal project that he had begun to put in motion in early 1995: the Internet 1996 World Exposition. Aided by longtime friend and internet pioneer Vint Cerf (today the vice president and chief internet evangelist for Google) along with a coalition of corporate sponsors and technologists, Malamud set out to create a virtual world's fair introducing the internet to anyone who had not yet discovered it.[99] Malamud had long preached that "there needs to be a public space on the internet," and he believed that that was what he was offering with IMS: civic media and services like news and cultural programming, unfiltered access to government, information databases equivalent to public libraries.[100] The Internet 1996 World Exposition took the idea of a virtual public sphere even more literally, creating a digital space that online visitors from around the world—an estimated forty thousand per day—could navigate through. *Time* magazine likened it to "a vast, bustling bazaar, a marketplace for the talents and offerings of thousands of individuals and small groups."[101] Anyone could participate as an exhibitor, not just corporations and nations, by creating a multimedia website. The whole thing revolved around a Central Park—"a public park for the global village"—a virtual representation of the types of public spaces on the internet about which Malamud long evangelized.[102] "We must prove that this technology works, we must challenge our engineers to make it better, we must educate our consumers," Malamud said about the world's fair at the Singapore World Wide Web Conference in May 1995. "We are all building a global village and now is when we decide what that global village is going to look like."[103] He packed the fair with some familiar names, such as the Global Schoolhouse Pavilion. The Internet Multicasting Service was incorporated into the Central Park in order to broadcast events, including "Concerts in the Park."[104] The virtual park survived through the year, but it was a quiet place, as its radio station went silent a mere four months into the fair's run.

Bring the Noise: College Radio Goes Online

Rewind to the fall of 1994. The Internet Multicasting Service had been in operation for more than a year, and the MBone, though still tightly controlled by the IETF, was increasingly being put to use for special live internet broadcasting events like the NPR *Science Friday* one-off, NASA space shuttle videos, and even a Rolling Stones concert.[105] Still, no terrestrial radio station had attempted to put its programming on the internet. If they were using the internet at all, most established radio producers were only just discovering electronic bulletin boards and email newsletters for networking with other industry insiders. At best, a few stations were beginning to develop websites and post their programming schedules online.[106] For the most part, the radio industry was adopting a "wait and see" stance when it came to the internet. "My fundamental belief is that radio is listened to away from a computer environment," stated one prominent commercial radio network executive, Scott Ginsburg.[107] As is the case with many well-established businesses, the radio industry—the commercial radio sector, in particular—was entrenched in its old ways and slow to adapt to change, a phenomenon economists call *path dependency*.[108]

This all began to change in November and December 1994 when a trio of student-run college radio stations put their signals online for live, 24-7 simulcast streaming: WXYC at the University of North Carolina at Chapel Hill (UNC), WREK at the Georgia Institute of Technology (Georgia Tech), and KJHK at the University of Kansas in Lawrence (KU). There is disagreement among them over which station was first to simulcast. WXYC and WREK both connected on November 7, though WREK did not "go official" with its achievement until months later.[109] KJHK began simulcasting nearly a month later, on December 3.[110] There was little press coverage of the events at the time. *The Guardian* and the radio industry trade publication *Gavin Report* cited WXYC as the first, while *Broadcasting & Cable* later identified KJHK.[111] The discrepancy between accounts largely boils down to specific technical feats or slight differences of definition. WXYC almost certainly started first, testing its system as early as August 1994.[112] WREK "did it ourselves" by developing its own custom software; the others both modified existing technology, namely the video conferencing software CU-SeeMe.[113] KJHK's claim is that it was the first to maintain a continuous live signal, the other stations experiencing problems keeping their signals online.[114] (WXYC refutes this claim, stating that its stream was continuous following the November 7 public announcement.)[115] The question of who was "first" carries curiosity value but less meaningful scholarly significance. This is a rather clear-cut case of *multiple discovery*, of several parties simultaneously developing a technological innovation independently.[116] The fact that multiple parties invented simulcasting technology at effectively the same time suggests that streaming audio was an important idea whose time had come. As such, the more interesting questions are how each group did it and what led them to pursue the innovation in the first place. Moreover, what was it about college radio stations, socially and culturally, that made them fertile ground for developing internet radio simulcasting?

Based in the underground music hotbed of Chapel Hill, North Carolina—among other notabilities, the location of Merge Records, one of the most prominent record labels of the 1990s indie rock scene—WXYC long had a reputation as an adventurous freeform radio station. The UNC campus was also known for innovation in technology development, including being home to a program known as SunSITE (Sun Software, Information, and Technology Exchange), a Sun Microsystems–sponsored program designed to promote the research and development of new internet technologies, as well as provide open access to software, government information, and other digital archives in the public domain.[117] It is perhaps unsurprising that the station's innovative programming mission and the university's academic spirit of discovery would converge. The idea for internet radio simulcasting came out of conversations between a pair of UNC information technology employees: David McConville, an audio engineer, digital media designer, and graduate student in the School of Journalism and Mass Communication who worked at SunSITE, and Michael Shoffner, a UNC computer systems administrator who worked alongside SunSITE and also happened to be a WXYC disc jockey. "David and I were talking about ideas for the station's web page," Shoffner explained. "It seemed like an obvious thing to do if we could: allow people to hear the broadcast over the web in real-time."[118] Under director Paul Jones, SunSITE employees, most of them UNC students, operated with a broad remit to pick and choose projects that explored the possibilities of the internet. "Paul just let the entire crew experiment," McConville said, adding, "which of course forced me to learn much more than I was learning in classes."[119] Doug Matthews, another SunSITE employee, explained that many of the program's projects were spawned from similar "'Wouldn't it be cool?' conversations."[120] The SunSITE team, led by McConville, began working on a solution to the problem of establishing a real-time streaming audio simulcast of WXYC's broadcast signal in August 1994, with Shoffner handling the logistics at the radio station.

A like-minded maverick engagement with technology prompted WREK's venture into simulcasting around the same time. Georgia Tech is a polytechnical university organized around advanced engineering and scientific research. In the fall of 1994, the radio station's general manager, a computer science student named John Selbie, was having dinner with some fellow staffers and describing to them the lab work he had been doing with TCP/IP protocols and the like when "someone suggested, 'Why don't we broadcast WREK over the internet?'" He proclaimed, "And the challenge was on at that moment."[121] Selbie, along with the station's operations manager, Eric Buckhalt, went about building the server and client programs from scratch. But it ultimately was not a desire to revolutionize radio that drove Selbie to experiment with simulcasting; rather, he explained, it was a determination to push his engineering skills and impress the other students he worked with in the university's Graphics, Visualization, and Usability (GVU) lab. "These PhD students seemed lights years ahead of me in terms of their programming skills and smarts. I wanted to be like them," he explained. "I love a challenge. I loved WREK and network programming. I was more interested in growing my own coding skills and being the inventor of this system than I was interested in actually getting the solution deployed."[122] Yet again in this story, radio as a medium became a platform for the pursuit of other personal, social, or civic ambitions, rather than the primary object of attention.

At KJHK, the lead computer programmer, Robert Burcham, was also a graduate student in computer science. KU's project grew out of a brainstorming session about future media in a journalism class. "When this becomes commonplace in the year 2004," said John Katich, the class's professor, "the students will look back and say, *Hey, we did that 10 years ago at KU*."[123] Burcham was also working with the University of Kansas Department of Special Education while in graduate school. The department was involved in a distance education program that utilized the CU-SeeMe video conferencing software, and thereby introduced Burcham to its potential for other applications like radio simulcasting.[124]

Testing one's own personal abilities while pushing the boundaries of possibility in media and technology: these passions that motivated both Selbie and the WXYC and KJHK teams are endemic to the hacker ethic. Tim Jordan describes hacking as the "pursuit of free creativity" and the desire "to create new things, to make alterations, to produce differences."[125] At its core, hacking—defined broadly as using technology to produce a change (and not the more illicit connotations the term provokes today)—boils down to an intellectual curiosity for tinkering with computers to find clever and unorthodox solutions to technical problems. In hacker communities, technical mastery is often its own greatest reward. Hackers care first and foremost about solving the problem, not profiting from or exploiting the solution. The social reward of bragging rights is perhaps the greatest achievement of all.

Notably, the groups at UNC, Georgia Tech, and KU were all aware of the MBone and at least some of the prior experiments with internet multicasting, and they took inspiration from these earlier exhibitions. SunSITE director Paul Jones, in particular, was actively involved in the IETF and had been friends with Carl Malamud for years. Yet the WXYC team is quick to point out that what they and the other college stations did was different from any of the previous efforts: it did not use FTP or the MBone, which in addition to necessitating a lot of cooperation between network administrators was only available on a small portion of the internet (many regional networks were not multicast enabled) and only facilitated temporary broadcasts, as opposed to a continuous stream.[126] The WXYC, WREK, and KJHK simulcasts streamed continuously and could be received on demand anywhere on the internet (with a few exceptions—at least initially, WXYC's and KJHK's streams were only on Macintosh computers, while WREK's stream was only compatible with UNIX operating systems). For his part, WREK's Selbie claims to have been unaware of the Internet Multicasting Service.[127]

Nonetheless, all three groups had been intimately familiar with early internet broadcasting, particularly through primitive internet audio and video conferencing tools like VAT and CU-SeeMe. VAT was the audio conferencing software used for the IETF meeting audiocasts. Selbie's internet broadcasting eureka moment came when he visited one of his teaching assistants at Georgia Tech to find the graduate student chatting with a colleague in California over the internet. "When I walked into his office, I saw what at that time was the most awesome thing ever," Selbie recalled. "Someone else's voice was coming over the speakers. I was floored."[128] CU-SeeMe, developed in 1992 by Tim Dorcey at Cornell University (originally in collaboration with the Global Schoolhouse Project), was internet video conferencing software that delivered small black-and-white real-time video to personal computers, albeit initially without audio and only to Macintosh computers. The silent picture was extremely low quality, but purposefully so: it was designed so as to be usable by the widest possible audience, including ordinary people with "low-end, widely available computing platforms."[129] Sending and receiving high-quality video would have required expensive hardware and the highest-speed internet connections, greatly limiting the potential number of senders and receivers. Fittingly, the program was made available for free, and reception required no special equipment apart from an internet connection and a camera and digitizer that cost approximately $100. In 1994, audio was added using a software client called Maven developed at the University of Illinois at Urbana-Champaign. It was also expanded to Windows computers. Although essentially a video chat technology in its person-to-person application, it could support multiparty conferencing (i.e., broadcasting) with the use of a reflector, software that replicated and redistributed the stream.[130] NASA used CU-SeeMe with a reflector to broadcast live space shuttle transmissions, astronaut interviews, and control room activity, which Selbie watched enthusiastically.[131] UNC's McConville spoke of watching blacked-out coverage of an IRA ceasefire in Northern Ireland through CU-SeeMe—witnessing the event in awe and thinking to himself, "It was all so new and strange and seemed to hold so much promise."[132]

CU-SeeMe may have been a crude technology, yet that is precisely what made it available to a wide audience and what also made it ripe for hands-on experimentation and add-on innovation by these computer science students. Also, they were working with minimal budgets and time constraints. "Once we had hit upon the idea, we decided, 'Hey, we better get this thing rolling. We better do it quickly because somebody else will do it.' It's something too good to let pass," said Shoffner.[133] Both WXYC and KJHK modified the CU-SeeMe video conferencing software to create their own internet radio simulcast technologies. (WREK's Selbie ambitiously designed his own client/server system, dubbed CyberRadio1, although he was nevertheless well acquainted with VAT and CU-SeeMe prior to writing his own code.) "The way we used CU-SeeMe was a total hack; it was not what that software was supposed to be used for," explained McConville. "Basically what we did was we said, 'Well, we've got this thing that can make sound go from Point A to Point B, and because it's a chatroom-type environment, lots of people can listen to it and all we have to do is be the only one talking all the time.'"[134] The way it basically worked was that, after converting WXYC's analog broadcast signal through a computer housed at SunSITE (using a two-dollar yard sale radio that once belonged to Michael Shoffner's sister), the digitized audio stream was fed to the CU-SeeMe reflector online. The reflector turned that one input signal into a broadcast by multiplying it for any computer signed in to the site. Listeners would then simply open the free CU-SeeMe software with the video turned off, log on to the reflector site, and receive the simulcast feed.[135] About their hack, Paul Jones added, "We took a lesson from Alexander Graham Bell in reverse. He saw the telephone as a way to bring music and news into homes, but ended up with what was used as a point-to-point device. We took a point-to-point conferencing software that had mods for group communication and turned it into a broadcast platform."[136]

This experimentation and ingenuity was taking place at college radio stations, as opposed to commercial stations or more professionalized public radio stations, precisely because of their smaller size and connections to educational institutions dedicated to cultivating learning and possibility. Without commercial mandates or large bureaucracies, student-run college radio stations have historically been able to be highly flexible, often adopting freeform programming formats and functioning as breeding grounds for new music to gain exposure. This reputation for being "raw and cutting edge" has meant that college stations have long been at the forefront of innovations in the radio industry.[137] In the case of KJHK, the radio station was attached to the school's journalism program, and putting the station on the internet was part of a commitment to "remain[ing] on the technical cutting edge of broadcasting," according to general manager and faculty advisor Gary Hawke. The simulcast also served as publicity for the journalism program and its students.[138] WXYC and WREK both operated as independent student organizations, and therefore were not bound by a particular department's mission. They only had to convince a handful of station management (mostly student volunteers) and advisory board members to green-light their plans. "We did more or less what we wanted," said WXYC's Shoffner, continuing, "I think this milieu made the innovation possible."[139] WREK'S Selbie pointed out, too, that conventional radio stations did not have technology staffs, and therefore would not have been in a position to develop internet simulcasting on their own.[140] On this point, Shoffner elaborated, "none of those groups"—not commercial radio stations or NPR public radio affiliates—"were in a position to see this confluence that we saw. Nobody mainstream even knew what the internet was at that time. But . . . I doubt they would have moved on it if they had seen it, for cultural reasons. Which is a general pattern you see in innovation."[141]

While the path dependency of established radio institutions made them risk adverse and slow to adopt the new technology, the culture at these academic institutions encouraged experimentation. In the case of WXYC, the SunSITE program existed precisely to spur on the type of unconventional, blue-skies research that fueled McConville and his colleagues.[142] "Our history at SunSITE [is] all about what is now called open-source software and open information. In particular, I had been an advocate of radical collaboration (with Tim Berners-Lee in the case of the WWW and Brewster Kahle in the case of WAIS) for almost a decade up to this particular event," explained Jones, referring to the World Wide Web and the Wide Area Information Server.[143] "At SunSITE, we were really being encouraged to experiment in all kinds of ways," McConville stated. "It really was an open-ended field dealing with early internet technologies. I doubt that we could have done it with a commercial station or a really well-funded station, because XYC by its very nature is very experimental, and I think they were willing to go with us and take some chances." McConville noted that Sun-SITE was also experimenting with early audio encoding around the same time, as well as putting some of the first record label websites and digital music archives online.[144] Likewise, at Georgia Tech, Selbie described the campus being "absolutely the right culture for this project. . . . Georgia Tech was very open to student innovations. The faculty and students there were really passionate about technology, tinkering, inventing, and doing things simply because 'it's cool.'"[145] To put it plainly, the multiple discovery of internet radio simulcasting was an instance of right place(s), right time: a combination of a few determined individuals (mostly students looking for ways to learn their craft) working with a technology (the internet) that was still relatively unproven and open, and in an environment that not only permitted but actively encouraged free creativity.

Notably, none of these students who worked on internet radio simulcasting pursued careers in radio after graduation. McConville, Shoffner, Selbie, and Burcham all went on to work in computer or information technology. Much as with IMS, these innovations in internet radio were happening from the internet side, not the radio side. Indeed, SunSITE likely would have launched WXYC's simulcast months earlier if it were not for the station management's anxiety over possible FCC violations. A number of WXYC staffers who were also law students were tasked with investigating any potential copyright law issues stemming from the internet rebroadcast; after some delay, they ultimately determined that they were in the clear since both WXYC and SunSITE were noncommercial entities and the station was already paying the legally required ASCAP/BMI performance royalties.[146] At Georgia Tech, Selbie claimed to have "never once thought about the FCC. . . . I felt it best not to ask questions—I had more of an 'act first, and ask for forgiveness later' approach."[147] Neither WXYC nor WREK received any contact from the FCC or the music recording and publishing representatives. It would be another few years before internet streaming royalties became a major legal issue, as discussed in chapter 5.

Despite the considerable innovations, these simulcasts were accessible by very small audiences of only one to two dozen people at a time. For instance, the KJHK simulcast could support twenty concurrent listeners (they hoped to eventually increase that number to sixty).[148] WREK hardcoded its system to limit the feed to between ten and twenty listeners at once.[149] If the simulcast was maxed out and a new listener signed on, the system would abruptly drop one of the existing listeners. The reason for these restrictions came down to network capacity. The audio was being unicast, not multicast, meaning it was transmitted in a point-to-point, one-to-one fashion; each individual listener required a dedicated stream rather than a single stream being delivered to multiple recipients, in true broadcast fashion. The number of listeners was therefore strictly limited by the amount of bandwidth available to the radio station. Even at the 64 kbit/s bitrate these stations were using—which was very low quality and about the lowest possible bitrate for streaming—twenty streams was an enormous amount of bandwidth in 1994. There was a good chance that any additional listeners could crash the network. WXYC actually claims not to have capped its simulcast audience; however, the students were concerned enough about damaging UNC's network that they ran special tests on the network load prior to launching.[150] Being college stations, the streaming audiences were not likely to ever exceed more than a few dozen students in the computer labs, and perhaps some nostalgic alumni or a DJ's doting parents. KJHK, for example, registered only two hundred listeners signing on to its reflector site during its first two months online.[151] In the spirit of both technical innovation and sociability, internet streaming was never about profit or marketing for these college stations: it was simply about exploring technology, expanding the reach of the station into the community, and connecting with listeners in new ways.

Still, as internet usage was expanding in the mid-1990s, there was a widespread feeling that much of the new content appearing online was frivolous or even dangerous.[152] As early as 1993, even before the dot-com boom was in full effect, critics were warning of an impending internet *infoglut*, a flood of unnecessary, valueless media content and analysis.[153] Many of the technologists who had been the longest users of the network argued that it should only be utilized for research and personal communication, such as email and data exchange. They felt that new uses like streaming radio bastardized the medium and, even worse, risked damaging its infrastructure. In response to an IETF listserv post about the WXYC simulcast by UNC's Paul Jones, software architect Brad Templeton (the founder of the first online newspaper, ClariNet, and years later, the chairman of the Electronic Frontier Foundation) lambasted the group behind the project. "I certainly hope this is NOT the start of a trend," Templeton wrote, contending that "multicasting is an interesting thing to play with . . . but the internet was meant to be point to point, and it would be a shame to see its bandwidth wasted by broadcasting. . . . Imagine if even a tiny fraction of the radio stations in the world tried to do this, or anybody who thought they would want to be a radio station. It would quickly swamp things."[154] He went on to argue that the broadcasting of radio and other traditional media content was best left to cable and satellite. WREK was also met with "lots of recrimination in the network community about what a flagrant waste of bandwidth this was."[155] At this stage, it was becoming clear that the internet *could* indeed be successfully used for broadcasting. However, there was not yet consensus that it *should* be used for these purposes. This was an existential moment for the internet, when the very question of what the internet was *for* remained uncertain and hotly contested.

Whatever the case, the severe listener capacity limitations meant that the simulcast technologies developed at WXYC, WREK, and KJHK were not ready for prime time. They were usable, but only just so—more prototypes than full-fledged inventions. Burcham openly admitted that "although functional, the current configuration of KJHK's [simulcast] is not very Internet friendly."[156] Moreover, none of the inventors actively sought to further develop their technologies for the marketplace. "It was clear that this was a whole new way of distributing live content of any sort, but we, or at least I, didn't have any plans to do anything else beyond what we did," admitted Shoffner.[157] They were working from the hacker ethic of curiosity and exploration; they were driven to build something simply to do it—to create something new and previously unknown—rather than to make money.[158] Burcham published a how-to guide for transmitting radio broadcasts online in July 1995, giving away his protocols to the rest of the software development community.[159] It must be noted, too, that they were mostly modifying existing technologies, namely the freeware CU-SeeMe—meaning that, for intellectual property reasons, they likely would not have been able to commercially market their software even if they tried. WREK's Selbie was the only one to write all his own code. Yet he ended up giving away his CyberRadio1 software for free. "Now anyone with the proper resources can start their own Internet radio station," he wrote in an October 1995 bulletin board message announcing the release, adding, "It is especially hoped that other computer hackers that work in college radio will attempt to use it and perhaps improve it."[160] Only about a dozen radio stations, including Stanford University's KZSU, are known to have put Selbie's software into use.[161] By that point, RealAudio had emerged; the college radio simulcasting technologies had a problem of scale that RealAudio and a number of other commercialized internet streaming media delivery systems were soon able to solve.

Conclusion

There are a number of reasons why most established radio stations and networks were hesitant about, or even outright hostile to, internet radio—not just in 1993–95 but throughout the 1990s. There was the issue of technology: existing radio institutions mostly stood on the sidelines waiting for an industry standard to emerge—RealAudio became that de facto standard, as will be discussed in chapter 2. Most importantly, the radio industry was not actively developing its own streaming audio technology. As Selbie pointed out, radio stations did not employ technology staff—definitely not internet technology staff, at least not in the mid-1990s. Another reason for this unresponsiveness on the part of the established radio companies may have been that the focus in the broadcasting industries in 1993–94 was on "interactive television." The widely held belief among broadcasters and marketers was that cable television set-top boxes, such as Time

Warner's planned Full Service Network, would be the point of access for the information superhighway, not the internet and online services.[162] The prediction was that the emergent digital radio, or *digital audio broadcasting* (DAB), would be delivered by satellite to these cable boxes.[163] In addition, the potential online audience was still relatively small, and thus there simply was not a huge incentive to invest in internet radio. In 1995, only 14 percent of adult Americans were using the internet, and a meager 3.5 percent, or roughly six million people, went online daily.[164] A mere 2 percent of all internet users possessed the high-speed internet connectivity that would make streaming online radio a realistic possibility. By the turn of the century, the number of Americans with internet access had increased to roughly half the adult population (46 percent), with three-quarters of them going online every day.[165] Still, that potential audience hardly touched terrestrial radio's nearly universal penetration rate—approximately 96 percent of American adults listened to the radio weekly in 2000.[166]

The audience to be gained via either original internet broadcasting or simulcasting was of negligible value to most radio stations in the mid-1990s. As much as the media were hyping the internet's ability to give stations a global reach, and programmers likewise bragged about their listenership in far-off places like South Africa and New Zealand, the fact of the matter is that the US commercial and public radio business model is based on local markets. It was unclear how listeners halfway around the world, or even the next state over, benefited a Dallas, Texas, station whose advertisers were mostly Dallas area merchants relying on reaching local customers who would walk into their doors to shop.[167] To that point, an internet simulcast could actually be seen as competition, if it drew listeners away from the broadcast feed where audiences were measured and advertising rates were set. It was not until 1999 that any form of online radio ratings were established, when Arbitron initiated its InfoStream webcast ratings service.[168] In other words, for an industry with a firmly established business model, there was little motivation to invest heavily in a new platform that had questionable commercial prospects and might even undermine its existing business. In comparison, internet start-ups like IMS and college radio stations like WXYC, WREK, and KJHK were nonprofit organizations, and mostly unburdened by the bureaucracy and conservative mindset that actively steered commercial radio away from innovation. As part of a larger technological or academic community, these online and college radio entities were sandboxes, not just for future broadcasters but also for budding technologists experimenting with the internet.

The story of internet radio's development in the late 1990s continues in chapter 2. Suffice it to say, few of the individuals and institutions profiled in here in chapter 1 remained involved in the field much past these initial few years. The Internet Multicasting Service ended in 1996, right on the cusp of internet radio's upswing. Its wonkish programming and nonprofit model did not fit well among the capitalist fervor of the dot-com boom. The college radio stations survived, carried on by new student volunteers: they kept on streaming, although they faced a number of challenges in the years to come, including copyright issues and the imposition of increased music royalty fees following the passage of the Digital Millennium Copyright Act (DMCA) in 1998.

In the introduction to this chapter, I laid out a few guiding questions, including asking who first decided to bring radio to the internet, and why. In both of these case studies, the individuals involved were keenly aware of the fact that internet radio, and the internet more broadly, had yet to be clearly defined as a technology or a cultural form. They sought to test out what internet radio could potentially be, and in doing so imposed their personal vision on the format as it solidified into a new medium. For instance, Malamud's combined efforts with the Internet Multicasting Service—the internet radio programming plus nonradio projects like the Global Schoolhouse and the SEC EDGAR database—were intentionally designed to demonstrate the potential of the internet as a general purpose communications system that could be made to do whatever people wanted. "Our goal is to see how much data we can put on the net at once," he said, adding, "The answer is: a lot."[169] Unlike how most mass media had come to be used for a single purpose—radio was used almost exclusively to transmit voice and music, and mostly professionally produced programming at that—the internet could carry almost any conceivable form of media and information. Moreover, through the open standards that Malamud advocated (along with Paul Jones, John Selbie, and the other college radio streamers), it was the audience who was in control. Any hacker could come up with an innovative new way to use the internet.

Internet radio's earliest pioneers were not especially committed to furthering radio as an art form or an industry, however. Radio was a means to an end. Malamud astutely realized that if the internet was going to grow into a mass medium, there needed to be original, high-quality content to attract audiences. Drawing an analogy to television, he pointed out that no one would have bought television sets unless there was programming to receive.[170] This was also very much true of early US network-era radio in the 1920s and 1930s, where programming was the carrot-on-the-stick used to get people to buy radio sets or consume paid advertising. In other words, IMS's internet radio programming was created first and foremost as an incentive to lure people online and demonstrate the value of the medium. Notably, Malamud had his particular ideas about how internet radio could and should be used to fulfill the utopian promise of a global village. Yet he was not dogmatic with his opinions; he welcomed entertainment as well as commercial uses of the technology. "This method of communicating with people is going to be major media," Malamud proclaimed in 1995. "Eventually we're going to have audiences of millions of people."[171] Yet he would not be the individual to take internet radio to the masses: RealAudio's Rob Glaser and AudioNet's Mark Cuban and Todd Wagner, profiled in chapter 2, would be the ones to do that. Nevertheless, reflecting on IMS in 2003, Malamud said, "We started a public radio station. We always said others should start commercial stations. Cuban and Glaser both did good work."[172] This neoliberal vision of internet radio as a largely unregulated space for private and pseudo-public cultural production remains intact to this day.

The fact that the *sound* of early internet radio was largely a facsimile of over-the-air radio broadcasting points to it being an extension of traditional radio, rather than a distinct medium. Internet radio's initial tinkerers may not have been outsiders to broadcasting, but when it comes to the structure and aesthetics of the programming they produced—the style, form, genres, and so on—they conformed to radio's preexisting norms and conventions. The college radio stations, of course, were straight simulcasting their broadcast signals; none of the stations profiled here produced original web content, at least not early on. Malamud and the IMS calculatedly copied NPR news and talk radio principles and techniques. And as IMS grew, it increasingly relied on third-party radio producers for additional content, in particular US and foreign public radio services. This was primarily programming made both for and by white, highly educated, upper-middle-class social groups (not to mention English-speaking, North American, and northern European). In other words, the technology and distribution models may have been innovative, but the content was more of the same. As a platform for cultural diversity, creative artistry, and sonic exploration, the internet remained an untapped resource for many years to come.

CHAPTER TWO

Radio Dot-Com

Internet Radio Goes Mainstream

When Mark Cuban went to bed on March 31, 1999, he was a freshly minted billionaire. His business partner Todd Wagner's net worth had also skyrocketed to just shy of a billion dollars over the course of that Wednesday. More than 300 of their 330 employees were suddenly millionaires, too—at least on paper.[1] The pair had signed an agreement to sell their four-year-old company Broadcast.com to web portal Yahoo!—then the second most trafficked website in the United States, second only to America Online (AOL)—in a stock swap deal worth $5.7 billion.[2] It was one of the most expensive transactions of the dot-com era, and Broadcast.com remains Yahoo!'s costliest acquisition of all time.[3] Even before the Yahoo! deal, Broadcast.com had firmly established itself as one of the hottest companies of what has come to be known as the internet dot-com boom: the period from roughly 1993 to 2000 during which the newly developed World Wide Web grew the internet into a commercial mass medium. (The latter few years of that period, from 1997 to 2000, represent the speculative economic boom-and-bust cycle of the dot-com bubble.) Back in July 1998, during its initial public offering (IPO), the firm's shares soared to nearly 250 percent above the offer price, resulting in a company valuation north of $1 billion—despite the fact that Broadcast.com was unprofitable and millions of dollars in debt at the time.[4] As the *New York Times* announced, the "frenzy" over Broadcast.com stock resulted in "the best opening day gain of any company in Wall Street history."[5]

What was Broadcast.com, and why was it a multi-billion-dollar company? By its own description, it was "the leading aggregator and broad-caster of streaming media programming on the web."[6] At the time of its sale to Yahoo! (when its name changed to Yahoo! Broadcast), the company's website featured content from 420 radio stations and 56 television stations, plus game broadcasts for more than 450 college and professional sports teams. It also provided coverage of a wide range of events such as politi-cal speeches, business conferences, and concerts.[7] In addition to its live webcasts, Broadcast.com offered more than 65,000 hours of on-demand content, including hundreds of audiobooks and nearly 2,500 full-length music albums in its "CD Jukebox."[8] And all this media content was avail-able to online audiences free of charge. Importantly, while Broadcast.com presented both audio and video content, the audio far outweighed the video. And the bulk of that audio consisted of simulcasts from terrestrial radio stations. The site had only commenced video programming in 1997, and livestreaming video was still "not ready for prime time" in 1999; due to a lack of widespread high-speed internet access and limited server tech-nology, the image quality was choppy and videos were mostly restricted to very short clips.[9] Indeed, as a sign of the significance of radio and other audio content to Broadcast.com's business, up until May 1998 the company was known under its original name: AudioNet.[10]

Unlike video, streaming audio on the web had been a viable technology for the better part of a half-decade when the new millennium hit. Most of that "audio" content consisted specifically of radio: some from new internet-only streamers native to the web, and a much larger segment of it arriving via *simulcasts* of traditional radio broadcasts. By the year 2000, an estimated three thousand radio stations were webcasting worldwide, with more than 90 percent of those so-called webcasters being established ter-restrial radio institutions simulcasting their signals. And that number was growing by at least a hundred stations each month.[11] Put simply, the bulk of internet radio in the 1990s consisted of recycled broadcast radio content. This "gold rush" of internet radio stations, as one industry commentator put it, meant that aggregator sites, internet radio directories, and fledgling online radio "networks" were plentiful.[12] AudioNet was hardly alone in its efforts to create a web radio portal that brought together a large num-ber of streams. It was joined during the 1990s by websites like NetRadio, Real Broadcast Network, Netcast, SonicNet, Lycos Radio, Spinner, and Live365, to name just a few. And in a burgeoning internet industry that highly valued *traffic* (or the number of visitors to a website and the number of pages they visited) along with interactive, multimedia web content (or *rich media*, in the parlance of the day), these internet radio websites became hot commodities for venture capitalists, stock market investors, and even legacy media conglomerates.

Following the small-scale, mostly DIY (do-it-yourself) efforts of the upstarts operating from 1993 to 1995 (chapter 1), this chapter traces internet radio's explosion into a mass media practice during the late 1990s. This diffusion from 1996 to 2000 required developments across technology, business, and content, and the case studies in this chapter provide a snapshot of the advancements in each of these areas. In terms of technology, RealAudio offered a standardized streaming audio software format, the convenience and affordability of which accelerated the spread of online radio in the 1990s. In terms of business, AudioNet introduced the content aggregation model to internet media distribution, its web portal creating a convenient access point for radio online and establishing the commercial potential for webcasting. And in terms of content, Pseudo created a slate of original online radio programming that brought fresh voices and perspectives to the radio medium, while its interactive "ChatRadio" format was a harbinger of the participatory culture of Web 2.0 and the modern social media logic. At the same time, these events were not without their drawbacks, including that much of the energy and enthusiasm for internet radio was redirected into web video at the start of new millennium, slowing internet radio's cultural momentum at the start of the 2000s.

Confluences of Sound: RealAudio

The simulcasting technology that chapter 1's small college radio stations developed worked for their specific purposes. Streaming audio could not diffuse into the broader media industries until a simpler, more uniform product came along, however; something that any radio station or listener could easily implement. Major media outlets were not going to invest their time and money in jerry-built software tools. But as early as 1994–95, a number of computer software companies were working on more refined, marketable systems. Next to web browsers, streaming technologies for the continuous delivery of media content became the internet's "killer apps" during the second half of the 1990s, with so many firms vying to create a superior product that talk of "streaming wars" pervaded the business and technology press year after year. Launching in April 1995, RealAudio, developed by Progressive Networks (later known as RealNetworks), was the first company to release its streaming audio software, and it maintained a significant market advantage throughout the late 1990s. Some estimates indicate that in the year 2000, 85 percent of all available online streaming content was in Real formats (identifiable by the .ra and .rm file extensions).[13] Still, it was challenged by a number of rival technologies, including StreamWorks from Xing Technology, iWave (Internet Wave) from VocalTec, VivoActive from Vivo Software (acquired by RealNetworks in 1997), Shockwave from Macromedia, Audioactive and QuickTime from Apple, and NetShow and MSAudio from Microsoft (later incorporated into Windows Media Audio). By the end of the millennium, there was no shortage of streaming audio options. This technological abundance gave way to a flood of simulcasts from established broadcasters, along with a wave of upstart internet-only radio stations and networks.

All these streaming audio developers were technology companies with no previous involvement in broadcasting, including Progressive Networks. The roots of Progressive and RealAudio lay in Microsoft, which was the dominant personal computer hardware and software firm in 1995 and would become Real's bitter rival in the streaming wars.[14] Progressive was the brainchild of Rob Glaser, a Yale graduate who joined Microsoft in 1983 and rose to become its vice president of multimedia systems before the age of thirty—at that time, the youngest-ever VP in Microsoft history. Glaser then promptly stepped away from Microsoft in 1993. It was clear that "Microsoft was fundamentally about a computer at every desk in every home," Glaser said about his decision to leave, "and I was much more interested in the intersection between computing, communication, and media."[15] He quickly founded Progressive Networks to fulfill these goals, albeit absent a clear business plan or product.

Glaser knew only that he was "determined to use technology to pursue a progressive agenda," that mission inspiring the company's original name.[16] The son of 1960s political activists, including a mother who was a New York City social worker, Glaser wrote a leftist political column for the *Yale Daily News* and organized the campus' draft resistance organization while attending the university during the late 1970s and early 1980s. Although he never worked professionally in the media industries, in addition to writing for his college newspaper, Glaser helped create a "radio station" at his Bronx high school using the building's intercom system.[17] Already a prominent philanthropist in his late twenties and early thirties, he joined the boards of a number of politically liberal nonprofit media organizations in the time between leaving Microsoft and launching RealAudio, including the digital rights group the Electronic Frontier Foundation and the Foundation for National Progress, the publisher of *Mother Jones* magazine. In 1993, working from a foothold in computing, Glaser was looking to bridge the worlds of information technology, media, and politics—but he still did not have an enterprise for Progressive Networks to undertake.

In trying to find a way to marry his technological and political interests, Glaser initially hit upon the idea of creating "a cable [television] channel focused on politics and culture."[18] Hence the name Progressive Networks, which sounds more like a media conglomerate than a computer software company. This was in mid to late 1993, when cable-delivered interactive television was the subject of much investment and hype in the US media industries, and initially Glaser planned to take that route with his embryonic broadcast network. Despite having built multimedia computer systems for Microsoft, these had not been network-based programs, and he was not all that well acquainted with developments in internet technology. By chance, Glaser was shown an alpha version of the web browser Mosaic before its official release in late 1993. It was a "total epiphany," Glaser stated.[19] From that point on, he was convinced that the World Wide Web would beat out cable as the preferred method of delivery for multimedia content.

Yet Mosaic and other early graphical web browsers only supported text, images, and hyperlinks and could not easily display audio-video content. "I thought, well, how do we do for audio and video what the web browser is doing for text?" recalled Glaser.[20] The idea was still to create a broadcast network that was "a version of a Fox or a Disney that would be more politically progressive and more culturally adventurous," according to Progressive's cofounder, David Halperin (who left the company early on and ended up a national security aide and speechwriter for President Clinton).[21] To achieve this through the web, they would need to develop a plug-in, an additional piece of software that could overcome the limitations of HTML to process multimedia audio-video content on a web page. It quickly became clear that investors were more interested in funding that software than an internet-based far-left political television network, and Glaser and company shifted from creating progressive content to creating the tools that others could potentially use to distribute their own innovative content—or at least, that was the rationale.[22] (Glaser would continue to funnel money into leftist media ventures, including being the largest financial backer—to the tune of $10 million—of the failed liberal talk radio network Air America Radio in the mid-2000s.)[23] Along the way, Progressive also realized that the network bandwidth capacity was not yet available to support streaming video on a large scale, and so it settled on streaming audio.

RealAudio officially launched on April 10, 1995. Symbolic of how the technology was positioned to integrate into the established radio industry, the launch announcement took place at the National Association of Broadcasters (NAB) annual trade show in Las Vegas.[24] There Glaser was joined on the NAB Show stage by National Public Radio president and CEO Del Lewis and Capital Cities / ABC vice president Katherine Dillon. Progressive had lined up NPR and ABC News as programming partners, astutely realizing that in order to get a mass of web users to download its audio players it had to provide content to lure them in.[25] Moreover, that content needed to possess relatively broad appeal—unlike the college stations, which, even if its streaming audio technology could scale, produced content that was only ever going to appeal to niche audience segments. NPR initially licensed content to Progressive from its flagship news program *All Things Considered*, as well as the popular *Morning Edition* and *Weekend Edition*. The RealAudio webcasts only featured edited clips of each show, not the entire live radio broadcasts. Meanwhile, ABC News produced special bulletin-style daily news segments, including a four-minute national news webcast, a minute each of entertainment and sports news, and a one-minute report called *Peter Jennings' Journal* (the latter of which was a commentary segment from the network anchorman that also played over the air each weekday on ABC Information Radio Network affiliates). Other traditional radio programming available at launch were Voice of America broadcasts, a talk radio show about cyberspace called *RadioNet*, and a selection of old-time radio programs, including the Groucho Marx quiz show *You Bet Your Life*, presented under the banner of "Radio Yesteryear."[26] Combined with Progressive's first-mover advantage, these strategic agreements with content producers were crucial to RealAudio's success, as the popular programming helped secure a sizable user base—over six hundred thousand RealAudio Players were downloaded in the first six months—that, in turn, demonstrated to other major broadcasters that Progressive's system was worth investing in.[27]

The RealAudio streaming audio delivery system presented a number of important technological breakthroughs, in addition to some shrewd business innovations. Although it was not the first streaming audio tool, it was remarkable for compressing the audio data stream, making internet radio available to a much wider audience than it ever had been previously. Listeners no longer had to wait hours for an audio file to download, nor did they need a state-of-the-art high-speed network connection for streaming. Within a few seconds of clicking play, the audio would start and stream continuously while the file continued to download. The RealAudio compression allowed for dependable streaming over a dial-up internet connection with a 28.8-Kbps or even 14.4-Kbps modem, which was the standard setup on most newer personal home computers in 1995.[28] In addition, since its player was either a plug-in or helper application (depending on the browser being used) and it worked with various web browsers, anyone with an internet-connected computer could listen in. It did not require access to special networks like the MBone, as was the case with many of the Internet Multicasting Service and IETF internet broadcasts, nor did it hog lots of bandwidth like the college radio simulcasts, since they utilized uncompressed audio.

Furthermore, RealAudio was a fully integrated system that combined an encoder, server, and player—a simple configuration that resulted in a seamless delivery from the broadcaster to the listener.[29] Like all streaming audio products, it used a codec (coder/decoder) that compressed the audio stream at the producer's end (the encoder)—shrinking the audio to match low-bandwidth internet connections—and then decompressed and played the stream at the listener's end (the player). The RealAudio Studio (renamed the RealAudio Encoder) and RealAudio Server were software products for audio producers. The encoder enabled producers to convert their audio into Progressive's standardized, proprietary RealAudio format. The server then distributed the files to listeners over the internet; it delivered the actual streams, in effect functioning like a transmitter in AM/FM radio. The player was the final component in the system; it was the client-based software listeners downloaded in order to play the streamed audio online (and Progressive made arrangements for subsequent versions to come preloaded in several web browsers, making playback even more user-friendly). For listeners, the player was uncomplicated and featured a timer, volume control, the ability to pause and resume clips as well as drag the status bar forward or backward (primitive fast-forward/rewind). It also displayed various other connectivity statistics and audio file information. For producers, the encoder-server combination was full-featured and easy to configure, plus most of its functions were fully automated. Therefore, once a radio station set up its server and encoder (a process that only took a couple hours), any mildly computer savvy staffer could operate the streams.[30] This is one of the reasons why the RealAudio system became so widely adopted, so quickly: it did not necessitate a dedicated technology staff. For enterprise software of the era, it was about as close as a broadcaster could get to plug-and-play technology.

Progressive's biggest business coup was that it opted for a *freeware* model, giving the player and encoder software away at no cost and only charging internet broadcasters for the server software.[31] This move enabled RealAudio to quickly capture the greatest market share. Especially in its earliest days, when there were few viable streaming audio alternatives, there was little reason for web users *not* to download the RealAudio Player, as it was the easiest way to experience the novelty of internet audio. People tend to be resistant to downloading multiple software applications for a single function like streaming audio, and RealAudio's early foothold gave it a significant advantage well into the future—despite many users having a love-hate relationship with the player.[32] (Influential *New York Times* technology columnist David Pogue summed up end users' frustrations over RealAudio when he called it "the world's most mercenary, obnoxious, and relentlessly tacky software.")[33] RealAudio Player downloads went from approximately six hundred in October 1995 to over four million in April 1996, twenty million in August 1997, and over ninety-five million by the end of the decade.[34]

Market domination required buy-in from both listeners and producers, and giving away the encoder lowered the risk for radio makers to adopt RealAudio. They only had to pay for the server, which was adjustably priced based on the size of a station's online audience. Circa 1996, base prices for the RealAudio Server ranged from $1,500 to over $10,000. For instance, a server supporting ten simultaneous streams cost $2,500. A server supporting one hundred simultaneous streams cost $14,000, including technical support and upgrades for a year, which were additional fee services.[35] These costs were not inconsequential, though they were economical compared to what it would cost a radio station to develop and maintain its own server system with custom-built software and hardware and high-skilled engineers on staff. As Microsoft, VocalTec, and others attempted to capture streaming audio market share away from RealAudio during the late 1990s, they offered their servers for free, in addition to the encoders and players.[36] In order to stay competitive, Progressive made some of its smaller servers available for free, recouping profits by selling premium versions of its player. While this business model helped make RealAudio (and later RealPlayer) nearly ubiquitous for many years, the fact that it was a proprietary format meant it was not compatible with other players. Real streams could only be played on a RealPlayer. (This was also true of other internet radio software like Windows Media Player.) By the early 2000s, many producers and users grew tired of these restrictions. RealAudio was passed over in favor of free and open-source formats with interoperability, specifically the MP3 audio standard.

The original RealAudio (Version 1.0) that was introduced in April 1995 arrived with some other significant limitations. Most notably, it only allowed for on-demand streaming, not livestreaming. Playback was instant, but it could not actually support the simulcasting of radio broadcasts in real time. This is one of the reasons why the initial RealAudio website only offered short segments from NPR and ABC News, as opposed to full-length programs. The digital audio files would be uploaded on the same day as the original radio broadcast, but they needed to be edited and reformatted for the web and thus appeared after some delay. The other major factor influencing the decision to use abridged clips was that shorter audio files were smaller in size, and thus streamed more easily without disruption on computers with slower modems and lower-speed network connections. Like the Internet Multicasting Service, RealAudio 1.0 only provided AM radio-quality sound—or worse. *The Guardian* described it as sounding "like AM with a lot of static."[37] Indeed, RealAudio's compression added extra distortion, meaning it was actually lower quality than some of the internet audio formats that preceded it. RealAudio playback, especially with larger files containing more data, was known to be "flaky, jerky, and low-fi," meaning the stream would frequently stutter and break up.[38] Even a few couple-second-long buffering pauses could make a radio program essentially unlistenable. For these reasons, the first two versions of RealAudio only adequately supported talk radio programming; more data-intensive programming like music was unfeasible. Still, many of these issues were quickly resolved; Progressive released a number of consequential updates between 1995 and 1997. With RealAudio Version 3.0 in September 1996 it became a truly viable option for music content, its sound quality on par with FM stereo radio.[39]

The streaming internet radio era began in proper in November 1995 when Progressive launched RealAudio Version 2.0, which added support for live "netcasting."[40] This meant that terrestrial radio stations could now simulcast broadcast feeds in real time, as opposed to posting previously aired program segments for time-shifted on-demand streaming, as had been the case with RealAudio 1.0. The player also became more ubiquitous by expanding from only computers running Windows operating systems to supporting Apple Macintosh and UNIX systems as well. It also bumped up the audio fidelity from AM radio-quality to FM radio mono-quality. (It was still best suited for talk radio content and not music, but a significant improvement nonetheless.)[41] Progressive debuted RealAudio's live webcasting capabilities with a broadcast of Bill Gates's opening remarks at the August 24 launch event for the much-hyped Windows 95 operating system, though the feed was only shared with a specially invited group of company friends and business partners. That demonstration was followed by a much more heavily publicized event: a broadcast of a Seattle Mariners–New York Yankees Major League Baseball game on September 5, 1995. (Glaser was a part owner of the Mariners.) The livestream went out through the ESPN SportsZone website, which itself soon emerged as a major sports webcaster.[42] ABC quickly followed with the launch of a dedicated internet news radio network using the livestreaming RealAudio technology, called ABC RadioNet.[43]

Hundreds more US broadcast radio stations and media conglomerates deployed RealAudio from 1996 to 2000, using it to deliver internet simulcasts or customized internet-only radio content. The rechristened RealNetworks corporation also led the way into streaming video in 1997 with the introduction of RealVideo.[44] Glaser never entirely gave up on the idea of being a media producer, either, creating the Real Broadcast Network (aka RealNetwork) in August 1997: a partnership with MCI that aggregated audio and video content from ABC News, ESPN SportsZone, Home & Garden Television (HGTV), Atlantic Records, and others.[45] Yet in this endeavor to form a wide-spanning internet broadcast network, RealNetworks had already been beat out by a number of players, not least of all AudioNet.

AudioNet and the Rise of Content Aggregation

Promoting itself as "*The* Broadcast Network on the Internet," AudioNet's business model revolved primarily around the redistribution of existing radio content for online audiences. (It also offered some live events, music, audiobooks, and eventually, video content.) Like so many technology companies from the late 1970s on, the origins of AudioNet began with two guys in a garage—or a spare bedroom in Dallas, to be exact. The story goes like this: a pair of sports fanatic Indiana University alumni living in Texas, Mark Cuban and Todd Wagner, want to listen to the basketball games of their beloved Hoosiers.[46] This is 1994, before the emergence of cable/satellite television regional sports networks, and thus there is no easy way to receive the local Indiana game broadcasts from 870 miles away. Cuban was such an avid Hoosiers fan that he claimed to call friends back in Bloomington and have them put the phone up to the radio so he could listen to games. The pair had some familiarity with the internet and together hit upon the idea of using the web to stream the sportscasts. Cuban, in particular, had worked in computer hardware and software resale/distribution since the 1980s, selling his company MicroSolutions to CompuServe for $6 million in 1990. He then spent the next few years increasing his personal fortune by day-trading computer technology stocks under the guise of a start-up named Radical Computing.[47] Cuban and Wagner were looking for a new business venture together circa 1994—Wagner, a corporate attorney, wanted to switch careers, while Cuban was eager to invest in something at the cutting edge of the emerging web technology. Assuming they were not the only homesick sports fans who wanted to hear games from far away, they went about developing a business plan for an internet broadcast network that would syndicate existing radio broadcasts from stations around the country, akin to a cable or satellite television service provider except for radio on the internet.

The part of the AudioNet story that always gets left out is that the basis for this idea of a web portal licensing the rights to rebroadcast radio on the internet came from another Dallas entrepreneur, Chris Jaeb. Years prior, Jaeb started Cameron Audio Networks, which centered around hand-held radios that would receive broadcasts inside sports venues, allowing fans to hear the play-by-play of the game they were attending. He eventually switched the distribution platform from hand-held radios to the web sometime in 1993–94, and although he had secured some rebroadcasting rights, he did not have the internet streaming technology in place. Jaeb consulted with Wagner, who then brought in Cuban as an investor. The pair took control of the company and bought out Jaeb in July 1995, though he continued to work for AudioNet for a few years, acquiring broadcast rights agreements.[48]

To establish a proof of concept for the AudioNet internet broadcasting scheme, Cuban and Wagner convinced a local Dallas news and talk radio station, KLIF 570 AM, to let them rebroadcast its programming. AudioNet piggybacked on the RealAudio streaming technology to build its business; it was among the first companies to purchase a RealAudio Server.[49] The rebroadcasts of KLIF started around July 1995, prior to the availability of RealAudio 2.0. As a result, it was not able to simulcast the radio station's signal in real time. Rather, Cuban and Wagner taped the live broadcasts off the radio and onto cassette, digitized the recordings, and posted them on the AudioNet website later.[50] Its first use of livestreaming radio took place on the evening of September 2, 1995, when Cuban, Wagner, and Jaeb employed an alpha version of the RealAudio 2.0 real-time encoder and server to simulcast KLIF's broadcast of a college football game between Southern Methodist University (SMU) and the University of Arkansas. AudioNet actually scooped Progressive's own public debut of the livestreaming technology with the Mariners-Yankees baseball game by a few days.[51]

Though internet radio was beginning to wade into the commercial mainstream, it was still makeshift technologically. The AudioNet setup in Cuban's house relied on about $5,000 worth of the latest sophisticated computing and networking equipment: a $3,000 Packard Bell personal computer, the RealAudio server, a high-speed ISDN line. Yet it was a $15 radio tuner plugged into the PC's sound card that actually enabled them to get the KLIF broadcast from analog to digital and out over the web.[52] For the bleeding edge of internet technology, this coupling of relatively low-tech sound equipment and high-tech computing equipment was something of a recurring trend in early streaming audio: the Internet Multicasting Service and Malamud's Denon CD Jukebox, SunSITE/WXYC and Shoffner's sister's yard sale boombox, and AudioNet's budget radio tuner. Wagner once tripped over the tuner on the way to the bathroom, disconnecting the KLIF feed and knocking AudioNet's entire operation offline. This makeshift technical setup continued for a number of months, even as AudioNet added dozens more radio stations to its "internet broadcast network," as they called it. Many of the simulcast feeds consisted of the radio station simply putting a phone next to a radio, and AudioNet digitizing that signal at the other end in real time and streaming it over the internet.[53]

Although AudioNet publicized KLIF as "the first radio station in the world" to continuously simulcast its over-the-air radio transmission on the internet, it was beaten not only by the numerous college stations but also a few other commercial radio stations. The first commercial station to internet simulcast was KPIG 107.5 FM in Santa Cruz, California, which began simulcasting on August 5, 1995, using Xing's StreamWorks.[54] Nevertheless, many radio stations were not willing or able to put simulcasts online in the mid to late 1990s, even with the assistance of a full-service streaming audio software company like Progressive/RealNetworks or Xing. More stations opted instead to partner with an aggregator like AudioNet. Of the 1,100 licensed radio stations in the United States that were broadcasting audio of any kind on the internet in 1999, about 420 of those—close to 40 percent—were carried through AudioNet (renamed Broadcast.com by that point).[55]

The reasons why a radio station might have chosen to opt with Audio-

Net are multiple. One is that building and running a professional website was expensive and highly involved. While setting up and maintaining the RealAudio encoder and server were relatively simple processes, these components were only software. They still needed to be employed in conjunction with a web server and various other hardware elements. (College radio stations had a distinct advantage here, as their streaming audio systems were built on top of robust existing university computer networks.) Server expenses were based on bandwidth usage and storage space, and hosting streaming audio required significant amounts of both. Plus, these requirements increase exponentially: the more listeners, the more bandwidth; the more programs and archives, the more storage space. Leasing bandwidth could cost as much as $50 per month per stream, meaning that providing support for even ten simultaneous internet radio listeners could cost a station $500 per month.[56] AudioNet took care of most of this networking infrastructure and its costs, pulling radio stations' broadcasts in via satellite feeds and other means, then hosting the stations' streams on its servers and providing quality control and technical support.[57]

Moreover, even a well-equipped radio station might be able to support only a few hundred simultaneous internet radio listeners. For any station wanting or needing to support larger audiences, multiple servers and top-tier internet access were required.[58] Although it started with an ad hoc configuration in Cuban's spare bedroom, AudioNet quickly developed an ultramodern broadcast center in Dallas that, by 1998, included twenty-five satellite receiving dishes, over one thousand multimedia streaming servers, and a staff of dozens of technicians. It also had direct connections to major internet backbone providers (e.g., UUNET, MCI, Sprint, GTEI, AT&T) and even ran its own private data network, which jointly bypassed congestion points on the public internet. In its first nine months, AudioNet could only support a maximum of about twelve thousand concurrent listeners. Six months later that capacity was up to twenty thousand. By 1997 it was at a half-million and continuously growing.[59] With the setup it had in 1998, AudioNet could support 650-plus simultaneous live broadcasts, and the website attracted around five hundred thousand daily users.[60] During the late 1990s, AudioNet regularly ranked among the top twenty most trafficked websites in the world.[61] In sum, AudioNet could offer established radio stations turnkey implementation of internet radio at a scale fitting of their over-the-air stature.

AudioNet operated as a *content aggregator*, meaning it pulled media content from various sources and made it accessible at one dedicated, easy-to-find location. As an internet broadcast network, AudioNet was first and foremost a service intermediary that connected established broadcasters to the internet. Apart from assisting with the coverage of specific live events like the Super Bowl, AudioNet did not produce any original programming—and its involvement in these productions was mostly in a technological capacity.[62] It also did not develop any proprietary hardware or software; rather it relied on third-party applications like RealAudio, as well as other preexisting server networking and satellite delivery technologies. Yet AudioNet's configuration of these existing technologies was not easily reproducible and required a significant amount of skill and ingenuity, as well as immense financial investment.[63] AudioNet's primary business advantage was that Cuban and Wagner were willing to take risks by investing heavily in the concept of webcasting, which in 1995 was a completely unproven business model within the media industries. Furthermore, the duo excelled at branding and sales, getting high-profile partners to sign up for their services: CNN, BBC, CNBC, Arista Records, Intel, CNET, the National Football League, the National Hockey League, Major League Baseball, NCAA March Madness. Wagner was especially adept at getting radio stations and other broadcasting rights holders to sign exclusive multiyear licensing deals with AudioNet early on, when few people realized the potential value of these rights.[64] "We licensed a lot of content for multiple years because nobody else cared about it," explained Wagner. "I mean, we were creating something that didn't even exist. Internet broadcast rights— what were those?"[65] To the point, AudioNet rarely paid a dime for these internet broadcasting rights; rather it traded its services.

In exchange for distributing a radio station's content on its website, along with all the services and support involved in the rebroadcasting process, AudioNet received free airtime for ads during the simulcast.[66] It generally received twenty to thirty sixty-second spots per week, and either resold that time to other advertisers or used it to self-promote AudioNet programming and services.[67] These included prestream ads as well as inserted commercials that would replace some of the ads in the over-the-air broadcast—or what in the contemporary podcasting industry would be called preroll and midroll ads.[68] Put another way, AudioNet did not pay the producers anything to carry the content, and potentially even profited on it. The company also placed banner ads and other graphical display ads on its web pages, which listeners would then see while the radio content played. It was willing to trade its costly services for these broadcasting rights because of the web traffic it might generate. In a few cases, such as with Major League Baseball, AudioNet even provided additional services to the rights holder in order to acquire especially popular programming. For instance, it created special MLB websites with exclusive content for select teams.[69]

All audio content was made available to web audiences for free. In this way, AudioNet was essentially using a radio/television broadcast model of giving programming away to audiences and then selling those audiences to advertisers. While popular stations were hooked into the network for free, some smaller producers, such as the internet-only music radio program *Ethiopian Online Radio*, had to pay AudioNet to lease server space on a per-listener basis.[70] Other larger businesses paid to broadcast web-exclusive content, such as when Playboy presented an audiocast "party" to promote its film *Girls of the Internet*.[71] An additional revenue source came from business services: it was contracted by corporations and organizations like the Cato Institute to webcast earnings conference calls, stockholder meetings, press conferences, trade shows, training sessions, seminars, and the like. These special webcasts were sometimes restricted to company intranets, while other times they were made public for anyone to hear. The site was full of live audio from events that typically would not be heard on traditional radio: the annual Microsoft CEO Summit, the Macworld Expo computer trade show, stock forecasting announcements by Price Waterhouse LLP, and conference keynote addresses from AT&T and Intel executives. In fact, the Business Services Group was AudioNet's most profitable division, accounting for up to 70 percent of the company's revenue.[72] After the 1999 sale to Yahoo!, the renamed Yahoo! Broadcast streaming media portal struggled and was mostly shut down in 2002. However, Yahoo! retained the business services components for years to come.[73]

AudioNet's business strategy was simply to aggregate as much audio content as possible, based on the belief that more content would bring in bigger audiences that could then be translated into more and more profitable ads. AudioNet was indiscriminate when it came to what radio programs and stations it carried: "Content's content's content," Cuban deadpanned. "The Broadcast.com law says the more content you have, the more valuable all the content gets, because you become the destination."[74] As noted in this chapter's introduction, by the time of the 1999 sale to Yahoo!, AudioNet/Broadcast.com was simulcasting 420 radio stations and game-day broadcasts for more than 450 college and professional sports teams. These partners ran the gamut from popular commercial news and talk radio stations like WOR and WCBS in New York City to the Bay

Area sports radio station KNBR to Top 40 music radio stations like KRBE in Houston. Public broadcasters like C-SPAN and the BBC World Service provided AudioNet with content, as did US government agencies like NASA. CNN presented a special web channel called Audioselect. Some of the most popular nationally syndicated talk radio shows used the portal, including *The Rush Limbaugh Show*, *The Dr. Laura Program*, and *Coast to Coast AM with Art Bell*. Nevertheless, AudioNet was indiscriminate, and it also hosted many small college and community radio stations, including San Francisco's KUSF and the cult favorite New York area freeform station WFMU. It even included some early internet-only radio stations and programs, such as the aforementioned *Ethiopian Online Radio* based out of Washington, D.C., plus a year-round Christmas music channel and a station called dAISYrADIO run by the alternative rock band Tripping Daisy. In addition to the various conferences and live events, AudioNet webcasted police scanner frequencies from New York City, Los Angeles, and Dallas, plus the communications from the Dallas / Fort Worth International Airport air traffic control.

Despite this wide assortment of content, sports and news/information made up the core of AudioNet's network. This was in part due to the portal's largest audience, consisting of daytime office workers. Cuban astutely realized that the largest audience segment with regular high-speed internet access in the late 1990s was white-collar office workers.[75] They were heavy consumers of sports and news programming, yet they did not typically have radios or televisions at their desks. Radio reception was also spotty in urban high-rise office buildings. Yet this was a highly desirable advertising demographic, making it worth the effort to reach them. Thus, AudioNet targeted the in-office audience, who would listen during the workday to financial news reports, stream music in the background, replay last night's game, and so on. The second major target audience remained the sports diaspora, like Cuban and Wagner themselves, who wanted to listen to their favorite hometown teams from far away.

Another reason radio stations might have preferred an aggregator like AudioNet was due to the industry discourse about how people accessed the internet and how businesses could best take advantage of the web. The late 1990s was the era of the *web portal*: specialized websites that brought together diverse media and information sources into one place.[76] The idea was that users needed a home base from which to navigate the web—and to this point, many portals were also search engines, such as AOL, Yahoo!, Lycos, and Excite. These were de facto gateways to the web, people often using them as their home pages. Some portals brought together broad swaths of content: news, weather, entertainment, shopping, a little bit of everything. Others like AudioNet focused narrowly on content from a specific market or niche. These were known as *vertical portals*. With few proven advertising models in the early years of the web, the industry mostly adopted a crude version of the network-era television ad model: attract the largest audience possible. Thus, advertisers sought sites with high traffic, and in particular those sites that were *sticky*, meaning they held web surfers' attention for long amounts of time.[77] The notion of stickiness valued websites that functioned the most like mass media. The "stickiness" online business model, as Henry Jenkins, Sam Ford, and Joshua Green have more recently described, "refers to centralizing the audience's presence in a particular online location to generate advertising revenue or sales." This is achieved by "placing material in an easily measured location and assessing how many people view it, how many times it is viewed, and how long visitors view it."[78] Websites that function most like older mass media are valued by this notion of stickiness, as they maintain attention by its monopoly. As the term "surfing the web" implies, though, web surfers were notoriously fickle, flitting from website to website. Portals were thereby attractive to advertisers because their large selections of content promised to keep the audience locked into one place the longest. Indeed, this logic was one of the main things that made Broadcast.com an attractive purchase for Yahoo!, as the company hoped Broadcast.com's wide array of audio-video content could help make Yahoo! the first, last, and only stop on the internet for large numbers of users.

For a radio station wanting to expand its reach as far and wide as possible, AudioNet offered a much bigger audience than it would likely ever attract to its own website. The portal could also generate traffic for the station's own website, as curious new listeners might click through to find out more information about programs, DJs, and events. These visitors were presented with ads in the process. Indeed, in the late 1990s many commercial stations valued their websites less as a new listening platform and more as a marketing opportunity.[79] They were primarily a "value-added marketing mechanism for special events and imaging."[80] Websites could offer up a new revenue stream through display ads (including ones from national advertisers and technology companies that typically would not purchase audio ads on a small local radio broadcast). They could also publicize community and station events, facilitate contests, direct-sell merchandise and concert tickets, enable communication with the audience, and so on. As late as 1999, 75 percent of US terrestrial radio stations with websites did not offer any audio content: the sites were simply repositories for programming schedules and marketing.[81] A number of scholars conducting radio station website content analyses in the late 1990s and early 2000s found that the majority of websites functioned only as promotional tools, and were underutilized in terms of offering access to programming.[82]

Whatever the case, AudioNet served as a beta tester for many radio stations and networks, offering them a relatively low-cost, low-risk opportunity to determine whether there was a market for their content online. "We're curious to see what it will yield," said one CBS executive after signing up with AudioNet in 1996, adding, "Everybody is still evaluating the value of webcasting."[83] Especially for major legacy media corporations like CBS, CNN, CNBC, the NFL, and the like, the cultural pressure to have an online presence was intense by the late 1990s, yet they did not have the technical know-how to make it happen. Historically, corporations like these are known to be risk-adverse and slow to take chances on new technologies and platforms that might fail commercially or, in the least, be costly and burdensome to manage. Thus, in addition to being a safe testing ground, AudioNet allowed a low-stakes way for these "old media" institutions to appear innovative in their self-promotion, as they could claim that they were on "the net" and embracing the future of media.

Ultimately, what distinguished both RealAudio and AudioNet was that they led webcasting beyond the niches and into the mainstream. Compared to the Internet Multicasting Service and the college radio stations (chapter 1), these were more commercial, business-minded tools and services that were well suited to the newly emerged commercial web. RealAudio made it possible for any website to easily incorporate streaming audio at a financial cost that was, while not inexpensive, still relatively affordable, especially for the businesses that were looking for any way to make their sites stand out on the rapidly expanding web. Indeed, so-called *interactive web design* was all the rage during the dot-com boom of the mid-1990s. Before Flash animation and streaming video became widespread, adding streaming audio to a website was the best way to build multimedia cachet.[84] "Audio is the new toy," remarked one radio industry consultant in June 1996.[85] Likewise, AudioNet brought radio content to the web that contained mass appeal—sports and mainstream media news, in particular—and intentionally targeted it at the cultural mainstream. AudioNet was, in effect, the closest thing the early web had to a YouTube: a one-stop web portal that brought together radio and other audio content (and some video) from diverse sources and all walks of life. Online radio aggregators like TuneIn follow essentially this same model today.

Neither a technology developer nor a content producer, AudioNet was a new type of media company for the internet era—and yet it was modeled explicitly on broadcasting precedents. Many early internet radio providers described themselves as *networks*, including of course AudioNet ("*The* Broadcast Network on the Internet") and Progressive Networks (also with its Real Broadcast Network). They were not perfect analogues of radio or television networks, however. (Perhaps the double entendre of a "network"—of a broadcast network and the internet network of networks—was too hard to resist.) In its classic sense, a broadcast network (e.g., NBC, CBS, ABC) is a chain of interconnected local affiliate stations that are centrally controlled and share a single identity along with a core set of programs.[86] The aggregation of programs and simulcasts from many different radio stations, as well as a hodgepodge of miscellaneous audio content (music albums, audiobooks, live event audiocasts, and so on), into a single broadcast platform is much more akin to a cable service provider. This similarity was not lost on Cuban and Wagner, who frequently used the discourse of cable television to describe AudioNet. They often alluded to cable television magnate Ted Turner in interviews.[87] They would refer to themselves as a "superstation," in the sense that they were not limited by the power of a radio transmitter and could deliver a uniform slate of programming to the entire country (and beyond). This was another imperfect analogy, since a superstation denotes only a single channel and AudioNet's vertical portal was offering multiplex delivery.[88] Nevertheless, elsewhere Wagner stated, "We're just cable on steroids. We're the next step. We're 50,000 channels."[89] In other words, AudioNet was not seeking to radically redefine the medium of radio for the internet. Rather, they were simply bringing to radio a mode of distribution—along with certain services and functions, like on-demand access—that analog radio broadcasting technology previously could not provide.

Stay Connected: Pseudo

Enthusiasm for the internet grew steadily during the first half of the decade, yet it was not until 1995 that "internet mania" truly took hold.[90] Few companies came to symbolize the dot-com era, in particular New York's Silicon Alley, more than Pseudo (alternately known as Pseudo.com, Pseudo Programs, and the Pseudo Online Network). Although best remembered today as a web television network—if remembered at all—Pseudo initially made its mark as an internet-only radio network built entirely around its own original content. However, its shows could not have been more different from the relatively staid, NPR-style news and cultural affairs programming of the Internet Multicasting Service. It also deviated considerably from the conventional radio broadcasting that made up the bulk of the content carried by aggregators like AudioNet. Pseudo was the unusual by-product of the collision of an emergent technology with the 1990s downtown New York youth culture and what US Federal Reserve chairman Alan Greenspan famously called the "irrational exuberance" of the dot-com era financial markets.[91] The newfound enthusiasm for the internet made strange bedfellows of an uninhibited, wildly creative subculture with Wall Street venture capitalists, who pumped tens of millions of dollars into the enterprise. Depending on the point of view, Pseudo was either at the forefront of hipster culture and avant-garde performance art, or it was the online equivalent of the worst of community radio and public access cable television around.[92]

Pseudo was created in 1994 by thirty-four-year-old Josh Harris, who started out in the technology industry in 1986 as founder of one of the first internet market research and consulting firms, Jupiter Communications.[93] Harris was an early booster of the commercial internet and what were then known as "online services": companies like AOL, CompuServe, and Prodigy that enabled ordinary consumers to connect their personal home computers to networked services and, eventually, the web. Circa 1993–94, he was adamant that the internet, not cable television set-top boxes, would become the preferred gateway to the "information superhighway."[94] Jupiter was well known for its bold predictions and its industry conferences, which brought together the top executives in high tech and often served as the site of heated insider debates about the future of media.[95] Personally, Harris built a reputation as an outlandish provocateur and a master public relations manipulator. Yet, despite his peculiarities, he was regarded as someone who uniquely understood the fast-moving, chaotic business of the internet. Even if his projections were fantastical, he appeared to possess knowledge and vision that eluded even many of the computer and media industries' top CEOs and financiers.[96] Harris left the day-to-day operations of Jupiter Communications in 1993, though he remained a part owner and board member (eventually netting him a fortune reported to be worth anywhere between $10 and $100 million, after Jupiter went public in 1999).[97] As a result of his early success with Jupiter, Harris was something of a wunderkind in the burgeoning dot-com business sector of the mid-1990s—and he was particularly adept at attracting press publicity alongside corporate backers and venture capital investments.

The origins of Pseudo lie in a computer chat software business that was initially called Jupiter Interactive. Following his departure from Jupiter Communications, Harris acquired a contract with Prodigy to develop its Internet Relay Chat (IRC) software and manage a number of its chat rooms. Prodigy had begun in 1984 as Trintex, a joint venture between CBS, IBM, and Sears (though CBS divested itself in 1986). It was, according to *Forbes*, "America's first cutting-edge online service," and, circa 1990, it was viewed as "the network of the future."[98] Prodigy was still the world's largest online services company in 1994; however, poor management stemming from a conservative corporate culture and in-fighting between IBM and Sears led to a dwindling subscriber base and a stodgy reputation. Chat was an extension of popular early network services, such as message boards and online communities (e.g., MUDs and MOOs, which also had text-based interfaces), and this form of real-time messaging was one of the most widely used features of preweb online services. Prodigy, though, came under fire for aggressively censoring chatter comments that it deemed offensive. It hired an editorial staff to closely monitor and even direct chat room activity.[99] Enter the brash, youthful Harris, who was appointed in an attempt to retool Prodigy's tarnished image. He and a small team of Pseudo chat proctors ran a specially branded area for Prodigy, called Pseudo Online! Chat, with edgy chat rooms like "Big Beautiful Women," "The Neighborhood Bar," "Domination and Submission," "The Fetish Lounge," and "Vampire Pub."[100] According to *New York* magazine, Harris—who claimed to present himself online as a woman named "softmissy"—"presides over the harshest chat rooms on Prodigy, which still has no idea what hit it, but doesn't mind the revenue."[101] Pseudo, in fact, made an arrangement that it would receive a percentage of the profits from the chat application (users paid per-minute or hourly fees), which turned out to be a sizable royalty. Harris claimed that his chat rooms were garnering a quarter of all of Prodigy's traffic, earning Pseudo as much as $150,000 per month.[102] Pseudo operated the Prodigy chat rooms from 1994 to 1996, at which point Prodigy was sold to new owners and retooled itself as an internet service provider, reducing many of its networked services like the chat rooms.[103]

Simultaneous with Pseudo's management of the Prodigy chat operations and various other bulletin boards and online forums, Harris was using the profits to build Pseudo into an internet radio platform. The company name was quickly changed in 1994 from Jupiter Interactive to Pseudo Programs (in part to avoid confusion with Jupiter Communications). According to Harris, "Pseudo" was meant to suggest "the many different faces you can have in the online world," while "Programs" reflected a postmodern awareness that "we are in the business of programming people's lives." He added, "We are the good side of Big Brother. We know that this is going to happen"—referring to people's increased involvement in social network sites—"and instead of saying it's scary, we embrace it."[104] In other words, the "programs" in Pseudo Programs was a verb rather than a noun; it did not refer to the media content itself but rather the effect the producers and their content were having on the audience.

The audience was the content for Pseudo. Harris was becoming increasingly involved in building interactive multimedia experiences. While chat software was primarily text-based, Pseudo Chat on Prodigy let users share images and audio in "an uncensored adult area."[105] The audio mostly consisted of short, prerecorded soundbite and sound effect-style clips, such as "hello" and "goodbye" or a bomb explosion noise. These audio recordings were "best used as a punch line."[106] Moreover, prior to Prodigy's retooling in 1996, it was rumored that Pseudo was helping it develop a three-dimensional online environment—a sort of interactive chat room that would have been akin to online virtual worlds like *Second Life* (which did not emerge until the early 2000s).[107] Notwithstanding, Pseudo's biggest move into interactive multimedia was through internet radio and a service it launched in 1995 called ChatRadio. Taking up a variety talk show format, programming was broadcast over the air one night a week on WEVD 1050 AM, a Manhattan-based brokered radio station on which Pseudo leased blocks of airtime. During the live broadcast, listeners would log on to the pseudo.com website and participate in the radio show in real time via chat software.[108] It was a makeshift pairing of radio and the internet, and the potential audience was limited to those people living in the local New York City area reached by the analog broadcast signal. However, when RealAudio Version 2.0 was released in October 1995—enabling live internet broadcasting—Pseudo moved ChatRadio entirely over to the web and the online radio network rapidly expanded.

The Pseudo offices and production studios were in an old industrial loft building at the intersection of Broadway and Houston in the expensive downtown Manhattan neighborhood of SoHo. (Among the building's other tenants was the pop artist Jeff Koons.) This ten-thousand-square-foot space was in the heart of what in 1995 was coming to be known as

Silicon Alley, the New York equivalent to California's Silicon Valley, where the city's high-tech sector was centered. Compared to Silicon Valley, where much of the technology (hardware and software) for the dot-com boom was invented, Silicon Alley was hailed as the content capital (web design, entertainment, marketing).[109] The company's location was strategic: it itself was a dot-com start-up, while it also created content aimed primarily at the hip, tech-savvy young professionals who were the backbone of the so-called New Economy—the term popularized by the likes of *Wired* magazine cofounder Kevin Kelly to describe the information-based economy of the 1990s being brought about by globally networked computing.[110] Pseudo's target audience was teen and twentysomething Generation Xers more generally, and the Silicon Alley workers in particular. (After all, in the mid-1990s, white-collar office workers and university students made up a considerable chunk of the users with the high-speed internet access needed to decently receive Pseudo's streaming internet radio programming.) This was a group of mostly white, highly educated, upper-middle-class young adults that a scathing 1996 *Village Voice* exposé nicknamed "cyberyuppies," and that Richard Florida would christen "the creative class" in a few years' time.[111] According to Florida, these were people whose job it was to fuel the New Economy by creating "new ideas, new technology, and new creative content." They worked within the traditional corporate, capitalist world, yet they embraced a nonconformist spirit that was informed by the Gen X counterculture (cyberpunk, hip-hop, the rave scene, video games, and so on). That is, their radical outward appearance belied the fact that their creative labor, as Andrew Ross has observed, was underwritten by venture capital and powered the mainstream financial markets.[112] Silicon Alley in the latter half of the 1990s was flush with big money flowing out of Manhattan's Financial District. However, its workforce consisted mostly of recent college graduates who lived and partied in the East Village and brought that nightlife culture and libertarian attitude into the workplace.

Pseudo encapsulated this conflicting Silicon Alley lifestyle in its programming: the euphoric spirit of the nascent web culture along with the hedonistic party atmosphere of downtown New York City. It rapidly grew from an original nine shows at the beginning of 1996 to thirty shows that July.[113] In total, during 1996 and 1997, it developed around forty original weekly radio series, all of them produced out of the Soho loft space.[114] Pseudo's strategy was to target very narrow micro-niche audiences with programming that was either too specialized or too risqué and controversial for conventional radio.[115] This dual focus on micro-niche audiences and alternative topics closely paralleled the cable television model. Among other things, being an internet-only broadcaster meant that, similar to premium cable, it was not subject to FCC regulations of obscenity, indecency, and profanity. There were a number of romance and relationship advice talk shows that championed the X-rated and kinky, including *Cherrybomb!* and *Nellie's Love Chat*. A handful of comedy shows foregrounded the raunchy and outlandish, including *Pseudo Psychics*, which was ostensibly a call-in talk radio program for advice and psychic readings but was really a parody of the familiar radio format. The show was hosted by actors who were explicitly hired to "play 'psychics' and 'astrologers' on internet radio."[116] A number of shows incorporated elements of high-concept performance art, including the "sock puppet theater" of *Bare Feet*, the socially conscious Off-Off-Broadway fringe theater performances of *Galinsky's Full Frontal Theatre*, and the retro 1940s swing era-themed music and improvisational comedy variety show *Chatterbox Lounge*. There was even a pro wrestling talk show called *And Justice for Brawl*. Pseudo's shows ran the spectrum from high to low culture, yet they were all topics that generally did not have a place on conventional radio.

A large portion of Pseudo's programs were topically focused on a particular type of media, entertainment, or fan community. *Illumination Gallery*, for instance, was about film and video with an emphasis on underground cinema and movie websites. *Art Dirt* covered the fine art world, in particular web-specific artwork and online galleries, through news, reviews, and artist interviews. It included a "no holds barred dissing segment" called Art Trash. *GO! Poetry* featured live spoken word performances from participants in the hip-hop-centric poetry slam movement popular during the 1990s. *Star Trek Books: The Novel Experience* was a talk show entirely about novelizations of the cult science-fiction franchise. Other shows centered on computers and the internet. *QuakeCast* was an hour-long show entirely about the online multiplayer video game *Quake*. For web developers and computer programmers, there was a show called *Hackerz*. There was even a news talk show about the New York high-tech business sector, *Silicon Alley Reporter*, an offshoot of the popular trade industry journal of the same name.

Pseudo's programs covered more than just the media industries and niche taste cultures: a number of shows consciously addressed gender, racial, and ethnic minority communities, attempting to give a platform for diverse multicultural voices and perspectives. *Minx* was a "sexy and sassy" women's "zine" in the spirit of the Riot Grrrl female punk movement.

Cindy Something was an all-female hosted comedic advice show with an outspoken feminist point of view. There was *Ten4*, "an interactive forum dedicated to exploring today's issues from a woman of color's perspective." *Street News Review* reported on "the soul and the spirit of the streets" of New York City, featuring interviews and stories from community members. Perhaps Pseudo's most elaborate production, *African-American Stories* was an anthology drama series exploring "the hopes and dreams, triumphs and tragedies of Americans of African descent." These series, while they might be similar to programs found elsewhere on local community or college radio, did not have many counterparts on mainstream commercial radio of the era. They were given equal billing on Pseudo (often in prime programming slots) and, through the internet, being distributed to a potentially worldwide audience.

While its shows certainly covered unusual topics and subject matter, Pseudo nevertheless adopted many conventions of radio broadcasting. It followed various programming norms, such as shows occupying weekly half-hour or one-hour slots. The daily schedule fluctuated between a peak of about thirteen hours during the midweek and only a few hours of live programming on the weekends, though past episodes were archived for on-demand access and Pseudo Chat ran 24-7. The network also utilized block programming, thematically grouping its evening primetime programs together: "Lifestyles & Relationship Monday," "Games Thursday," "Digital Worlds Friday." Pseudo relied heavily on traditional radio formats and programming practices, especially when it came to its music radio programming, dividing up series by recognizable popular music genres: *freQ* (rave and electronic dance music with live DJ mixes); *T-BO Power Hour* (techno and electronica); *Desert Flower Indie Hour* (indie rock/pop); *Rock n' Roll Hangover* (classic rock); *Kool Out* (soul and funk); *Sucker* (garage rock and punk); *J'Open Mike* (eclectic alternative music show with live guest performances and interviews based around chat questions from listeners); *88 Hip-Hop* (hip-hop culture with emphasis on talk and live in-studio performances showcasing young talent). Even as Pseudo presented itself as a cutting-edge take on broadcasting, it never broke away from radio's norms of use entirely.

Nevertheless, the similarities to standard radio production methods mostly began and ended with programming conventions. Notably, no one who worked at Pseudo came from careers in professional radio. The staff consisted largely of web designers and computer engineers—many of whom were not formally trained and learned on the job. A majority of the on-air talent had no previous broadcasting experience, either. They were luminaries of their particular field, pulled by Harris from the New York City arts and culture scene. For instance, the *CBGB's* music show was hosted by the owner of the legendary punk rock venue Hilly Krystal. *Infinity Factory*, a talk show about the occult and the supernatural, was hosted by renowned conspiracy theorist Richard Metzger. One of Pseudo's earliest programs was *The Church of Billy*, a daily faux sermon of the anticapitalist performance artist Bill "Reverend Billy" Talen (a prominent figure in the culture-jamming movement who has since been the subject of multiple documentary films).[117] Perhaps the most recognizable early Pseudo on-air personality was Taylor Mead (*The Convertible Taylor Mead*), an actor and member of Andy Warhol's Factory in the 1960s. While most of these individuals had experience as entertainers, they were not trained broadcasters.

Much of the Pseudo talent consisted of ordinary people who Harris simply thought were interesting. Harris and Pseudo were notorious for their parties: the *New York Times* called the Pseudo loft "the Warhol Factory of 1995."[118] These parties were a stage for Harris—fancying himself a conceptual artist—to perform on; he often appeared at the gatherings and on Pseudo as his alter ego, a demented, childlike, cross-dressing clown name Luvvy.[119] The bashes only got bigger as the decade went on, each drawing more than two thousand people, with the featured events ranging from elaborate art installations by New York art world luminaries to indoor boxing matches and debauched drug-fueled sex orgies.[120] The parties were massive spectacles that generated considerable publicity for Pseudo, though the downside was that the press rarely recognized Pseudo for anything other than these wild events. For his part, Harris described the Pseudo loft as "almost a commune," elaborating that "anyone in the East Village with any semblance of talent winds up here at some point."[121] Indeed, for him, the main purpose of the parties was to scout new talent for Pseudo's programming.[122] Rather than hire away personnel from traditional radio stations, Harris created a new paradigm for radio that drew from the spirit of naive art, celebrating an amateurish, anything-goes sensibility. This paradigm also heavily emphasized sociability and affect through its visceral, often confessional content and hands-on, communal atmosphere.

In addition to the atypical subjects and hosts, Pseudo's entire ChatRadio approach—marrying the live internet radio broadcast to real-time messaging—upended traditional radio production methods. Pseudo emphasized interactive content, or what it called "user-participatory" programming.[123] There was a lot of industry buzz about "interactive" entertainment media in the late 1990s, although it was usually doublespeak for direct marketing or data tracking. At most, audience participation meant on-demand playback, user-generated playlists, and real-time audience polls. Mark Cuban, for instance, regularly spoke about AudioNet as an interactive service, yet the site did not contain any chat rooms or messages boards. Rather, he meant that AudioNet could do things like insert custom ads into streams based on a listener's demographics, or that listeners could hear an ad and instantaneously click through to purchase the product, rather than having to be sent elsewhere to shop.[124] This notion of interactivity was quite a far cry from the two-way dialogue between producers and audiences that was being enabled by Pseudo. With only a few exceptions, all of Pseudo's programs were talk based—or, in the network's parlance, chat based. The basic listening experience for any Pseudo show consisted of the audience member looking at a web page with the RealAudio Player embedded in one image window and the live chat window embedded beside it. The chat window was the focal point. (In addition to the chat rooms, listeners could email the hosts, and shows had message boards for non-real-time public messaging.) Hosts were called "chat jockeys," as opposed to disc jockeys.[125] Indeed, their job was roughly equivalent to a radio DJ whose job was to introduce and play a rotating selection of music recordings on the air, except with online chat conversations instead of records. In other words, the listener-driven online chat *was* the show, as opposed to carefully prepared segments from the Pseudo hosts and guests.

Unlike traditional talk radio where audience members usually only participate one at a time via phone calls, real-time online chat enabled a party-line-like media environment. The entire audience could be actively participating in the program, interacting both with the Pseudo hosts and studio guests as well as with the other listeners. Audience questions dictated the direction shows took, which, according to a *New York Times* reporter who shadowed the chat jockeys in the studio, meant the broadcasts were "unpredictable and sometimes off on a tangent."[126] Harris compared the ChatRadio setting to a local bar where people could talk with friends casually while also snooping in on those around them, listening in to strangers and potentially getting to meet them. The talk show host functioned like the bar's music DJ, simply providing background sounds to facilitate the socializing, and stepping in to influence behavior only if the atmosphere got dull.[127] Much of the pleasure of listening came from hearing the host(s) field the audience's comments or questions. Pseudo audiences were small in 1996–97: the network attracted a few thousand listeners per week, and radio shows averaged about a hundred live listeners, with twenty-five or so participating in the chat rooms.[128] (The exhibitionist nature of the chat rooms clearly was not for everyone, but many people seemingly got pleasure from voyeuristically eavesdropping or lurking.) Listening to a Pseudo broadcast, then, truly was an intimate, participatory, and sociable experience, as there was a high likelihood that audience members who wanted to connect with the host or guest and get their opinion voiced on air actually could do so.

This notion of a virtual bar, or virtual party, where audiences could hang out and mingle also extended into the studio itself. Described as "antioffices" by the *New York Times*, the Pseudo loft—where Harris also lived for a time—featured rooms full of networked computer games, a pet bearded dragon and turtle, and graffiti-covered walls.[129] The scene in the loft often resembled a social gathering (including the alcohol, drugs, and sex) with the microphones turned on more than it did a radio studio. There were no rules and little regard for standard broadcasting conventions; it was essentially anything goes. Often, there would be an audience of dozens of people—other Pseudo employees, friends, celebrities, and Silicon Alley insiders who randomly stopped by—who would lounge around the studio while a live broadcast was happening.[130] The scene for broadcasts of the rave music show *freQ* resembled an actual rave, with live DJs performing in the studio to a crowd of musicians, break-dancers, graffiti artists, and assorted hangers-on.[131] According to one former Pseudo host, Mike Rinzel of *J'Open Mike*, they realized early on that with webcasting there was no need to approach radio as a formal, scripted performance. "We can just turn the microphones on for hours and just do crazy stuff into them," he described. "Some things were produced, some things were just hangouts. Some things it was like a party going on outside the glass booth, sometimes the party would go into the glass booth, sometimes it wouldn't." With regards to the chat jockeys, Rinzel explained that "they were just nutty people that we thought would be fun to have online."[132] Unscripted, amateurish media productions based around interesting personalities and narrowcast to micro-niche audiences: this would become the basic model for a large segment of podcasting a decade later, as well as user-driven social network sites like MySpace, YouTube, and Instagram.

At the start of 1998, Pseudo switched formats from internet radio to internet television.[133] Not unlike the radio-to-television transition in the 1940s and 1950s, the network simply switched much of its internet radio programs over to the new internet television platform. This changeover happened almost literally overnight. As Rinzel explained, the production setup and approach did not alter much: "It was still a radio studio [but] with cameras now."[134] The programming continued to be chat based, except now it was transmitted via RealVideo instead of RealAudio and the audience was watching the ChatRadio hosts in the studio. (In this way, it was quite similar to cable television broadcasts of live talk radio shows, such as Howard Stern, Don Imus, and *Mike & Mike*—a growing format during the 1990s.) Eventually, Pseudo moved to a more tightly focused setup with curated "channels" and preproduced video programming.

The fact of the matter is that Pseudo had been experimenting with online video for years and always had television in its sights. In 1993, even before officially launching Pseudo, Harris spent a considerable amount of time and money working with a team of programmers to create a short three-dimensional computer-animated film titled *Launder My Head*. With PC computer monitor-headed people chanting cryptic refrains like "Conform with me / Come form with me" and "I am your conscience / I am not conscious," the video art piece is simultaneously a social critique and a celebration of a hypermediated society.[135] During the early internet radio years of the Pseudo Online Network, it promoted *Launder My Head* and a few other "NetMovies" on its website. A handful of its radio shows, including *Quakecast*, also started webcasting video versions in 1997. Nevertheless, the move into web television was, first and foremost, the result of influxes of investment funding. "Investment money wasn't that interested in audio," Rinzel explained; financiers were gripped by the prospect of internet television.[136] Web television, as *USA Today* reported in August 1998, had become "THE red-hot Net application."[137] The allure of web video and television obscures the fact that the entire Pseudo template was drawn from its initial hybridization of radio and the internet.

As money started to flood into Pseudo, the online network's programming became more professionalized and increasingly conformed to recognizable broadcasting conventions.[138] This new funding came from Intel Corporation and established media companies like the Tribune Media Company, one of the largest US newspaper publishers and owner of multiple broadcasting properties including the WGN superstation. Rinzel described Pseudo as "a bunch of young, idealistic artist-type people learning how to use technology." He continued, "And then the company [got] formal investment . . . and then had to somehow legitimize," which meant developing more conventional, "professional-looking" television programming.[139] With much of the newer programming being prerecorded, the interactive chat element diminished. Audiences grew from a few thousand listeners to ten million viewers per month, and the network started to produce branded content for corporate sponsors like Sprite and Levi's.[140] It licensed its programming to mainstream media platforms, including web portal Excite and satellite service provider EchoStar (owner of Dish Network).[141] Pseudo covered more mainstream topics, such as professional sports, for instance partnering with the NFL Quarterback Club, a group of retired star football players.[142] It produced high-profile coverage of the 2000 Republican National Convention in Philadelphia.[143] The company also brought in executives from established media conglomerates, including a new CEO to replace Harris, David Bohrman, formerly of CNN and ABC News.[144] As the financial investment increased, the experimentation that initially defined Pseudo dramatically decreased.

The year 1999 was when the internet fully broke through into the mainstream as a platform for media entertainment and business. However, much of the enthusiasm (and financial investment) surrounding dot-com companies like Pseudo quickly evaporated on April 14, 2000, when the Dow Jones industrial average dropped 5.66 percent in one of the worst single-day stock market crashes in history.[145] It was the beginning of the dot-com bubble burst. When Pseudo shut down in September 2000, it had raised in excess of $32 million in investments, yet it was running up operating expenses of $6 million per month. According to its executives, it had plenty of advertiser interest, yet the site was still not attracting a critical mass of viewers.[146] Whereas the goal had once been to be a sort of antiradio or antitelevision, suddenly Pseudo was seeking to supplant the broadcasting establishment (along with a slew of competing webcasting networks like Pop.com and the Digital Entertainment Network). In February 2000, Harris infamously bragged to legendary CBS News correspondent Bob Simon in a *60 Minutes II* report, "The new boy is in town; the new boy is taking over as king of media. . . . I'm in a race to take CBS out of business."[147] Many onlookers, even Harris himself in other moments, argued that web television was ahead of its time; the broadband user-base for high-quality streaming media was still a few years off into the future.[148] Harris's reputation as a dot-com icon was cemented with a December 1999 performance art project/sociological experiment called "QUIET," in which he locked a hundred "guests" in a custom-built bunker for a month and put them under 24-7 video surveillance. The $1.8 million dollar event turned into an epic bacchanal. He then proceeded to rig his own apartment with state-of-the-art video equipment for another project titled "We Live in Public," live broadcasting his and his girlfriend's every move on the internet.[149] Retroactively viewed as prescient warnings about erosion of privacy and reality television's influence on culture, the pair of events was turned into a 2009 Sundance Film Festival Grand Jury Prize–winning documentary film, *We Live in Public*.[150]

In the film, as well as in many of the posthumous newspaper and magazine articles written about Pseudo, the company is referred to only as an online television network—its roots in internet radio erased from historical memory. Yet Pseudo was among the first major producers of internet-only radio, and it defined a bolder, more iconoclastic vision for what radio on the internet might accomplish. It was perhaps a bolder vision than anything attempted in the nearly twenty years since. Furthermore, that radio vision—particularly the ChatRadio format—was directly copied over to Pseudo's web television programming, which in turn significantly influenced the webcam-centric, reality-based web video style from the early 2000s on.[151] This interactive, colloquial, deeply personal style of web video content is highly sociable and affect laden, referred to as *videos of affinity* by Patricia Lange due to how they generate feelings of "communicative connection" between people.[152] Ultimately, there is a distinct logic of radio—talk or "chat" radio, specifically, with its emphasis on connectivity and sharing—deeply embedded in the now-dominant style of loose, inclusive, charisma-driven first-person YouTube videos.

Conclusion: Internet Radio at the End of the Twentieth Century

In the span of just a few years, from 1993 to 1995, the internet went from a small, silent outpost used mostly by technologists and researchers to a bustling frontier town teeming with sound and a wide-eyed crop of young webcasters. Then, from 1996 to 2000, the frontier exploded into a city, as the masses began to inhabit the web and streaming radio flooded the network of networks. But as the old adage goes, pioneers get scalped while settlers prosper. Few of the internet radio companies that emerged in the 1990s made it out of the decade alive. The simple fact is that they were ahead of their time; the broadband internet connections needed for seamless, high-quality streaming audio were still miniscule in 2000 and would not become widespread until the latter part of the 2000s. The bursting of the dot-com bubble did its part to prematurely wipe out many streaming media ventures, starving them of advertising and investment capital while also slowing the general population's internet usage overall.[153] Pseudo died abruptly in the fall of 2000 and Yahoo! Broadcast (né AudioNet) slowly withered in 2000–2001 until Yahoo! pulled the plug on it in 2002—though both companies had mostly abandoned streaming audio for video a few years prior, anyway. There were many other internet radio start-ups that either disappeared before the millennium or straggled into the new century before breaking down: NetRadio, eYada, TalkNetRadio.com, GRIT, SonicNet, Qradio, Imagine Radio, to name but a few.

The major internet radio services that most readers will recognize today—Pandora, Spotify, iHeartRadio, TuneIn, Stitcher, Slacker, Apple iTunes Radio and Podcasts—did not exist in 2000. Yet these contemporary services—the settlers—reap benefits from their deceased pioneers. Most plain to see, the aggregators like TuneIn and Stitcher, plus Apple Podcasts to a degree, follow in the footsteps of AudioNet. There are also elements of AudioNet's raw, unabridged coverage of live events, such as conference proceedings and public speeches, found in modern livestreaming audio-video platforms and apps like Livestream, Ustream (now IBM Cloud Video), Periscope, and Meerkat. There are traces of Pseudo's ChatRadio framework in social web radio platforms like BlogTalkRadio and Spreaker that let individuals host their own live multiparticipant broadcasts. Moreover, the entire practice of amateur podcasting—particularly comedy and pop culture podcasts that feature a group of friends chatting around a microphone—bears at least a passing resemblance to Pseudo's approach in the mid-1990s.

If there is one point that stands out most of all, it is that few of internet radio's early pioneers were experienced radio producers or traditional broadcasting institutions. The college radio stations from chapter 1 are a notable exception, though they existed at the very fringes of the radio establishment, plus the people most involved in developing their simulcast streams were computer engineers. There certainly were numerous radio stations and networks that took to the internet during the 1990s—NPR, among others—although few of these stations produced new, original content for the net. They largely relied on more forward-thinking companies like the Progressive Networks or AudioNet for online distribution. There were many more radio stations, however, that ignored or even scorned the web during this period, including major conglomerates like Clear Channel (now iHeartMedia). Clear Channel did not undertake a major online radio initiative until late 1999, when it launched KIIS-FMi. It was nearly another decade before the nation's largest radio conglomerate adopted a cohesive internet radio strategy with its iHeartRadio platform.[154]

Most of the early adopters were technologists or entrepreneurs coming from completely outside the broadcasting field. This scenario is very much in line with the development of computer and internet-related technologies and businesses since the late 1970s. As Thomas Streeter has observed, the "two guys in a garage" motif of small, upstart technology companies have been the dominant entrepreneurial narrative of the mythic American Dream during the US economy of the past thirty-plus years.[155] While this narrative is almost always an oversimplification, in the case of internet radio ventures like AudioNet and Pseudo it is not entirely inappropriate either. The innovations in internet radio did indeed come from small groups of self-starters operating mostly outside of the broadcasting establishment. The major players in the radio industry did not foresee the demand for internet radio, as they were oblivious to the new technology, caught up in the security of their existing business models, or sidelined by an overly cautious wait-and-see attitude. This left the field open for internet entrepreneurs like Glaser, Harris, and Cuban and Wagner (and individuals like Carl Malamud before them).

In terms of style and form, there was remarkably little innovation in the content of radio during the 1990s, as I highlighted in the conclusion to chapter 1. Most internet start-ups merely reproduced principles and techniques from terrestrial radio. Aggregators like AudioNet primarily pulled content directly from broadcast radio. Therefore, while Cuban and Wagner may not have been broadcasting insiders themselves, at the level of content they were mainly repurposing the output of the existing "old media" radio industry. Even Pseudo, which experimented the most wildly with the established aesthetics and practices of radio, still relied on traditional radio formats and programming practices. Moreover, investment money and pressures to conform to the business logics of venture capitalists and traditional media corporations forced Pseudo to eventually adopt more traditional and conservative broadcasting conventions.

Still, by the late 1990s, internet radio deviated in marked ways from broadcast radio of the era. Radio in the United States became hyperspecialized during the 1970s and 1980s, divided up by increasingly narrow music and talk formats. Satellite syndication combined with ownership consolidation resulted in programming schedules and music playlists that were highly standardized. By the early 1990s, many radio stations across the country sounded remarkably the same, and there was little room on the airwaves for content that broke the mold. Frustrated listeners were hearing what sounded like "assembly-line radio," as Michael Keith and Christopher Sterling put it.[156] In contrast, internet radio sounded bold and fresh, if only because many of the first webcasters were pulling from the margins of college radio, public radio, foreign broadcasting, and the like. They were airing programming that did not fit the rigid formatting of 1990s corporate radio. AudioNet, for example, featured a wealth of live-event coverage from conferences, trade shows, business meetings, political proceedings, and the like. These broadcasts defied normal radio scheduling conventions, running for unspecified lengths of time, often multiple hours. Additionally, many of the events—computer engineering conferences, financial analysis seminars, university lectures—were too niche and arcane to ever find a home on traditional radio, even in a severely edited form.

Yet, while unusual for 1990s broadcasting, this type of content was typical of radio in the 1910s and 1920s, when amateurs and a few ingenious entrepreneurs were inventing broadcasting. College lectures, religious sermons, political speeches, and various types of unformatted informational reports, often scheduled erratically, were among the first "programs" ever broadcast on radio. Many of the first livestreaming audio broadcasts, including AudioNet's Mustangs–Razorbacks college football game and Progressive Network's Mariners–Yankees baseball game, were also sporting events. A number of terrestrial radio's most prominent early broadcasts featured sports as well, including KDKA's 1921 coverage of a Johnny Dundee–Johnny Ray boxing match. Live music performances were also popular subjects for much-hyped internet radio broadcasts in the 1990s, as they were for early experimenters with radio broadcasting in the 1920s. Even Pseudo's audience-driven ChatRadio harkens back to the earliest days of analog radio, before the one-to-many broadcasting system took hold, when the airwaves were full of amateurs chattering away in what many perceived as disorderly chaos.[157] Thus, while portions of internet radio seemed unusual and even confusing when compared to the heavily formatted radio spectrum of the 1990s, it must have sounded remarkably like terrestrial radio in the 1910s and early 1920s before the network era of broadcasting firmly set in. There was a particularly strong emphasis on liveness and copresent intimacy in this diverse and highly ephemeral 1990s internet radio programming, which are lasting trademarks of radio's sociability.

During the 1990s, the prevailing popular discourse about internet radio, and about the internet more generally, was that it was going to be a democratizing force on cultural production. The basic idea was that since the network is open to anyone with a personal computer, a modem, and a phone line, media production was going to circumvent the broadcasting networks, publishing houses, and other traditional media gatekeepers. This is the same discourse that currently surrounds newer radio practices like podcasting. In chapter 1, Carl Malamud spoke of an oncoming "desktop broadcasting" revolution.[158] Similarly, Glaser declared that with the Real-Audio software, "We want to jump start a self-publishing movement."[159] Both were drawing parallels to the desktop publishing developments of the 1980s that allowed anyone to create professional-looking printed documents on a basic personal computer, and provided the impetus for the rise of zines and other alternative press publications.[160] Cuban, too, proclaimed in a keynote speech to the NAB that with the web and new technologies, "everyone is a broadcaster."[161] They all subscribed to the basic idea that anybody could now be a radio broadcaster with little more than a basic personal computer and an internet connection.

Lost in this enthusiasm for the new technology was the reality that putting radio on the internet was a relatively expensive endeavor in the early years of the web. (It is still more costly today than most people presume.) Many of the operations costs, from streaming audio software to web servers, have been outlined elsewhere in chapters 1 and 2. But a few more figures: the Internet Multicasting Service required approximately $100,000 in start-up costs.[162] Malamud paid for at least $40,000 of that out of his own pockets, and the rest came from his corporate donors.[163] Another early internet radio network, NetRadio, spent $250,000 just to open its doors.[164] In 1999, Cuban estimated that simulcast streaming expenses alone could run a radio station anywhere between $5,000 and $100,000 per month. While these prices were a bargain compared to what it would cost to start a broadcast radio station from scratch—acquiring an FCC license, constructing a radio transmitter and antenna, and so forth—monthly expenses in the thousands of dollars still would have put desktop broadcasting out of the reach of the average citizen. It would be another ten years before desktop broadcasting, in the guise of podcasting, would become an affordable reality for most Americans. These high costs meant it was primarily current stakeholders in the broadcasting industry who could afford to use the technology. At the same time, many of the industry mainstays were unwilling to take the risks involved in internet radio broadcasting. As a result, it took entrepreneurs like Glaser, Harris, Cuban, and Wagner, unburdened by the strictures of a broadcast business model and self-preservation concerns, to prove that radio on the internet could succeed. The college radio stations from chapter 1, in particular, draw many parallels to the prenetwork era of radio in the 1910s and 1920s, when it was amateur hobbyists—many of them young boys and men in their teens and twenties—who drove the nascent medium of radio forward through a desire to tinker and explore.[165]

Another curious parallel between 1990s internet radio and 1910s–1920s broadcast radio lies in listening practices. During the first few years of radio on the internet, many listeners marveled at the abundance of new radio options available to them online, and in particular the ability to hear radio broadcasts from far-off places. As one *Washington Post* reporter wrote in 1993, "The typical Internet user often is someone who enjoys the hunt as much as the catch."[166] The experience of early internet radio was "listening in the rough," as another columnist put it; however, for many listeners that only added to the unpredictable, explorer-like atmosphere that was part of its joy.[167] Mark Cuban remarked that much of the pleasure of listening to internet radio in the mid-1990s came from browsing the random assortment of unfamiliar options. "It's the fun of walking into the New York Public Library," he said. "So much stuff, you are awed and you just walk through the stacks till you find what you want."[168] Indeed, not only did the global nature of the internet eradicate the restrictions created by analog broadcast signals, but the digital format and graphical interface of the web meant that audiences could search radio via metadata (e.g., written descriptions of radio stations/programs), making browsing and discovery easier. A batch of specialized web directories sprouted up, such as the MIT List of Radio Stations on the Internet, which provided comprehensive lists of internet radio stations with hyperlinks and details about the station's geographic location, programming formats, and so on. A *New York Times* article reported that a third of AudioNet's audience in 1996 was "listening from afar."[169] Listeners loved to call in to radio shows from far-off places or send emails and post message board comments about how they were listening all the way from Australia, Denmark, Malaysia, and so on. And radio hosts relished putting these foreigners on the air or reading their messages aloud. This "listening from afar" was almost identical to the *distance listening*, or DXing, that was the main focus of much radio transmission and reception in the 1910s and early 1920s.[170] The sociability of radio has always entailed an audible yearning for companionship, with listeners turning on their radios as a means of reaching out to distant others and bringing themselves closer to the outside world. Internet radio in the 1990s not only gave this old practice renewed meaning, it also vastly expanded the world of voices with whom listeners could commune.

Internet radio's relationship with terrestrial broadcast radio was quite contradictory during its first years of existence from 1993 to 2000. It was widely understood as an extension of radio rather than a completely new medium—even by those entrepreneurs who had no prior experience working in radio. Yet it was also envisioned as a means to fix all manner of perceived shortcomings or flaws with contemporary radio broadcasting, whether it be a lack of openness and a failure to serve diverse identities and tastes, or simply an inability to be consumed on the listener's schedule and replayed at will. Nevertheless, internet radio ended up being merely a stopgap measure for many of its earliest producers and supporters, a temporary pursuit on the way to streaming video. In many ways, radio and streaming audio were a proof of concept, particularly in the first few years of the web, when the network did not yet have the carrying capacity for video. In another case of history repeating, internet radio was passed over for internet television in the late 1990s much the same way radio was passed over for television in the 1940s–1950s. This, along with the disinvestment following the dot-com bubble and the imposition of restrictive copyright and royalty regulations governing the use of music on internet radio, served to stunt internet radio's growth for the better part of a decade, at least at the level of large-scale commercial platforms. It would not be until the mid-2000s, with the arrival of podcasting and the maturation of streaming music services, that internet radio really began to take off in a meaningful way again. On the upside, much like broadcast radio had to redefine itself in the mid-twentieth century, the major media corporations turning their backs on internet radio in the early 2000s opened it up again for new players and new ideas, as the following chapters explore.

CHAPTER THREE

Everybody Speaks

Audioblogging and the Birth of Podcasting

Guglielmo Marconi, one of the inventors of radio technology at the turn of the twentieth century, late in his life reportedly came to believe that sound never dies. All sound waves continue to radiate eternally, he theorized, they just become too quiet for human ears to detect—and so he envisioned something akin to a super radio that would allow him to hear every sound ever made.[1] John Cheever's 1947 short story "The Enormous Radio" tells of a New York City husband and wife, Jim and Irene Westcott, who buy an expensive new radio set that, to their surprise, allows them to overhear the conversations of their neighbors: nannies singing to children, the sound of lively house parties alongside people listening to music alone, children complaining and couples bickering, even violent domestic arguments.[2] These are but two examples of how generations of thinkers have imagined the radio as a device for listening in to the voices of the world around us, to every manifestation of sound from the mundane to the spectacular. The association of radio and everyday life is deeply rooted in our cultural understanding of the medium. Radio has long been perceived as perpetually present, which creates affective connections to our daily routines and makes the medium inextricably tangled with our understanding of reality. Sound may be the most evanescent of our sensory experiences, and yet it is also ubiquitous. It is always all around us, and our auditory senses are perceptually first to be activated.[3] It is not unusual, for instance, for us to hear that which we cannot see, smell, or touch. Radio signals appear to us as virtually omnipresent, saturating the atmosphere and forming what Jeffrey Sconce has called a blanketing presence.[4] This omnipresence partly accounts for the medium's quality of liveness. And yet, following a brief period in the 1910s–1920s during which the airwaves were dominated by the unstructured chatter of amateur radio operators, the institution-alization of broadcasting in the 1920s turned US radio primarily into a system of distribution for the messages and voices of the professionalized media networks.

The most ubiquitous form of sound in modernity, apart from noise, is *talk*. And among the mass media, radio talk is distinctive for its (seeming) spontaneity and expressiveness.[5] Despite a clear self-consciousness, when compared to television, film, or writing, radio talk most closely resembles the "ordinary talk" of everyday face-to-face communication. This conver-sational style of talk helps give radio its sense of directness and presence, which is fundamental to the "reality effect" and radio's liveness and socia-bility. For most of the twentieth century, however, the nearly ubiquitous talk heard on the radio was that of the major broadcast networks and their representatives. If the general public ever gained access to the radio air-waves it was typically as a soundbite or a carefully chosen voice on a call-in talk radio program. In other words, radio may blanket our culture and give off the appearance of access to everyday life and to a plethora of voices and perspectives, but for most of its existence who and what was heard on the radio has been tightly controlled.

The oft-repeated promise of the internet, especially since the main-streaming of the web in the 1990s, has been to democratize the mass media and enhance speech and civic participation. This utopian rhetoric is per-haps no more pronounced than in Web 2.0's "architecture of participa-tion" and other related ideas of user-generated content, citizen journalism, crowdsourcing, collective intelligence, prosumption, and produsage that took hold in the early to mid-2000s.[6] The underlying premise of all these concepts is that the open architecture of the web enables and encourages the participation of all people, not merely a few chosen experts or authori-ties. And there is perhaps no form of content more representative of Web 2.0 than the *blog*: personal websites, often single authored and diary-like or essayistic in style, that were seen as opening up print publishing to regu-lar people. In the words of Dave Winer, a software developer who helped develop blogging technology and who also wrote one of the earliest blogs (*Scripting News*), blogging in its purest form consisted of "the unedited voice of a person," by which he meant "one voice, unedited, not determined by group-think."[7]

The mention of blogging is significant here because *podcasting*—the practice that has most transformed radio in the internet age—actually began as an outgrowth of blogging. The name "podcasting" did not even exist for the first few years the practice was in place. Rather, the activity of ordinary web users creating their own radio—or radio-like—programs for download was most commonly referred to as *audioblogging*. Webcasting in the 1990s opened up radio production and distribution to a much wider range of individuals than terrestrial radio ever could. It was nevertheless still primarily the domain of experts—if not radio professionals, then computing experts and technology start-ups—due its high costs and technological complexity. It was not until after the turn of the millennium when a confluence of events, including the development of web publishing tools designed for nontechnical users, made podcasting an easily accessible form of internet radio. The idea of a multitude of ordinary voices blanketing the landscape suddenly became more of a reality . . . again. That is, a century ago, radio broadcasting first emerged as a participatory communications medium, with individuals—most of them amateurs and hobbyists—filling the airwaves with chitchat and all manner of assorted content from prerecorded phonograph music and live piano recitals to political speeches, religious sermons, and informational reports.[8] Thus, as Jonathan Sterne, Jeremy Morris, Michael Brendan Baker, and Ariana Moscote Freire have suggested, podcasting is best viewed not as an alternative to broadcasting but rather as a realization of broadcasting's original democratic potential, prior to its 1920s institutionalization into a professionalized, corporatized form.[9] In addition to enabling the production of alternative forms of radio, the rapid expansion of social media tools since the early 2000s has opened up new ways for the audience to participate in traditional broadcast radio, altering both listeners' experience of radio and producers' creation of it. Talk radio is a key site where radio's eventfulness has been transformed through the sharedness of online social media, these additional communication channels giving workaday talk radio broadcasts event-like qualities that amplify and expand the presentness and sociability of traditional radio.

This chapter addresses talk radio online during the post-dot-com boom decade of the 2000s—highlighting the ways in which the internet has opened up radio production to the general public, in the process changing what it means to be a radio producer and challenging expertise and who is allowed to exert cultural authority through radio. The central focus of this chapter is a history of podcasting and its roots in the nascent Web 2.0 culture (2000–2005), specifically as an audible version of blogging called audioblogging. Podcasting in its earliest stages developed an unscripted style of talk centered around personal experience and self-expression. This informal, self-reflexive, often mundane conversational style—which was actually a convergence of not only radio and the internet but also blog/diary writing and the telephone—carried over into many areas of modern podcasting, and even made an indelible mark on the social media discourse that today pervades platforms like Twitter, which itself started as an audioblogging service.

Internet Talk Radio and the Transition to Web 2.0

High-speed internet service, despite rapid expansion around the turn of the century, was still relatively scarce in the United States during the first few years of the new millennium. This was especially true in residential homes where a mere 1 percent of Americans had broadband access in 2000. Even by 2005 that number had increased to only about 35 percent.[10] Also, the 200 kilobits per second (kbps) speeds that then qualified as an "advanced service" would be considered painfully slow even a half-decade later.[11] Many internet audio streams were transmitted in mono, since it was the most stable and widely accessible format—few listeners had the bandwidth, computer memory, or updated player software to support stereo streams—the point being that in the early 2000s streaming audio was still a rather primitive-sounding technology with a limited reach. Although plenty of individuals and groups experimented with music programming in the late 1990s and early 2000s, internet radio in these years remained best suited for talk formats. Again, the Internet Multicasting Service started with technology industry talk programs in 1993, and the first major commercial internet radio ventures, including Pseudo, were talk-centric.

Sports talk radio was very popular in the mid to late 1990s, partly because it held strong appeal for the young male and white-collar, office deskbound audiences who were the primary users of the internet during this era. Moreover, liveness and immediacy prevail in sports, and many listeners were willing to exchange poor audio quality for direct access to the latest action, scores, and analysis.[12] The cable sports television network ESPN took to the web in April 1995 with its ESPNET SportsZone website, which was developed in partnership with Microsoft cofounder Paul Allen's Starwave Corp.[13] Initially it simply offered short audio clips of games and athlete interviews alongside text and graphics (box scores, statistics, print news stories). By 1996, however, it was providing live audio coverage of numerous games from professional sports leagues including the NBA and MLB.[14] The radio and television broadcaster CBS entered into sports radio online in 1997 through a partnership with the upstart webcaster SportsLine USA (forming CBS SportsLine).[15] Legacy media conglomerates NBC (via its NBC Sports and MSNBC venture MSNBCSports), News Corp (Fox Sports Online), and Time Warner (with CNN/SI, a partnership between its CNN television and *Sports Illustrated* magazine properties) also launched major online sports websites in the late 1990s that streamed radio broadcasts.

Liveness and immediacy are also integral to breaking news, and news talk radio was another area of heavy investment online in the 1990s. ABC Radio Networks partnered with RealAudio to launch ABC RadioNet in August 1995.[16] NPR and CNN, among other broadcast news networks, dabbled in internet radio early on. They typically only simulcast terrestrial radio feeds or created short "news flash"-style audio segments, however. These were then distributed to other web services such as AudioNet or Progressive Networks' RealNetwork. The relationships between internet radio and legacy media institutions, in particular broadcast news outlets, were plentiful in the heady days of the dot-com boom. Nevertheless, as early as 1997—well before the dot-com bubble burst of 2000—legacy media companies were already scaling back on their online ventures, in part because the advertising models were unstable and internet users were slow to adopt the web for shopping.[17] The result of the existing mainstream media's hesitancy to invest heavily in internet radio (and internet media, more generally) was that it remained largely a space for amateurs and independent start-ups to experiment.

Moreover, during the late 1990s countless major online entertainment ventures stalled; the web proved most successful as a medium for ordinary communication (chat, email, etc.) and other utilitarian, "dull data" functions: weather forecasts, maps with driving directions, phone number directories, travel information, job listings, and so on.[18] Put simply, audiences were most attracted to websites and services that were built around rather basic communication and information functions, and which easily fit into the rhythms of their everyday lives. As John Durham Peters points out, the internet most effectively (and profoundly) serves logistical functions of tracking and orientation that help navigate us through our daily lives.[19] Entertainment content proved slow to catch on. Unlike web television and multimedia "webzine" initiatives that demanded considerable audience attention (as well as relied on bandwidth-hogging video and motion graphic technologies), the everydayness of most talk and news radio content made these formats uniquely suited to the mundane internet. To evoke Pseudo's ChatRadio concept (chapter 2), a more inclusive, open approach to online talk radio was something of a natural evolution of the chat rooms and real-time messaging that helped popularize the internet during the 1980s and 1990s.

Other internet radio services like eYada built on the (semi)professional media network approach of Pseudo, including its emphasis on alternative lifestyles and sensational content. The brainchild of radio veteran Bob Meyrowitz, who made his mark in the 1970s with the syndicated rock and roll radio show *King Biscuit Flower Hour*, eYada launched in 1999 as a webcast-only all-talk station focused on gossip and "the most outrageous talk on the internet."[20] Flush with tens of millions of dollars in investor capital, eYada employed "all-star" hosts such as *New York Post* Page Six editor Richard Johnson, *New York Daily News* gossip columnists George Rush and Joanna Molloy, CourtTV personality Lionel (né Michael William Lebron), raunchy comedians Dan Schulz and Scott Wirkus (*The Dan & Scott Show*), and Bob Berkowitz, former host of the CNBC sex advice series *Real Personal*.[21] One of the online network's more prominent hosts was the iconoclastic Sex Pistols singer John Lydon (aka Johnny Rotten), who broadcast a weekly freeform talk radio program from his Los Angeles living room. The ad-supported original programming was webcast live and then later made available for on-demand streaming. Eventually eYada expanded from its original celebrity gossip and entertainment news channel to include "Sports & Fitness" and "Sex & Comedy" channels, producing as much as 150 hours of programming per week. Yet eYada's run was short-lived: it shut down in July 2001, another victim of the withdrawal of online advertising dollars that followed the dot-com bubble burst.[22]

One of the weaknesses of eYada's approach was that, despite encouraging audience interactivity through live chat rooms plus emails and phone call-ins, it was essentially replicating a traditional radio model of heavily curated, professionalized, "quality" content.[23] The online network adopted familiar talk show formats like entertainment news, comedy, and call-in advice. And these were hosted by established personalities, most of whom were drawn from old media: rock stars, television hosts, print journalists. The content might have been "edgier" than broadcast radio: "sex, drugs and lesbians . . . oh yeah, and celebrity interviews" proclaimed the description for *The Chaunce Hayden Show*; "talk about sex, drugs, girls, rock & roll, celebrities, girls, girls, girls" was how *The Big Show with Adam Paul* was summarized.[24] Nevertheless, eYada was merely trying to map an old-media business model onto a new-media culture.

The years 1999–2001 coincided with the emergence of Web 2.0 and an internet culture that was increasingly geared toward audience interaction and collaboration via user-generated content and networked technologies—what by the mid-2000s would alternately come to be known as *social media*. The emphasis in Web 2.0 was on media built on the contributions of ordinary web users who were engaged in networked digital environments (think sites like MySpace, Flickr, or Wikipedia and activities like remixing, tagging, linking, and sharing). "The tenets of Web 2.0 entice audience members to join in the building and customizing of services and messages rather than to expect companies to present complete and fully formed experiences," write Henry Jenkins, Sam Ford, and Joshua Green.[25] Ideally, content was created by amateurs or, more to the point, through the collaborative work of multiple authors—which could mean groups of amateurs but also professionals coordinating with amateurs (hence, hybrid terms like "pro-am," i.e., professional-amateur). It is not as though this was a culture completely devoid of celebrities or authoritative voices. However, Web 2.0 audiences were drawn to commercial media that actively encouraged fan participation. They also preferred grassroots internet stars over ones imported from old media. eYada and its ilk, then, were out of step with the times, still trying to force a top-down media production model on a culture that was embracing bottom-up modes of authorship and active consumption.

Mere months before eYada debuted, back in January 1999, another all-talk internet radio start-up launched with a contrastingly "do it yourself" business model.[26] TalkNetRadio (which soon changed its name to GiveMeTalk!) positioned itself as a platform for user-created radio: "TalkNetRadio's mission is to allow any user to create and broadcast original programming to listeners worldwide," stated the Silicon Valley–based company's founder, internet marketing executive Ted Ganchiff.[27] Rather than adhere to standardized topics or formats, GiveMeTalk! espoused an anything-goes programming philosophy (hardcore pornography and hate speech were among the few content restrictions imposed on contributors). "For individuals, this means creating programs about hobbies, special interests, foreign-language programming, sports, anything they can come up with," Ganchiff explained.[28] The service "opens up the world of internet radio to anyone," he further proclaimed.[29] "Everybody has wanted a soapbox. This is a way to get your voice out."[30] Ganchiff and his colleagues were quick to connect this approach to a free speech rhetoric. "We are the place on the internet where free speech is really free," announced CEO William Gross.[31] GiveMeTalk! worked by giving registered users access to free, downloadable GiveMeTalk! Home Studio software (developed under contract by RealNetworks), which when combined with a microphone enabled individuals to record themselves offline on their PC and then upload it to the service's website for on-demand streaming or download. There was no preset broadcast schedule, although to preserve bandwidth the streaming audience for any given episode was capped at thirty listeners at a time.[32] GiveMeTalk! unceremoniously ceased operations in late 2000. Its basic talk radio hosting platform model, however, would be repeated by other companies—including the popular BlogTalkRadio, which began in 2006 and is still in operation today. Furthermore, its emphasis on original, user-generated internet radio content (as well as time-shifted recording and listening) signaled the shifting tides from Web 1.0 to Web 2.0, and predated podcasting by a couple years.

Podcasting History and Its Misconceptions

Most histories of podcasting begin around 2005, when the practice hit the cultural mainstream and spawned hundreds, if not thousands, of enthusiastic news stories about an "audible revolution" and "the people's radio."[33] There was a feverish response to this presumably new method of distributing "audio broadcasts" via the internet: as one example, the prominent tech blogger Doc Searls began tracking the number of Google search engine hits for the term "podcasts," finding a mere twenty-four search results in late September 2004; that number rocketed to over seventy-three thousand a month later and nearly fifty-seven *million* a year later, in September 2005.[34] 2005 was declared "the year of the podcast" and, also that same year, *New Oxford American Dictionary* named "podcast" its Word of the Year.[35] Likewise, a wave of articles in 2015 celebrated podcasting's tenth anniversary, reinforcing the popular conception of 2005 as podcasting's birthdate. The fact of the matter is that podcasting originated years before 2005—at least a few years and as much as a decade earlier, depending on one's definition of a podcast. Indeed, the real history of podcasting is in large part obscured by the very name *podcasting*.

Apart from the rush of mainstream media attention that brought podcasting into the popular consciousness, there are a number of reasons why 2005 is routinely misidentified as podcasting's origin date. First, the term "podcasting" did not exist until 2004, and without a recognizable and widely agreed-upon name it was impossible for the emergent practice to be legible to anyone outside the small community of technologists and bloggers who were developing it. What coalesced as podcasting had for a few years prior been called a whole range of different things by different people: audioblogging, DIY radio, MP3 messages, to name a few. This made it hard for journalists and potential listeners, as well as media-makers, to identify and comprehend. Second, blogging had been the internet media phenomenon of 2004: "blog" was picked as Merriam-Webster's Word of the Year in 2004 and the previously niche web publishing practice had begun to make a sizable impact on the mainstream media, most notably during the 2004 US presidential election.[36] Media and technology pundits were eager to spot the next big thing in/after blogging, and podcasting easily fit the bill. Even though "podcasting" had beat out the term "audioblogging," nearly every cultural commentator in 2005 made the direct connection between blogging and podcasting. Third, the Apple iPod was *the* must-have consumer electronics gadget of the 2004 holiday season: iPods made up 90 percent of the rapidly expanding portable media player market. Apple had sold some ten million iPods by the end of 2004, and though the device had debuted back in 2001, nearly half of those iPod sales were made in the final few months of 2004 alone.[37] The fad for personalized mobile media devices was in full effect, and all of those new iPod owners were searching for content to fill their hard drives. While many in the podcasting industry today complain about the arcane name "podcasting"—including its associations with a single company (Apple) and a device (the iPod) that is now nearly obsolete in the smartphone era—the fact of the matter is that, circa 2005, these close associations with the iPod helped significantly in bringing the media practice to prominence. Fourth, and relatedly, Apple integrated podcasts into its iTunes digital music software with the release of iTunes 4.9 in June 2015.[38] While other podcast directories and "podcatcher" software already existed for aggregating, managing, and transferring podcast content to portable media players, including iPods, the presence of simple podcast features in the world's largest online digital media store made podcast listening easily accessible to an immensely larger audience than before. In other words, 2005 did in fact mark the point when podcasting entered the cultural mainstream. Nevertheless, its earlier history is significant, both in order to understand what podcasting eventually became and what else it could have potentially been.

There are *three waves of podcasting*: the *first wave* runs from 2000 to

2005, the *second wave* from 2005 to 2010, and the *third wave* from 2010 to the present.[39] This periodization contrasts with that offered by other media scholars, such as Tiziano Bonini, who have suggested that we are currently in "the second age of podcasting"—the "first age" existing from 2004 to 2011 and the "second age" from 2012 to the present.[40] Such models do not account for the crucial first wave from 2000 to 2005, in which a yet-to-be-named podcasting developed alongside blogging as an unconventional form of personal media. The second wave, running from 2005 to 2010, contained the practice's first surge in popularity, marked by a flood of repurposed terrestrial radio content and a still relatively small yet consequential group of podcasters who explored the potentialities of the form, particularly in the interview/roundtable talk show and comedy formats. The ongoing third wave started in 2010 and includes the post-2014 surge in popularity marked by *Serial*. (The popular narrative carried out in the press tends to imply that *Serial* somehow created modern podcasting as we know it—though it must be noted that most of the podcasts that rose to popularity soon after *Serial* had roots going back at least a few years prior. *Serial* merely shone a massive spotlight on work that was already underway.)

These three waves of podcasting overlap, and there are many continuities between them. Nevertheless, there are distinct attributes to each wave. Unlike the first two waves that were dominated by talk-centric programs, this third wave is marked by a shift toward more professionalized and produced "storytelling" radio feature-documentary programs (as explored in chapters 6 and 7). There is also a prehistory of podcasting from 1993 to 2000, in which numerous iterations of internet radio emerged that, while they are not usually referred to as podcasts, in retrospect clearly have a lot in common with podcasts. The remainder of this chapter focuses on the first wave and pieces of that prehistory.

Proto-Podcasting in the 1990s

The pre-2005 history of podcasting is routinely glossed over, if acknowledged at all. If there is a generally accepted story that encompasses this first wave, though, it goes something like this: Adam Curry, a radio host and former MTV VJ who remade himself as an internet entrepreneur, wanted to find a way to deliver audio files over the internet without the wait involved in either on-demand streaming or downloading. After discussing this conundrum with Curry in late 2000, software developer Dave

Winer published RSS 0.92 in January 2001, a new version of the RSS (Rich Site Summary or Really Simple Syndication) web syndication format—one of the backbones of blogs and newsfeed aggregators—that enabled digital audio files to be delivered in RSS feeds.[41] Over the next few years, Winer, Curry, and a handful of other bloggers and techies—many of whom were associated with Harvard University's Berkman Center for Internet & Society (now the Berkman Klein Center) or Winer's BloggerCon conference series—experimented with carrying audio files in RSS feeds. It was not until 2004, however, that podcasting emerged as a viable technology. And crucially, the term "podcasting" was coined haphazardly by *Guardian* writer Ben Hammersley in a February 2004 newspaper article.[42] Many cite as watershed moments in its path to widespread use Curry's release of an RSS-to-iPod "podcatcher" client, iPodder, along with his launch of podcasting's first breakout program, *Daily Source Code*.[43]

While everything in this basic story is factual, it should perhaps come as little surprise that this narrative omits a number of integral people and events. It also does not help that the figures involved have subsequently diverged on and feuded over some of the central facts. (This discord partly stems from Curry being caught anonymously editing the Wikipedia entry on podcasting to emphasize his own contribution.)[44] This tale is overly focused on the technology, though, so much so that it is almost completely devoid of agency or motive. *Why* were these individuals, most of whom had no prior connection to radio broadcasting or webcasting, interested in internet radio in the first place? And why were they not satisfied with the plentiful online radio options that were already in place?

There were numerous versions of internet radio prior to the year 2000, some that even enabled on-demand, time-shifted listening (as opposed to livestreaming), and most of which accommodated downloading and mobile listening too. These are the essential features of podcasting according to practically any definition of the term: digitally downloadable radio programs available for personalized consumption—the audience being in control of what they listen to and when, where, and how they listen to it. Many definitions of podcasting pinpoint RSS (Rich Site Summary or Really Simple Syndication) web syndication technology as its distinguishing factor. However, the significance of RSS has greatly diminished in the internet environment of the past decade, as there are now various ways to distribute digital media files over the web without utilizing syndication.[45] Furthermore, going back to the very beginning of internet radio in 1993, the Internet Multicasting Service's original programming was prerecorded and distributed online via downloadable files. IMS was producing, in founder Carl Malamud's words, "asynchronous radio"—or what he alternately called "random access radio"—referring to the fact that the delivery was time-shifted and that the audience was in control of the listening experience. It was even mobile: recall the stories of people copying the audio files onto their laptops and playing them through their car stereos. Technologically speaking, that covers most any modern definition of a "podcast": digital, internet-distributed, downloadable, mobile. But in existence more than a decade before podcasting's oft-acknowledged birthdate of 2005. GiveMeTalk!, described earlier this chapter, was also facilitating something remarkably similar to podcasting in 1999–2000.

There were yet more proto-podcasting technologies and services. In 1997, nearly a half-decade before Apple's iPod, a Silicon Valley company called Audio Highway developed the first portable personal digital media device, the Listen Up Player. Audio content could be downloaded free to a desktop computer using the company's AudioWiz software (a precursor to Apple's iTunes). Anyone could listen to the content there on their PC or, if they wanted on-the-go mobile playback, they could transfer it to the Listen Up Player, which retailed for $300 (later dropped to $200). Audio Highway carried some music downloads but the bulk of the offered content consisted of audiobooks and syndicated talk/news radio programming from NPR (*Morning Edition*, *All Things Considered*, *Talk of the Nation*) as well as CNET, *Newsweek*, and the Associated Press (all of which were at that time creating original news radio segments for the internet).[46] The digital audiobook company Audible also produced an Audible Mobile Player in 1998–99; among the content that subscribers could port to the device were talk radio programs *Fresh Air* and *Car Talk*.[47] There were other portable "MP3 players" before the iPod, too: Saehan/Eiger Lab's MPMan and Diamond's Rio devices were both released in 1998.[48] SHOUTcast and i2GO were additional podcasting precedents that enabled personalized audio-on-demand listening experiences in the late 1990s.[49] Put simply, there was plentiful on-demand as well as portable internet radio content years before podcasting.

Furthermore, by the mid to late 1990s, NPR and many of its member stations, along with numerous other terrestrial stations and online-only radio outfits, were supplementing their livestreaming webcasts with archives that could be accessed at any time. Aggregators like AudioNet/Broadcast.com also archived much of the radio programming these outlets carried. It can be misleading, however, to call these "archives," as they were remarkably limited in scope and short-lived. Few individual stations had the storage or bandwidth capacity to preserve all of their programming online in perpetuity. Rather, they would make available the output of only their flagship shows and/or present abridged broadcasts and other select audio clips. Typically, these files would then be cycled off the website every few weeks for storage reasons. Nevertheless, these saved files could be listened to in a time-shifted manner—later in the day, or days or even weeks later (depending on how long the provider maintained them).

As a result, on-demand listening was a widely available feature of internet radio from the very start. It was not without its complications or restrictions, though. Depending on the website's choice of streaming audio software, playback might have been restricted only to on-demand streaming in a web browser. Most would have allowed for offline listening via download, although if proprietary streaming audio software like RealAudio was being deployed then listening would have been restricted to a desktop or laptop computer with the RealPlayer application installed. In fact, this predominance of proprietary software and walled garden platforms restricted the free movement of internet radio content in the 1990s, and the podcasting community's embrace of open-source software and formats like MP3 is one of the factors that set it apart and allowed it to flourish. Indeed, though there was plenty of podcasting-like content and numerous technologies available in the 1990s, none of these predecessors ever coalesced into an identifiable practice or cultural movement like podcasting would in the mid-2000s.

The phenomenon of podcasting as it is known today grew out of the blogging-inspired "audioblogging" community that Dave Winer and Adam Curry helped spearhead. The mere existence of precedents does not mean the prior technologies and platforms were culturally successful. Media historian Brian Winston has developed a model of change and development in media technology that he has applied to every medium from the telegraph to the internet.[50] In each instance, he shows that there is a pattern of what he calls scientific competence and ideation, prototype and "invention," and socially driven diffusion and suppression. Within this model, he outlines three stages of technological transformation that any technology must go through in order to become socially useful and accepted. The first transformation is "ideation," where basic scientific competence or know-how—which is socially determined to begin with—is used by technologists to envision a device that is eventually developed into a prototype. A prototype, though, does not become an invention until there is a widespread social necessity for it—which is why Winston puts "invention" in quotes, arguing that a device is not actually acknowledged as an invention until it achieves general cultural acceptance. (Thus, a lot of prototypes never achieve the status of "inventions.") This process is dependent upon the second transformation: "supervening social necessities" or social circumstances, ranging from changing cultural tastes to business or military demands. These social factors move a prototype from the lab into the world at large. Winston's notions of invention and social necessity suggest that technologies must always "fit" into preexisting social patterns, thus doing away with naive ideas of technological determinism and spontaneous invention. Here, citing Fernand Braudel, Winston inserts into his model the phenomena of historical "brakes" and "accelerators": social forces that both create the demand for, but also slow down or otherwise impede, the introduction of new technologies.[51]

These proto-podcasting initiatives obviously made it into production. Audio files were distributed on the internet and listened to. Listen Up and Audible devices were manufactured and sold. In retrospect, however, they are what we could call *failed technologies*. They were pushed to market but there was not enough of a social necessity for them to achieve broad cultural acceptance and become diffused widely. Services such as Audio-Net, though still niche when compared to over-the-air broadcast radio, were very popular for the internet at that time. All available accounts suggest, however, that their livestreaming functions were much more utilized than their on-demand archives. The marketing of AudioNet and all other webcasters of the late 1990s also heavily promoted live listening. Radio broadcasting has historically been most closely associated with live, real-time listening, and early webcasters primarily focused on exploiting this established expectation of liveness. Time-shifted listening was something of a novelty, a feature that was offered but little utilized. Likewise, the early handheld digital media players may have offered the potential of personalized mobile listening; however, the demand for it was not yet there in 1997–1999. For instance, Audible only sold about four thousand of its mobile players before discontinuing it.[52] In many ways, the ideas for the content and devices were simply ahead of the technology itself: the Listen Up Player, for example, could hold only one hour of audio (compared to the first iPod in 2001, which could hold ten hours). Also, USB ports did not become standard on desktop computers until the mid-2000s, and connecting devices like the Listen Up Player via a parallel port could be cumbersome and slow.[53] Using Winston's term, there simply was not a supervening social necessity for a podcasting-like practice in the 1990s. The question, then, is what about this particular technology and, more importantly, the culture surrounding it that made podcasting viable in the early 2000s?

"A Weblog for the Ears":
Blogging with Audio in the Early 2000s

The roots of blogging date back to the mid-1990s soon after the World Wide Web went public. Blogs are generally defined as publicly accessible personal websites: typically single-authored online publications featuring periodic posts that are organized in reverse chronological order, thereby emphasizing *frequency* and *recency*.[54] The term "blog" can be rather ambiguous, and many types of blogs exist, including multiauthored professional websites published by mainstream media corporations. Nevertheless, the discourse around blogging, particularly in the early 2000s, privileges those sites that emphasize personal opinions, plus a look and feel (tone and structure) that is more informal than traditional print journalism.[55]

The origins of blogs lie in online diaries and email mailing lists. The notion of blogging as a form of online diary writing or journaling has certainly dominated the popular discourse. For example, in its first article on the blogging phenomenon, *Time* magazine described a blog as a "spontaneous online public journal" filled "with whatever ramblings come to mind—what [the blogger] had for dinner, how their grandparents are getting along, their 10 favorite songs of the year—all sprinkled with links to cool Web pages they have discovered."[56] While not representative of all styles and genres of blogs, this was the prevailing conception of blogging established in the early 2000s web culture. Many scholars point to the emergence of online diaries in 1994 as laying the groundwork for blogging. One of the pioneers of the web-based diary was an undergraduate student at Pennsylvania's Swarthmore College, Justin Hall, who published a website called *Justin's Links*.[57] Initially a labyrinth of hyperlinks that told Hall's life story, he started writing more traditional diary entries—in a suitably rambling style—in 1996. Web 1.0 websites were mainly conceived of as fixed, completed products that did not often change. Those sites that were regularly updated, such as news sites or entertainment portals, visually and structurally replicated daily newspapers. Therefore, minimalist looking websites containing frequently updated diary-like discussions of an individual's personal life stood in stark contrast to the rest of the commercial web. These proto-blogs needed to be hand-coded, however, which meant that access to publishing was limited to individuals with computer coding skills. That began to change in late 1998 and 1999, when a number of easy-to-use blog publishing services, including Pyra Labs' Blogger (later acquired by Google) and early blog-like social networks such as LiveJournal, made online journaling simple and pervasive.[58]

Also back in 1994, computer software and hardware developers were looking for a way to discuss news in the tech industry. One of them was software developer Dave Winer, then already a fifteen-year industry veteran with a reputation as a provocateur, who felt that the trade press was corrupt and either ignored or poorly covered computer industry news.[59] He sought a way to circumvent the media establishment and converse directly with his colleagues. Email would not become mainstream until the late 1990s, but in the computer business email accounts were already widespread by 1994. The well-connected Winer assembled a mailing list of all the influential industry insiders he knew and began circulating his short essays to everyone on it. The messages were unpolished, usually only a couple hundred words long, and read "less like a trade-magazine column than like a late-night discussion among friends."[60] DaveNet, as it was called, did not allow subscribers to reply to the whole list; however, Winer replied to feedback and selectively quoted from some of the response emails he received. This gave the mailing list a public forum-like quality. He also archived the DaveNet email messages online.[61]

Winer continued distributing longer essays through email but, in April 1997, he sought another venue for voicing shorter, more time-sensitive messages. That is when he created *Scripting News*—one of the first widely read blogs—by expanding on a "news and updates" webpage he had started a year prior for his scripting software company Frontier.[62] Winer would post information about new products, link to articles, and share casual observations along with inspirational quotes and other random ephemera. The visual appearance was austere: bare black text on a white background, blue-underlined hyperlinks with little to no commentary, no images or graphics, and a "Community Directory" sidebar. Winer would publish almost daily, although most of the posts were only a few lines long and consisted of little more than hyperlinks to other web sources.[63] Indeed, many of the earliest blogs were made up almost entirely of lists of links to other websites of interest, accompanied sometimes (but not always) by brief commentary. The name "blog," as a matter of a fact, is an abbreviation of the term "web log," which when it was first coined in 1997 was literally meant to represent a log, or chronicle, of websites.[64]

From the very start, then, blogging functioned as a sort of metacommentary. The practice was diary-like, in the sense that it was a record of one person's thoughts and experiences. But more to the point, many of the first blogs were personalized explanations of the web itself, the authors directing their readers to what was going on elsewhere in the online world. To put it another way, blogs were websites about websites—or, websites about web culture, more accurately.

In addition to simple and free publishing platforms like Blogger, one of the tools that helped propagate blogging circa 2000–2005 was the web syndication format RSS, which provides metadata information about websites and allows content to be shared between sites. Most importantly, it allows people to subscribe to updates from their favorite websites. Anyone could publish a blog, but this democratization of content was only exacerbating the infoglut. Keeping up with all of the web's real-time content was becoming nearly impossible in the early 2000s, as the blogosphere was expanding and daily or even hourly posts were piling up. With RSS, web users did not need to visit each and every website separately; instead they subscribed to a website's RSS feed and then new posts were delivered automatically to aggregator software downloaded to the users' computer, known as a feed reader (e.g., Google Reader, Feedreader, Feedly). This software configuration (RSS feed plus RSS feed reader) effectively turned blogging (and later, podcasting) from a *pull media*, where each piece of content sat waiting for the audience to find and "pull" it, into a *push media*, where the latest subscribed content was immediately delivered or "pushed" directly to the self-selecting audience.

Dave Winer was one of the inventors of RSS technology in the late 1990s and, as previously mentioned, it was a desire to expand web syndication to include multimedia content that first brought Winer and Adam Curry together in 2000. This partnership eventually led to the development of one strain of audioblogging, aka podcasting.[65] Curry's idea was to get around the still-persistent problem of slow, poor-quality audio-video streaming (the "click-wait" problem, in Winer's words) by allowing large digital files to be "dripped in" overnight, or at other off-peak times when people's computers were not in use.[66] He called this "the last yard": taking static audio-video files posted to a website and removing the inconvenience of on-demand streaming and the then-inescapable problem of buffering by scheduling downloads of the content in advance, making it available for time-shifted instantaneous playback whenever the individual audience member desired.[67] It was Winer who suggested using RSS to achieve these ends, and in January 2001 he modified the text-based RSS to also deliver "multimedia payloads," also known as "enclosures."[68] (Others had previously proposed the idea of using RSS feeds to deliver digital audio files, although Winer appears to be the first to have actually executed it.)[69] Most feed aggregators of the era would not support multimedia, although Winer's Radio UserLand desktop aggregation software enabled audio-video playback as early as 2001.[70] The practice did not really take off until July 2004, however, when Curry released a stand-alone audio feed receiver, iPodder, which not only aggregated audio content but also connected that content to an MP3 player. The free, open-source iPodder was what has come to be known as a "podcatcher," and it was quickly followed by a number of other software apps, including iPodderX (later called Transistr), Doppler Radio, and eventually, iTunes.[71] It was this software that put the "pod" in podcasting, enabling mobile listening by shifting the audio files from the desktop computer to the portable media player.

The fact that podcasts circa 2000–2005 were downloadable distracts from the overwhelming similarities between podcasting and broadcasting. These are similarities that were obvious to the technology's earliest developers and users at the time, but which have been discursively buried in the years since. Popular discourse over the last two decades, particularly surrounding digital music, has established a starkly oppositional relationship between streaming and downloading. Again, the initial need to make podcasts downloadable and deliverable via RSS was a solution to a historically situated technological problem—that is, the low bandwidth in the early 2000s that made listening to streaming audio on the internet troublesome. Winer, Curry, and their colleagues were, in fact, resorting to downloadable audio because they felt it *most* resembled the broadcast experience. The "click-wait" lag of streaming audio disrupted the long-established expectations of audiences accustomed to broadcasting, who anticipated that when they switched on a radio the content would stream flawlessly and continuously. In other words, livestreaming online radio in the early 2000s actually lacked liveness, since the transmission (the "now" of output) was routinely disrupted by buffering. Their solution was that if the content was automatically delivered and downloaded on your computer hard drive, then the next time you logged onto your machine you could hit play on the audio file(s) and "the wait is zero."[72] Your continuous flow of "live" content would again be intact. You are, in effect, creating your own customized radio station featuring exclusively the content you are interested in.

It was not lost on these early adopters that the technology shifted power to the audience and also opened up opportunities for more democratized radio production. Nevertheless, the most important point early on was that RSS with enclosures improved the audience's experience, making it resemble traditional modes of broadcast media consumption as closely as possible. Significantly, radio's liveness is *not* reliant on simultaneity (i.e., real-time production) but rather it is defined through the experience of the listener. The implementation of RSS was aimed at erasing the sense of disconnection that primitive streaming technology created, the time-shifted listening actually making the audio signal once again endlessly copresent. This actually enhanced the perception of liveness, as counterintuitive as that might initially seem.

There are, of course, some significant differences between live broadcast radio and podcasting, most notably that podcast listening is time-shifted and the individual listener, not the centralized broadcaster, determines what to listen to and when. However, podcasting early on was largely understood as an extension of traditional radio rather than an alternative to it. For instance, countless articles compared podcasting to a "TiVo for radio," referring to the digital video recorder (DVR) brand that was transforming the broadcast television industry in the early 2000s.[73] The prevailing idea was that with RSS's capacity for scheduling and filtering content downloads, you could stop at a convenient time and catch up on that morning's or the previous day's internet radio content interspersed with print news stories and video. "Then when I arrive in the morning," wrote Dave Winer when first describing the software, "there are fresh bits, news clips, a song of the day, whatever, provided by all kinds of content providers, from big TV networks like CNN and MSNBC, to a Dutch school where kids are taking a film class using inexpensive video recorders and iMacs."[74] It was always understood that the technology would support existing media practices and institutions as much as it might create challenges to them. Whatever the case, the relationship between podcasting and downloading was something of a temporary means to an end, and the current era of on-demand audio streamed via podcast platforms/apps on high-speed smartphones is arguably more true to the original spirit of the idea.

Radio and broadcasting were understood as metaphors not only for podcasting but also for blogging. Although blogging and RSS evoked many features of print media (e.g., publishing, subscription, diary writing), there were also frequent allusions made to broadcasting. For instance, RSS feeds could be compiled into "channels"; individual users merging different feeds within their feed reader software into customizable groups ("sports," "news," "gossip," etc.). The basic idea being that, based on your mood or taste, you could tune to these the way you would a radio or television channel: if you wanted the latest news headlines you would switch to your "news" feed, the way you might choose CNN on TV; if you wanted the latest baseball scores you would flip to the "sports" feed, the way you might choose ESPN on TV; and so on. In fact, this was the single most impactful innovation of RSS technology: it turned previously dispersed and isolated audio files on the web into a sequenced channel. Similarly, Winer's aggregator software was designed first and foremost for online print news, yet he named it Radio UserLand—the "radio" being a metaphor for how it pushed (i.e., broadcast) content from multiple sites to one source, much the way a radio set pulls together an array of audio signals into one device. Furthermore, the real-time function of RSS feeds recalled radio's liveness. As prominent blogger Andrew Sullivan observed in 2002, "Poised between media, blogs can be as nuanced and well-sourced as traditional journalism, but they have the immediacy of talk radio."[75] New-media scholar danah boyd also argues that blog writing blurs textuality and orality.[76] In other words, if blogging is to be regarded as a medium distinct from traditional modes of textual production, then one of the components that distinguishes it most is the fixation on newness and recency. Attention is focused on the newest posts and the most current information, and this immediacy is tied to a writing structure and style that is chatty, affable, and personality driven—qualities very much akin to talk radio.

Podcasting's early history is best understood through an examination of its content and form. The discussion of podcasting too often gets obfuscated through an overemphasis on its technology (RSS, downloading, etc.). More attention must be paid to the *who* and *what* of podcasting (the people making the podcasts and the shape of that content), not just the *how* (the technology). As we have seen, Winer and Curry conceived of podcasting in 2000 and put the basic technological tools for the practice in place by early 2001. The fact of the matter, however, is that hardly anyone used it at first. Curry was the experienced radio personality, and yet he did not actually produce anything resembling a podcast until August 2004. (Moreover, despite his eventual reputation as "the podfather," Curry was initially interested in a subscription-like system for delivering digital *video* content, not audio.)[77] Winer was a technologist and writer who had no experience with radio broadcasting whatsoever; his primary interest was in pushing the technology of RSS forward. He tested out his system for RSS with enclosures not by creating original internet radio content but rather by delivering a different Grateful Dead song each day to the RSS feed's subscribers.[78] The problem was that very few people subscribed. "We threw it out there, we evangelized it. It was like, 'If you build it, they will come.' But they didn't come," said Winer.[79] He went back to blog writing and other pursuits, and the idea of multimedia blogging lay mostly dormant until the summer of 2003.

Committed content creators were needed in order to transform podcasting from a prototype into an actual "invention" with social utility. Podcasting began in earnest in mid-2003 when Winer was awarded a fellowship at the Berkman Center for Internet & Society at Harvard University. He was there "to get Harvard blogging"; however, he soon met another Berkman fellow, Christopher Lydon, and the two started collaborating on self-produced radio segments for the web, or audio blogs.[80] Lydon was a journalist and popular Boston public radio host who had been fired from WBUR in 2001, after Lydon attempted to take co-ownership over his weekday call-in program *The Connection*.[81] Ousted by the media establishment, Lydon became enamored with the self-publishing power of the web. He quickly transitioned to producing a "live streaming audio webcast" through his personal website, christopherlydon.org. The program was streamed weekly on Tuesdays at 10:00 a.m. (EST), as well as archived online afterward. It was also syndicated by a few terrestrial public radio stations. The webcast featured typical NPR news magazine-style fare: political discussion and debate, interviews with authors, and so on.[82]

At Berkman in 2003, Lydon was studying blogs alongside Winer when he penned an essay praising the nineteenth-century transcendentalist philosopher Ralph Waldo Emerson as "a god for bloggers." He wrote, "In the booming energy of blog world, we are glimpsing the fulfillment of an Emersonian vision: this democracy of outspoken individuals," to which he added, "I want to embrace bloggers in general as Essential Emersonians: radical democrats and individualists, sick unto death of our imprisonment by mass media, mass emotion, the retribalization in our time by mass labeling, mass marketing, mass following, sheepish mass everything. That modern Emersonian is nonetheless cheerfully banking on the reawakening of individual conscience, individual ambition, individual possibility."[83] Inspired by his colleague's words, Winer encouraged Lydon to record the essay as an "audio blog," which he subsequently posted to *DaveNet* for the world to hear.[84] Winer and another Cambridge-based computer engineer, Bob Doyle of the Desktop Video Group, set Lydon up with a digital recording studio and protocols for posting and distributing audio with an associated RSS feed.[85] Lydon was a veteran radio producer who had all the knowledge and skills needed to record audio. Nevertheless, despite all the rhetoric in the 1990s and early 2000s about how easy it was to become an internet radio star—many variations of "All you need is a computer, a microphone, and an internet connection!"—recording audio onto a standard desktop PC and formatting it for internet distribution was actually a rather difficult task. Without skilled technologists like Winer and Doyle at his aide, it is unlikely that someone like Lydon would have been able to pull off a full-blown internet radio program on his own.

Beginning in July 2003, Lydon embarked on a weekly series of audio interviews, dubbed *Christopher Lydon Interviews*. This audioblog—or "weblog for the ears," as Winer lyrically declared it—featured interviews between Lydon and various intellectuals, such as Harold Bloom and Cornel West, plus conversations with prominent bloggers and technologists, like Doc Searls, Julie Powell, and The Daily Kos.[86] These were among the first-ever podcasts, although the production quality was quite high and the overall style and tone was very NPR-esque: friendly and conversational but buttoned-up, casual but still very conventional. In other words, there was little to distinguish these audioblog interview segments (or Lydon's previous webcasts, for that matter) from traditional talk radio programming, other than that they were never broadcast/webcast live and instead circulated only via RSS.

In the fall of 2003 and into 2004, a number of other individuals directly or indirectly affiliated with the Winer-Curry circle of bloggers and technologists began experimenting with podcasts. Many of these individuals were associated with the BloggerCon conference, which Winer organized and which was held at Harvard's Berkman Center—the first one taking place in October 2003. Their contributions may be underacknowledged but their names are still known, among them Stephen Downes (*Ed Radio*), Dave Slusher (*Evil Genius Chronicles*), Doug Kaye (*IT Conversations*), and Steve Gillmor (*The Gillmor Gang*).[87] Notably, much like the Internet Multicasting Service in the 1990s, these earliest programs were micro-niche talk radio shows mainly about computer technology and the techie culture. Dave Winer started his own "audio weblog" series, *Morning Coffee Notes*, in June 2004.[88] Adam Curry debuted his daily talk program *Daily Source Code* in August 2004. Winer and Curry also collaborated on a short-lived series called *Trade Secrets*.[89]

Despite not being the first podcast (a frequent misattribution), Curry's *Daily Source Code* became the catalyst for a much more widespread phenomenon of podcasting—"podcasting" being the new name that got embraced by the audioblogging community in September 2004.[90] As its title suggests, Curry's podcast was still tech-centric (he talked a lot about his ongoing work as an internet entrepreneur), yet it also mixed in a considerable amount of personal musings, diary-like admissions, and random miscellany, such as music from his favorite up-and-coming recording artists. He frequently recorded episodes while driving in his car, and the finished product featured a single take with little or no editing. For instance, in his very first episode, Curry encounters a horse and buggy along the road he is driving on in the Belgian countryside; rather than stop the recording and pick it up again later, he keeps the audio rolling while he navigates around the traffic obstacle, narrating the entire diversion in real time to the would-be listener, whom he addresses in a convivial first person.[91] This sort of free-flowing, conversational style had a "Hey, I can do that" accessibility that inspired newcomers to join the podcasting ranks.[92] It stood in contrast to more professional-sounding audiobloggers like Lydon, whose "NPR-like thing," Winer admits, did not come across to the average listener as something they could do themselves.[93] And yet Curry's approach was not unfocused rambling either: a professional radio and television host, he was comfortable extemporizing in front of a microphone. As a result, his podcasts had the off-the-cuff vibe that so many listeners today construe as "intimate" and "authentic," yet it was something of a controlled chaos that was compelling and actually entertaining. Such a balance requires considerable training and skill. Curry's celebrity also attracted listeners, who otherwise would not have paid attention to the insider tech talk of most other early podcasters.

All of the podcasts that rose to popularity in 2004–5, along with most of the followers that then copied them and created the world of podcasting as we know it today, were *series* based on the broadcast radio model of episodic programming. This adoption of the basic structure and form of broadcasting is a key reason why I contend that podcasting is not an entirely new medium but rather a new radio practice. The shift from the *audioblog post* to the *podcast episode* as the fundamental unit of podcast culture represents a deviation from some of the core principles of blogging. That is, there is more to the transition from audioblogging to podcasting than a simple name change. The relatively short-lived and little-used practice of audioblogging, in place from about 2002 to 2005, represents an unrealized future for podcasting—one that could have potentially developed into a unique and stand-alone medium (although, as I emphasize in this book's conclusion, a more audience-centered participatory audio medium like this is something theorists have envisioned for radio since its very beginnings). And it is a history that involves a whole set of characters outside the Winer-Curry community who are barely remembered today.

The *post* is the primary unit in blogging, compared to the page in non-digital print culture. A blog, at its most basic level, is "a collection of posts (each with its own permalink), appearing in reverse chronological order, time-stamped, and archived," according to Jodi Dean.[94] (The other fundamental structure of blogging is the *link*—since blogs tend to be self-published and decentralized, it is through the networked practice of linking to sources and to other bloggers that communities of bloggers form and interact. An RSS feed is also a link, and thus at a technical level linking makes up the primary method of distribution for a blog—or a podcast, for that matter.) This emphasis on the post gives blog writing its unique character, creating a situation where frequency, brevity, and personality are prioritized above all else.[95] There is an expectation that blog posts are relatively short in length but issued in quantity, and that stylistically they are highly subjective and written in the first person. (Even for blogs that are written by a group of contributors and that adopt a more journalistic approach, posts will be clearly opinionated and eschew the attempted objectivity of traditional journalism.)[96] This centrality of the post has clearly carried over from blogging to social networking websites, in particular "microblogging" platforms like Twitter and its 280-character messages. The post—or its stand-in, the *message*—has become the primary unit of all digital culture in the past decade-plus.

The audio equivalent of a blog post, an *audioblog post* would be a short, personal audio recording. Blogs are reflexive and reflect on every aspect of contemporary life: a blogger might post about political views one minute, comment on careers the next minute, and share a dinner recipe later in the day.[97] Whereas a written blog post on average runs a couple hundred words, the audio equivalent contains only a couple minutes of an individual monologizing on a topic. In totality, an audioblog features a succession of these brief posts, most of which are one-off dispatches that do not directly connect to each other apart from the fact that they come from the mind of a single individual (or a closely tied group of individuals). The main thing uniting the posts is the audioblogger's personality, not a set of formal or structural conventions. An audioblog post, if not formless (since to lack any form is, theoretically, itself a form), comes as close as possible to lacking a formal structure or shape. It is, in effect, the expression of a single thought from start to finish, unscripted and unedited, warts and all. In this way, it would be akin to a telephone answering machine message or a voice memo that is broadcast via the internet. Another corollary would be a ham radio or CB radio message. If in print culture a blog post is juxtaposed with the more formally constructed and periodic unit of the page, then in audio culture the audioblog post stands in contrast to the radio episode: a formally constructed and longer block of programming. (Even an open-ended call-in talk radio show adheres to standardized conventions, such a sign-on/sign-off and multiple segments. In contemporary practice, a radio episode usually lasts thirty minutes, one hour, or two hours) An audioblog post is more like a sound clip or segment—one small piece of a longer episode. Moreover, an episode is one installment of a regularly occurring *series* of programs that continues the same story (in fictional programming) or explores the same predefined topics (in nonfictional programming), typically appearing at a set day and time. Even if the content was produced by amateurs and sounded the part, the series became the preferred form for podcasts after 2004. However, earlier attempts at audioblogging rejected these conventionalities.

Think It, Speak It: Audblog and Phone-to-Blog Services

Despite the anybody-can-do-it rhetoric of "broadcast your own show" and "be your own shock jock," starting and maintaining even an amateur podcast series required more of a commitment than most ordinary web users were willing to take on.[98] Recording a few quick personal audio posts every often so, on the other hand, was a much less daunting and more accessible prospect. If maximum participation in the mediated public sphere is the desired goal, then audioblogging in its nascent form seems a much more practical solution to get the audience speaking than does podcasting in its mature form. Beginning in 2002, at least a handful of individuals, including Garth Kidd and Harold Gilchrist, were tinkering with tools like Radio UserLand to share personal audio posts.[99] Most of their posts were only one to two minutes long. (As it happens, Radio UserLand had a 1 MB file size limit, a technical constraint that forced audioblog posts to be two minutes and under.) Since they were members of the Winer-Curry blogging tech community, much of Kidd's and Gilchrist's commentary revolved around the fledgling practice of audioblogging itself: what it was, how it worked, suggestions for its future. Again, blogging in its earliest incarnations functioned as a metacommentary on blogging and the web, and thus it is perhaps unsurprising that the earliest audioblogs were preoccupied with audioblogging itself. The audiobloggers were frequently in dialogue with one another, too: Gilchrist would issue a call for audioblogging standards, Kidd would post a response comment, and Adam Curry would reply to both. The posts were time-shifted—not unlike friends playing voicemail phone tag—but when linked together took on the shape of a call-in talk radio forum.

There were still other audiobloggers who were completely unconnected to the Winer-Curry community. A San Francisco scientist and blogger named Jish Mukerji began audioblogging as early as August 2001.[100] Novelist Chuck Palahniuk was also an early audioblogger, beginning around July 2003. The subject of a prolific fan-run "official" website (called *The Cult*, chuckpalahniuk.net), the well-known *Fight Club* author maintained a close relationship with his readers. He contributed audioblog posts every few days to communicate with these loyal supporters.[101] Palahniuk's posts were only a minute or two long each and consisted primarily of phone messages recorded while traveling on a book publicity tour for his novel *Diary*. Always beginning cordially with "Hey, this is Chuck," Palahniuk adopted a friendly, chatty tone. He spent most of his time announcing his upcoming public appearances and giving rundowns of previous events, thanking his fans for showing up at the readings, encouraging them to introduce themselves in person or to mail letters. On the tour, he was routinely reading from "Guts," a disturbing horror story about masturbation, and it became something of a running joke in his fan community as to how many audience members passed out at the reading each evening. Palahniuk would consistently update the "body count" of fainters through his audioblogs. "Boy, I am less and less proud of the body count," he announced in a September 14, 2003, post following a New York City reading, "but two more people passed during the reading of the short story, both of them men and the second man it was very upsetting, he blacked out, he shouted unconscious and he vomited. So, in addition to that, it cost 400 dollars to take him . . . no, 500 dollars to take him four blocks to the hospital and his folks had to come from outside the city to come to collect him. So, boy, I'm not really proud of that, but I'm going to start warning people before each event, just how much has happened around this story."[102] In another late September 2003 audioblog post, Palahniuk broke from the tour diary format to speak out about his personal life. Up until then a closeted gay man, he feared that an *Entertainment Weekly* reporter was about to out him. Rather than have the press divulge his personal information, he recorded a frenetic rant for his fans in which he lashed out at the reporter and, in the process, revealed he was in a decade-long relationship with another man. At the request of Palahniuk and his publisher, the audioblog post was quickly deleted by *The Cult*'s webmaster, Dennis Widmyer.[103] Nonetheless, the Palahniuk example displays one way in which the direct, do-it-yourself practice of audioblogging forged intimate social relationships between the speaker and his audience, in this case between a famous author and his fans.

Many audioblog posts such as Palahniuk's not only sounded like telephone voicemail messages, they actually *were* voicemail messages. Recording, formatting, and uploading digital audio for the web was still a burdensome, if not entirely infeasible, undertaking for most average internet users in the early 2000s. As a result, a number of phone-to-blog services sprung up that would allow users to call up a phone number and record an audioblog post over the phone. The service would automatically convert the recording into an MP3 that would then be uploaded to the user's blog and made available for on-demand streaming through a "play this audio post" link. The most widely used of these services was Audblog (stylized in all lowercase lettering as audblog), which Palahniuk also used. There were multiple such services, though, including PhoneBlogger, Voice Blogging, and Audioblog.com. BlogTalkRadio, still in business today, used similar phone-to-blog recording technology when it started out in 2006.

The San Francisco–based Audblog launched in February 2003, promoting itself as a community-building, free speech-enhancing participatory media service that was the next evolutionary step in blogging. Podcasting had yet to be named, though Audblog's take on the emerging personal media practice used the same language of time-shifting and mobility: "Audblog enables audio posts to your blog from any phone, anytime, anywhere," the company's inaugural press release announced.[104]

This version of audioblogging resembled telephone technology in multiple ways. Services like Audblog did not provide any editing tools: the one-take, real-time voicemail recording was posted as-is. The length of recordings was also restricted to a maximum of two to five minutes, meaning that at the level of form and structure they were brief, one-way voice messages that really were quite similar to personal voicemail recordings. It was common for bloggers to call in from mobile phones while driving or waiting somewhere for a few minutes—interstitial moments associated more with making a quick phone call than recording a professional radio broadcast. Moreover, these audioblog posts *sounded* like voicemail messages, in that they were short, unscripted, low-quality spoken-word communiqués delivered over the phone. Yet they were still in essence radio because they were intended for public consumption, and they addressed an audience (real or imagined) rather than an individual. And the content of audioblogs consisted almost entirely of *talk*—ordinary, everyday, banal talk, which is also the principal stuff of broadcast talk radio. In sum, audioblogs crafted through services like Audblog represent not only a convergence of radio and the internet but also a convergence of radio, the internet, and the telephone. Infrastructurally, the internet and telephone networks are closely integrated, and thus it should perhaps come as little surprise that the new media of audioblogging/podcasting relied directly on the old media of the telephone early on.

The popular blog publishing service Blogger promised "push-button publishing" for text-based blogging, and Audblog and its competitors were attempting to replicate that strategy for audioblogging. Initially, Audblog was a subscription service: users would pay $3 per month for a dozen posts, and additional blocks of twelve posts could be purchased for $3. However, it quickly teamed with Blogger to offer a co-branded service, audioBLOG-GER, which was eventually made free to Blogger users—the cost being subsidized by Blogger's new corporate parent, Google.[105] Blogger even claimed that audioblogging was "simpler than publishing a text post."[106] Now widely available as a free extension through the world's most utilized blogging platform (more than a million users in early 2003), audioblogging nonetheless never developed beyond novelty status.[107] It was experimented with by a few high-profile web personalities like Palahniuk and the actor Wil Wheaton, along with various Silicon Valley tech entrepreneurs.[108] But these short audio *posts* never took off the way more formal episodic podcast series eventually did.

Nevertheless, audioblogging was a significant transition point between blogging and podcasting in other ways. Audblog was the brainchild of Noah Glass, a young software developer who, in 2002, happened to live next door to Evan Williams. The then-thirty-year-old Williams was an internet entrepreneur who cocreated Blogger, which he sold to Google in February 2003 for undisclosed millions. Williams briefly stayed on at Google, including brokering the audioBLOGGER licensing deal, but by the fall of 2004 he was looking to branch out and start another company. Between spring 2003 and fall 2004, audioblogging had given way to podcasting, which had begun to diffuse rapidly—in a little over a year the number of podcast series jumped from a handful to as many as a couple thousand, and dozens more were emerging each day.[109] In December 2004, Williams partnered with Glass on a new company, Odeo, which aimed to create a "one-source solution for finding, subscribing to, and publishing audio content."[110] Right away, they attracted more than $5 million in investment capital.[111] The existing Audblog/audioBLOGGER service was Odeo's main product, although they surrounded the phone-to-blog software with a set of audio recording and editing tools (Odeo Studio software) plus a podcast directory, customizable playlist capabilities, and other applications that enabled users to easily find and manage online audio content.[112] The problem is that many free or inexpensive hardware and software options were quickly arriving to aid podcast producers, while Apple's inclusion of podcatching tools in the June 2005 release of iTunes 4.9 became the go-to platform for podcast listeners. The market for an end-to-end podcast solution evaporated, and the barely year-old Odeo was forced to "pivot" and reinvent itself. Switching from a simple blogging-like tool for audio messaging to a simple blogging-like tool for text messaging, Odeo became Twitter.[113]

The link between Odeo as a podcasting company and Twitter as a microblogging platform is typically seen as trivial or inconsequential, if it is acknowledged at all. The roots of Twitter are described as being in SMS (Short Message Service) text messaging. As the story goes, when the company was forced to reboot, Odeo employees split up for brainstorming sessions, during which computer programmer Jack Dorsey proposed an old idea of his based around the notion of status, which allowed individuals to use SMS to update a small group of people about what they were doing—basically group texting but web-based.[114] Twitter (originally twttr) was born and the rest is history. Except that this idea of broadcasting one's status in a short personalized message to a close group of friends and followers is essentially what Audblog/audioBLOGGER was already doing, only with audio and spread across individual websites rather than together on a single shared site. Moreover, even before Dorsey's "stat.us" idea was prototyped, the Odeo team was working on a "mobile listening post" function that would turn audioBLOGGER into "a kind of voicemail dispatch service where people could both post and listen to ongoing conversations," according to one of the company's engineers, Dom Sagolla.[115] There were competing prototypes being developed in late 2005 and early 2006: there were actually three versions that shared status via voice and only the one that shared status via text.[116] The text version eventually won out, and the alpha version of Twitter was launched in March 2006.

In the ensuing decade, Twitter became emblematic of the social media age and the blurring of lines between producers and audiences. All

Twitter users are also active contributors to the online networked media culture—continuously liking, posting, and linking to or replying to other users' content. Ordinary individuals are engaged in what Alice Marwick has called *lifestreaming*, or "the ongoing sharing of personal information to a networked audience, the creation of a digital portrait of one's actions and thoughts."[117] Marwick and other new-media scholars like Jodi Dean are ultimately critical of these participatory media practices, pointing to loss of privacy, commercial exploitation, and a cultural preoccupation with status and publicity that they argue undermines progressive politics.[118] Nevertheless, despite these legitimate concerns, many theorists still see decentralized media technologies like Twitter as "free public spaces" that are a boon for democracy. They enable ordinary individuals to represent themselves and their interests, as well as engage directly with other citizens and communities, all done relatively independent of government, the mass media, or other gatekeepers.[119] Twitter's construction of a continuous, never-ending flow of informal and mostly banal content closely resembles talk radio, producing a very similar sociability.

Conclusion

The forerunner of today's podcasting—audioblogging—presented a potentially very different way forward for internet radio. Audblog's tagline was "Think It, Speak It." Media theorists often talk metaphorically about how text-based social media like Twitter can "give voice" to ordinary individuals, in particular marginalized or underrepresented groups.[120] Yet audioblogging sought to literally give voice to everyone, people sharing their thoughts with their own speech—sound bringing with it a wider range of expression and affective dimensions than text alone can capture (accent, tone, inflection, etc.). Twitter came very close to creating a world in which everyone communicated through, say, twenty-eight-second audible tweets instead of 280-character written ones.[121] Indeed, audioblogging pointed to a unique short-form style of talk radio, one that was entirely raw aesthetically and content-wise centered on personal experience and self-expression. This merged public and private speech.

Again, the close relationship in audioblogging between the radio and the telephone is instructive. Radio speech *sounds* mundane and casual, mimicking private conversation, but it is in actuality intended for a public audience and (usually) heavily premeditated.[122] Telephone speech, in comparison, is typically conceived of as private, informal, and domestic—it is actually spontaneous and intimate, whereas radio speech only presents that illusion. Audioblogging narrowed that gap considerably, creating a highly reflexive hybrid mode of public-private speech. Moreover, although audioblogging circa 2002–5 was time-shifted, it shared with broadcast radio a desire for immediacy and newness: like diary writing, posts were short and to the point, focused on what speakers were doing in the present moment (or what they had just recently done or were just about to do in the near future). And posts were updated frequently, many audiobloggers leaving them every day or multiple times per day. This conversationality, frequency, and focus on content concerning the living reality of everyday existence evoked radio's liveness and sociability. Furthermore, its existence in a continuously connected social media environment generated a sharedness that transcended the fact that audioblog posts were time-shifted: listeners' comments and subsequent audioblog response posts gave the original post an extended life and a heightened sense of immediacy and presence.

The innovation of audioblogging, then, was not so much the technology (though elements of audioblogging/podcasting technology certainly were crucial) as it was a new form of mediated speech that was a convergence of talk radio, blog writing, and telephone conversation. Audioblogging brought together broadcast talk, everyday telephone speech, and diary-like blog writing into a hybrid speech form that was familiar but also new—more personal, mundane, and self-reflexive than most institutional talk radio speech. This highly reflexive style of talk was carried over into the practice of podcasting as it coalesced circa 2005, and it has stuck with it through to today. This affective audio aesthetic and subjective conversational style is integral to the perceived "intimacy" of modern podcasting.

A large portion of the many thousands of podcasts that have come and gone over the past decade have been nonprofessional talk radio-style productions created by individuals as a means for self-expression, social/political activism, or simply fun. Stylistically, most of these nonprofessional podcasts would fit the "amateurish" tag: unplanned, unedited, and ultimately appealing only to a micro-niche of likeminded individuals.[123] Nevertheless, even the most unpolished podcast series is more conventional than an audioblog post: podcast episodes tend to be less spontaneous and less focused on immediacy, and the series format carries with it higher expectations. More to the point, audioblogging came closest to envisioning a true "people's radio"—a world in which everyone was a radio broadcaster.[124] Despite the proliferation of podcasts in recent years, most people still remain radio listeners rather than radio producers.

In the fifteen years since podcasting's first wave ceased, the practice of podcasting has largely developed in ways that are complimentary to, rather than challenging of, traditional radio broadcasting. This is not to suggest that there are no podcasts today giving voice to the voiceless or addressing oppositional politics marginalized within the mainstream media. Nevertheless, broadcast radio and podcasting share more similarities than differences, especially when it comes to their content and form. If anything, podcasting has revived talk radio. Talk and news radio formats are among the most popular in podcasting today. For instance, look at the recent explosion of digest-style daily news podcasts like the *New York Times' The Daily* and lengthy political talk podcasts like *Pod Save America*. Yet audioblogging today is not only a forgotten, alternative history of podcasting—it is also an unfulfilled promise, a road not traveled.

CHAPTER FOUR

On the Line and Online

Talk Radio Meets the Internet

Google the term "podcasting" and a majority of the top search results will offer some version of "start your own podcast." Populist notions of democratization and self-empowerment remain intricately bound up in a discourse of user-generated content and DIY (do-it-yourself) media practice. Often, the idea is that to find a platform for one's voice individuals must create and control their own "message"—by, say, starting a podcast—rather than work through existing platforms in the mediasphere. But perhaps not everyone wants, or can easily facilitate, his or her own podcast. Or perhaps traditional mass media like talk radio can work equally well, if not better, as a platform for community engagement and social change.

Up to now, *Sound Streams* has primarily examined newer forms of radio that emerged *on* the internet in the 1990s and 2000s, such as streaming audio and audioblogging/podcasting. Remediation, however, is a two-way process: "new media" adapt older media, while "old media" adapt newer media in return.[1] That is, radio-internet convergence involves much more than just the "old" medium of broadcast radio being integrated into the "new" medium of the internet. To be sure, as Jay David Bolter and Richard Grusin's theory of remediation informs us, newer media forms must always define themselves in relationship to earlier media.[2] For podcasting, that meant adopting criteria from broadcast radio, as well as from the telephone and diary writing/blogging. However, radio-internet convergence has also caused older media like terrestrial talk radio to remediate and become a renewed version of itself in light of modern media forms—in this case podcasting, but even more consequentially, social media, online chat, and collaborative software tools for crowdsourcing and knowledge sharing.

In other words, the internet has done much more than facilitate alternatives to terrestrial radio. It has also drastically changed the way traditional radio producers and institutions make their programming for broadcast radio. This chapter explores the ways in which talk and news radio producers harnessed the internet, especially social networking tools and services, to make their programming more open and inclusive of the audience during the past twenty years. This emerging environment of *sharedness* enhanced broadcast radio's already very strong capacity for liveness, allowing even more participants to feel a sense of "being there" in the broadcast moment (even after the initial live transmission passed). This is especially true of the already participatory media format of the call-in (or phone-in) radio program.

Talk and news radio now operate in a digital environment of *convergent journalism.*[3] Sometimes alternately referred to as multiplatform, cross-platform, or integrated journalism, the basic premise is that stories are told through more than one medium and more than one outlet or platform. For instance, a radio producer might write and record an audio version of a story for broadcast and then create a text version of the story for the web—possibly with photos, hyperlinks, and even video added—all in the same day. As makers of a "multimedia" production, contemporary radio producers must be skilled in various digital media disciplines. In addition to audio reporting and production skills, the radio producer in this example would also need at least a basic level of expertise in print journalism, multimedia design and web programming, and photo/video editing. The forms and logics of these different media are converging in the labor of these cultural workers.

In the case of traditional call-in talk radio, the on-air talk style has not changed markedly in the past twenty-odd years. On a program like *On Point* or *The Tech Guy with Leo Laporte*, the calls still adhere to a specific topical agenda, rather than meander as they might in an ordinary phone conversation. Furthermore, the participants fall into given speaker identities ("host," "guest," "caller") who mostly talk in turns.[4] However, the very form of *interactive broadcast talk* has become the norm across networked digital media. From newspaper comments sections to Rotten Tomatoes movie ratings to Twitter backchannels during live television broadcasts, the expectation in the Web 2.0 era is that all media be interactive and provide channels for instantaneous feedback from the audience. This conversationality and responsiveness are fundamental traits of talk radio. In the modern social media era, audiences expect to always have the ability to "talk back" to the media they are participating in.

Not surprisingly, *talkback* is another name for call-in radio.[5] The basic formula of direct audience participation in radio broadcasting dates to the 1930s and series like *Vox Pop* that sought to capture "the voice of the people."[6] Since the mid-1980s, the popularity of "shock jocks" like Howard Stern, Don Imus, and Rush Limbaugh has revitalized a populist approach to talk radio.[7] Celebratory accounts of user-generated content, cocreation, and the active audience overstate the novelty of user production in today's digital media environment: audiences have been contributing firsthand to mass media for as long as it has existed.[8] I do not intend to suggest that talk radio alone is the progenitor of today's multifarious forms of user-generated content. Nevertheless, the "phatic and metalingual" function of call-in radio—meaning that producers acknowledge the audience as "present" and capable of contributing to the program—has become everywhere common in contemporary media.[9] This situation makes call-in talk radio a particularly suitable candidate for cross-platform success on the internet.

This chapter further explores talk and news radio, building on chapter 3's examination of podcasting and the ways in which the internet opened up radio production to ordinary people. Here, though, attention turns to how radio-internet convergence made broadcast talk radio more accessible and inclusive of the public in the 2000s and 2010s. Through a case study of the New York City weekday public radio program *The Brian Lehrer Show*, produced by NPR member station WNYC, I exhibit how call-in (or phone-in) broadcast radio was itself already an interactive format and, in the hands of an ambitious production team, it was combined with online social networking tools to create an even more robust and effective public forum. Contemplating notions of crowdsourcing and citizen journalism, I point to how the internet opened up new channels and platforms for the audience to speak, and for public service-oriented talk radio producers like Lehrer to listen to them and elevate their voices.

WNYC and US Public Radio's Adoption of the Internet

Describing itself as "America's most listened to public radio station," WNYC consists of a pair of NPR member stations in New York City (820 AM and 93.9 FM) that provide a mix of news, talk, music, and entertainment programming.[10] Beginning with its AM channel in 1924, WNYC is one of the oldest broadcast radio stations in the United States. Renowned during the 1930s–1960s, especially, for its educational as well as cultural and public affairs programming, WNYC was municipally owned by the City of New York up until 1997, at which point a long and contentious public struggle resulted in the stations being sold to an independent, non-profit foundation.[11] Today, the network's 1.7 million weekly radio listeners make it the most popular talk radio outlet in New York City and also give it the largest public radio audience in the United States.[12] In addition to producing the original local weekday programs *The Brian Lehrer Show* (previously named *On the Line*) and *Midday on WNYC* (the successor to *The Leonard Lopate Show / New York & Company*), WNYC is among the largest producers of nationally syndicated public radio programming, creating the weekday news series *The Takeaway* (distributed by Public Radio International) as well as the weekly shows *On the Media, Radiolab, Studio 360, Snap Judgment,* and *Freakonomics Radio.* In recent years, WNYC has invested so heavily in original, talk-centric "podcast first" productions—*Death, Sex + Money, Here's the Thing with Alec Baldwin, Note to Self, Sooo Many White Guys, Nancy, Caught, The New Yorker Radio Hour, The Stakes,* and *Only Human* among them—that it has also set up a dedicated podcast division, WNYC Studios.[13] Despite experiencing its share of the usual budgetary woes faced by public media institutions in the neoliberal era, WNYC's location in the media capital of New York City and its relatively considerable financial resources have meant that it has been an innovation leader in terms of bridging traditional broadcast radio and the internet, not just among the US public radio sector but within the terrestrial radio industry writ large.

NPR and many of its seven hundred-plus local affiliates (or "member stations") retroactively like to boast that they were pioneers in the use of internet radio.[14] No doubt, compared to commercial radio stations and networks—many of which were slow to adapt to the internet or even actively resisted it—NPR in the 1990s and early 2000s was relatively proactive about venturing onto the web. Select NPR national programs had makeshift web pages in place as early as 1993, and NPR launched its first network website in 1995.[15] Yet most of NPR's earliest online ventures were merely content partnership deals in which other start-ups, such as Progressive Networks or Audio Highway, were the ones developing the technology and driving the creative vision. (College and other smaller local nonprofit stations tended to be more aggressive in their approach to the internet than did the larger and more institutionally conservative NPR affiliates, as examined in chapter 1.) Even at the turn of the century, most member stations operated websites that consisted only of programming schedules.[16] As Richard Dean, the founder of NPR.org and the first employee of the network's New Media department, describes, "To say that NPR resisted the move into the online world would be a bit of an understatement."[17]

WNYC's initial foray onto the internet was in January 1998 when the network launched its first website (www.wnyc.org). It was hardly the first terrestrial radio station online, though its website was two years in the making and quickly emerged as one of the more ambitious radio station domains. Initially, the site featured program schedules, news updates, music playlists, membership information, and some audience-centric novelties like a quiz page. There was an "archive," although it did not contain any audio, only transcripts of select interviews or news stories (usually one or two per day pulled from programs like *Morning Edition*). There was also occasional web-exclusive content like written theater and food reviews.[18] Within months, WNYC.org linked to RealAudio streams of NPR newscasts and then, between May and August 1999, it began offering on-demand streams and downloads of a handful of its original programs, including *On the Media*, *On the Line*, and *New York & Company*, plus local news reports and other audio "selections."[19] Due to storage and bandwidth costs, only the ten most recent episodes of a program were made available for streaming. In other words, these were not permanent archives—online listening, even if it was not live, still emphasized recency, the "archive" primarily being a means to quickly catch up with yesterday's or this morning's broadcast that you might have missed.

24-7 livestreaming of WNYC's AM and FM channels was implemented in January 2000.[20] Nevertheless, beyond the simulcast streams and select downloads, at this stage the majority of the web content barely connected to the stories and conversations heard on the radio. The WNYC program guide announcing the website described the material as "elements that animate the studios and sounds of live radio."[21] These news updates, program schedules, donor links, and the like were effectively just marketing materials, the World Wide Web equivalent of a promotional brochure.

In fact, in the late 1990s and early 2000s, most radio station websites were the products of press and public relations departments (or "audience services" in public media jargon). Editorial personnel and on-air talent were seldom, if ever, involved with the website. Even within a major radio organization, it was common for a website to emerge as essentially the pet project of an entrepreneurial lower-level employee. This was the case at NPR, where Richard Dean, then an associate producer on *Week-*

end *All Things Considered* and former employee in NPR's Audience Services department, started the first *Weekend ATC* webpage in 1993. He did so in relative secret, only "with the quiet approval of my show's producer."[22] He saw it as a way to communicate information to the audience, such as recipe requests, that previously had to be exchanged through the inefficient one-to-one medium of paper mail. At WNYC, the first website was developed by a publicist, Virginia Prescott, and an intern, Hunter College High School student Luke Stein (the child of board member Howard Stein), "in their spare time."[23] Indeed, from its launch up until 2001, wnyc.org was overseen by Prescott, the station's first "new media manager," whose other job responsibilities included being the editor of *The WNYC Program Guide*—a bimonthly (later quarterly) promotional paper publication mailed to donors. In 2000, the website and an email listserv replaced the printed program guide entirely.[24] Well into the decade of the 2000s, the vast majority of radio station websites functioned primarily as radio tuners (i.e., digital receivers for the simulcast of the live broadcast signal) and as resources for general program and operational information.

Most of the "new" features on radio station websites of the 1990s merely duplicated functions that previously took place by mail or by phone. For instance, listeners could use email and message boards to voice approval or disapproval about a previously aired segment, correct an announcer's grammar and pronunciation, suggest a future topic or guest, and make a donation or buy a product. These were predominantly one-way exchanges, in that they only allowed audience members to reach out to stations like WNYC and "leave us messages."[25] Sending an email did not guarantee a response and rarely resulted in a substantive public dialogue. It is possible that a listener comment might prompt an on-air correction or encourage producers to tackle a new topic some days or weeks later, the same way a written letter might do. Regardless, across the terrestrial radio landscape, the audience's actions online infrequently fed back into the content heard on the radio. At best, chat rooms fostered a small community of aggressively opinionated listeners who conversed regularly with one another, forming a backchannel that reacted to the topics and discussions heard on air. Yet few radio hosts, producers, or guests directly participated in this secondary conversation before the mid-2000s. Discursively, the radio and the internet remained rather separate media.

In the late 1990s, WNYC nevertheless began to experiment with "interactive" web features that would eventually become more pronounced in the mid-2000s as social networking tools and practices spread throughout the culture. These initiatives, such as message boards, made audience activity on the website more directly connected to the radio broadcast. Seeking to establish "a radio cyber-dialogue," the network created a number of chat rooms on wnyc.org between 1998 and 2000.[26] Included were Lopate's Place (a discussion board for the Leonard Lopate–hosted *New York & Company*), Resonating Chamber (a forum for music discussion), and Public Conversation (a series of live web chats with public figures such as Frank Serpico and Ira Glass—a precursor to the Reddit AMA).

One especially notable exception to the previously stated rule was the *On the Line* discussion board, The Soapbox. It was wnyc.org's first truly interactive internet feature and also a harbinger of the social media world to come.[27] Launching in October 1998, The Soapbox actively sought listener's opinions on topics *before* the daily broadcast, not merely accumulating reactions afterward or entertaining tangential debates. This way, *On the Line* and its host, Brian Lehrer, could drum up enthusiasm for the broadcast as well as gauge public opinion on the issues. And they could use that feedback to shape the forthcoming on-air discussion. Unlike most public radio producers in the 1990s—whom NPR's Richard Dean described as "decidedly analog," still using typewriters and refusing to have email addresses—Lehrer himself actively participated on The Soapbox board.[28] In fact, Lehrer had been an early adopter of the internet, taking part in The WELL's online forums in the 1980s and creating a bulletin board for his radio show all the way back in the early 1990s.[29]

The Brian Lehrer Show and Call-In Radio 2.0

"It's your neighborhood, your city, your country, your world, and now your website" was *On the Line*'s motto in 1998.[30] The series was—and remains to this day—WNYC's public affairs call-in show on its AM channel (now also simulcast on its FM channel), featuring "real talk" about local and national politics, the economy, crime and the law, education and housing policy, the city's social and cultural life, and other issues of everyday concern to New Yorkers. Hosted by Lehrer since its inception in 1989, the program (which changed its name to *The Brian Lehrer Show* in 2002) airs weekdays from 10:00 a.m. until noon with Lehrer moderating wide-ranging discussions between politicians, experts and others "newsmakers," and listeners. (Lehrer was also the second host of the nationally syndicated media criticism show *On the Media*, from 1997 to 2001, before it relaunched with current cohosts Bob Garfield and Brooke Gladstone.)[31] A radio DJ and reporter since his college days in the 1970s at the State University of New York at Albany, the Queens-born Lehrer has been called everything from a "radio mensch" (he is Jewish) to "the Walter Cronkite of New York City broadcasting" and "the Voice of New York."[32] When bestowing Lehrer and his eponymous program with a George Foster Peabody Award in 2007, the judges proclaimed that he "reunit[es] the estranged terms 'civil' and 'discourse' five mornings a week like no other show on the air" and that his show "is a wide open yet shrewdly managed forum in which every sort of political, social and cultural issue is considered and where New Yorkers, in all their diversity, can get to know each other."[33] As he has proven time and again over a career of more than thirty years, Lehrer strongly believes in the power of talk radio "to build community rather than divide" and "to bring Americans together, to get Americans to listen to one another, rather than just get us all riled up into our little echo chambers, which is frankly the easy way out."[34]

At first listen, the structure of Lehrer's talk show has changed seemingly little over its three-decade long run, as it has always remained organized around the no-frills formula of levelheaded one-on-one guest interviews interspersed with plentiful questions and comments from listeners. It is a quintessential live news-talk radio show in its focus on the day's current events—there is little preproduced material included in the broadcast and the production team frequently has to "turn on a dime" in order to attend to breaking news events.[35] (Lehrer's deep knowledge of New York City paired with his adeptness at managing a live broadcast under pressure has meant that he has become WNYC's go-to anchor during nearly all the city's major crises including 9/11 and Hurricane Sandy.) Ultimately, the meat and bones of *The Brian Lehrer Show* boils down to "people sharing their different experiences and points of view."[36]

This public forum-style broadcasting format has dwindled since the 1980s and the emergence of pundit-driven talk radio. In the contemporary media landscape (and throughout the entire history of Western civilization), speaking is heavily privileged over listening.[37] Lehrer himself has observed that conservative talk radio host Rush Limbaugh often repeats a joke in which he brags, "This show is not what you think, it's about what *I* think." To which Lehrer retorts, "For me [*The Brian Lehrer Show*] is *not* about what I think. It's about what I think and about what everybody thinks."[38] He adds, "I try [and make] the show an act of openness, outreach to people who agree or disagree with me."[39] More to the point, Lehrer's program is about *listening*: inviting in a diversity of opinions and treating all his callers with respect, hearing each and every one of them out even when they occasionally get agitated, all in an attempt to "humanize each other" and "try to turn the debate inside out and find out what we have in common as Americans."[40] Lehrer talks regularly of his desire to listen to ordinary people and their stories of everyday experience: "The most satisfying on-air moment keeps happening," he states, explaining, "It's when we get a new caller, with a story about life in New York that we haven't heard before."[41] This type of *listening out*, as Kate Lacey calls it, is a form of "radical openness" that Habermasian public sphere scholars regard as fundamental to democratic political debate and deliberation.[42]

Upon first listen the twenty-first-century, new-media-era version of *The Brian Lehrer Show* closely resembles the analog twentieth-century version of the show. The live broadcast remains paramount, and each episode is still dominated by phone-in or face-to-face interviews with guests and listener phone calls.[43] Yet, upon closer inspection, the internet has impacted the show considerably—most of all by opening up new channels and platforms for the audience to speak, and for Lehrer and his team to listen to them. Indeed, Lehrer himself astutely recognizes the intrinsic similarities between the "old media" of broadcast talk radio (particularly local call-in radio) and the "new media" of social networking, in terms of both mission and basic practice. "As a call-in show, the program has always had interactivity and community-building as part of its core," he observes, further explaining, "Streaming and on-demand web listening have drastically expanded our audience base, and the explosion of ways to connect with people beyond the telephone—from Facebook and Twitter to Instagram, Reddit and Vine—allows us to generate greater engagement, hear more perspectives, and tell more stories."[44] Rather than resist internet media and technology, as many veteran radio broadcasters have done, Lehrer sees them as opportunities to further his and WNYC's broader goal "to provide information and a politically diverse public forum."[45]

These public service objectives of providing information, fostering civil discourse, and building community were embedded in *On the Line*'s The Soapbox message board from its beginning back in 1998. Lehrer and his producers posted and commented on the board regularly. It was never a case of a publicist down the hall generating the content. Most mornings they would announce in advance the main topics of discussion for that day's broadcast, starting a topic thread for listeners to comment on. For instance, "Should Jack Kevorkian be convicted of first degree murder?" or "What do you think about the Clinton investigation?" In addition, the producers linked to news stories and other background reading materials, many of which were written by the guests who would be appearing on the radio show. Thus, listeners could give reasoned feedback on the message board that might be incorporated into the forthcoming on-air conversation. Or they could prepare for making a phone call during the live broadcast. Or they could simply take in the broader context of the discussion happening on air as a more informed listener. (Today, the show's producers routinely post "What we're reading" links to Facebook and Twitter, as a means of opening a window into their story development process and simultaneously gauging audience interest in the topics.)

While it may seem fairly trivial nowadays, this convergence of radio and print media via the internet was not an inconsequential shift. First, it offered the equivalent of a works cited section for radio (or to use a sound media metaphor: album liner notes). Second, the boundaries of the live broadcast were shifting: the daily radio show was still the primary text, but the conversation now started online before the broadcast and continued online afterward. Third, relationships could develop between listeners, as they interacted regularly with each other via the online message board. Producers allowed participants to "post a message on any topic," and it was common for listeners to initiate unsolicited threads as well as carry out conversations with fellow audience members completely independent of the producers. The *Lehrer* producers say they aim to generate "engagement" online—a commonly uttered buzzword in web marketing speak.[46] Yet here they are genuinely dedicated to creating a forum for conversation and then only moderating the conversations that ensue, as opposed to constantly responding to or dictating the exchange, which is the approach of most corporate social media managers. Once again, listening as much or more than speaking. Lehrer has personally praised these networked media practices for bypassing the one-way or two-way conversations of traditional mass media, creating instead "a three-way street: me-to-you, you-to-me, you-to-you." He argues this is ultimately "very good for democracy—much better than any media model we have had before."[47]

The Soapbox was eventually phased out as Facebook, Twitter, Instagram, and other social networking sites replaced chat rooms in the mid-2000s. The *Lehrer* producers experimented with each new service, though, in an attempt to engage listeners wherever they might be online. If an online tool could help facilitate this "three-way" exchange of views, they tried it out.

Listen to any *Brian Lehrer Show* broadcast from the past decade and you will hear a mix of listener responses coming in via phone calls, over Twitter and Facebook, through comments on the wnyc.org show page. While researching *Sound Streams*, I personally had the opportunity to observe a *Lehrer* live broadcast from inside the studio control room.[48] The production team for the show included an executive producer, four assistant/associate producers, and two interns. During the live broadcast, one assistant producer remained upstairs actively overseeing social media (mainly Twitter), while the others were in the control room answering phones, logging callers into PhoneBOX call-routing software, and monitoring social media (but not actively participating). The executive producer, Megan Ryan, was positioned in the middle of the control room with a direct view to Lehrer in the booth and the ability to feed information to his earpiece. In the booth, Lehrer had a desktop computer in front of him with multiple windows open, including the call log, plus a separate tablet computer and printed notes. He could also view the multiple television monitors in the studio tuned to CNN, MSNBC, and the local New York City 24-hour news channel NY1. This enabled Lehrer and his producers to track any breaking news. For instance, on the day I observed in January 2015, suspects in the *Charlie Hebdo* terrorist attack in Paris, which had occurred two days prior, were killed during the broadcast—news that Lehrer was quick to report to his listeners. Even though the show eventually winds up archived online and podcast, this liveness remains paramount in daily talk-news programs like *Lehrer*.

The real-time "continued connectedness" of the internet influences radio broadcasting by enabling producers to always be attuned to the world outside the bubble of the studio.[49] Lehrer himself constantly monitored social media on both his studio desktop and tablet computers. For the most part, Lehrer personally chose all the callers and social media comments to include on air, selecting them on the fly rather than having them fed to him by the producers. His staff only entered the caller's name, location, and a brief description into the call log, and Lehrer used his expertise to pick the callers or online commenters that he judged would most diversify the conversation. (The caller's location is important because, as Ryan explained, Lehrer likes to include a mixture of neighborhoods in most discussions—and listeners enjoy hearing their neighborhoods called out on air.)

Not all listeners are willing or able to call in, either because they have anxiety about speaking on air or they simply are not in a position to wait on hold and talk. Social media and the like can therefore bring in a greater number of voices and, ideally, a greater variety of perspectives. Still, a program like *Lehrer* continued to attract a bounty of traditional phone calls: on average, I witnessed about ten callers parked on hold at any given time, and invariably most of them did not get on air. When a segment ended, the producers urged the remaining callers to go to the website and comment. On air Lehrer also encouraged listeners to take to social media to continue the conversation. This was not mere lip service: those comments could potentially become fodder for future broadcasts. If a vibrant discussion continues online, that is an indicator that the topic resonates with listeners and potentially warrants further attention in the program.

The People Formerly Known as the Audience: Citizen Journalism Radio

Furthermore, the networked culture of the internet enables broadcast journalists to crowdsource reporting. Lehrer has been a vocal proponent of *citizen journalism*, a broad term referring to ordinary citizens without any professional journalism training who use networked digital media tools (blogs, web video, Twitter, etc.) to expand or fact-check established media reporting.[50] The traditional format of call-in talk radio already shares much in common with this and other forms of participatory culture that are characteristic of Web 2.0. Call-in shows are all about tapping into the *collective intelligence* of the listening public, whether it is to reinforce a narrow ideological agenda (as is the case with the siloing of much contemporary political talk radio) or to open up public debate about governance and social justice (as is the case with more progressive-minded public affairs talk radio like *Lehrer*). This practice of soliciting contributions from the audience in order to obtain ideas and create content is specifically known as *crowdsourcing*, and it is a fundamental principle of modern social media.[51] When talk radio hosts put a call out seeking perspectives on a particular topic or hold an open-lines segment inviting listeners to discuss any topic of their choosing, it is a form of interactive broadcasting that features all the hallmarks of crowdsourcing in online communities. Thus, it was a natural fit in the late 1990s and early 2000s for the *Lehrer* program to take up online crowdsourcing initiatives that could be integrated into its on-air broadcasts.

One of *Lehrer*'s first major online crowdsourcing projects came in 2007 when the show invited its listeners to report on a series of rather mundane items and events taking place in their own neighborhoods. As oil prices were rising to record high prices in the summer of 2007, Lehrer, assisted by *Wired* magazine writer and "crowdsourcing" evangelist Jeff Howe, asked his audience to determine "How many SUVs are there in the New York City area?" To do so, they tasked listeners to go outside their homes and count how many of the gas-guzzling vehicles were on their block, and then post their findings to the show's comments webpage.[52] After a week, the producers analyzed the nearly 450 listener reports, sharing the results in a thirty-five-minute segment as well as through an interactive data map on wnyc.org.[53] (They determined that approximately one-third of all cars in the New York City region were SUVs.) Though the SUV conclusions were instructive, their next crowdsourcing, or "group journalism," project resulted in findings with considerably deeper social significance. In September 2007, Lehrer and producer Jim Colgan instructed listeners to gather grocery prices (lettuce, milk, beer) from their neighborhood bodega or supermarket.[54] The more than 350 audiences submissions culminated in a thirty-minute radio segment with *New York Times* writer David Leonhardt and NYU journalism professor Jay Rosen, revealing considerable price disparities across the city.[55] On their own, one or a few sets of these prices revealed little more than an anecdotal talking point, but when compiled with hundreds of samples, they highlighted patterns of price gouging that correlated with socioeconomic inequality.

These were investigative reports that WNYC would not have had the resources to carry out on its own (although journalists did verify some results). Moreover, the interactivity fueled listener engagement and also garnered considerable public attention.[56] In the case of the grocery pricing project, in addition to data maps of the results—which listeners could look up online to determine whether they were unsuspecting victims of price gouging—the producers also posted the phone number for the New York State Department of Agriculture and Markets, to whom listeners could report the illegal overcharging. As such, Lehrer's crowdsourcing initiatives did more than just create audience enthusiasm for marketing purposes, they also undeniably served the public good. Although citizen journalism in its purest form entails activists operating entirely independent of established media institutions, numerous media scholars have pointed out that in practice citizen journalism often necessitates that ordinary citizens collaborate with professional journalists and institutions in order to make their reporting both feasible and accessible.[57] Large-scale cooperative efforts like *Lehrer*'s grocery-pricing survey—especially initiatives that involve big data—require considerable editorial vision and expertise to both create and translate, in addition to substantial financial and technological resources to execute and get in front of audiences. Therefore, intermediaries like Lehrer and WNYC who have the cultural and economic capital are often needed for citizen journalism to be successful and impactful.

The Brian Lehrer Show has continued to issue similar crowdsourcing assignments to its audience over the past decade, in addition to undertaking a number of large-scale political projects. For instance, in 2008 the show invited its listeners to pore through an eleven-thousand-page data dump of then–New York senator and presidential candidate Hillary Clinton's schedule from her years as First Lady, with the goal of the audience helping identify items of interest for future on-air discussion and analysis.[58] In 2009, producers set up a wiki for listeners to help structure the conversation regarding urban development issues in New York and New Jersey during the show's recurring "30 Issues in 30 Days" election cycle series. In effect, the producers were crowdsourcing their research process—the goal of the wiki being "to shape a segment that addresses the questions and concerns of all interested parties and the public at large."[59] Between 2010 and 2012, the Lehrer team also spearheaded a political web series for WNYC, *It's a Free Country*, that combined crowdsourcing investigations and social networking with original print-based online reporting. These actions fed into broadcast radio segments and a stand-alone podcast.[60] "The point is for listeners to collaborate in producing the series by adding topics, guests, questions, sound bites and research material—even intro copy for me to read," explained Lehrer about these online endeavors. "We're a Web 2.0 radio show that does a lot of multimedia and interactivity," he further clarified.[61] In terms of convergence, these various projects utilized the internet to gather data for developing original radio reporting while also then extending that reporting beyond the life of the live broadcast—through the maps, hotline information, discussion forums, and so on at wnyc.org. The archived audio could also continue to be a resource for listeners days, months, or even years later.

Separate from podcasts and audio archives, the internet extends a radio show like *Lehrer* far beyond its daily two-hour time slot. Today, Lehrer and his producers often source story ideas from Twitter, checking the trending topics and generally using the social networking platform as a sort of barometer for what topics are resonating within New York City and their community of listeners.[62] They will also occasionally publish print articles to the web first, using these work-in-progress story ideas to gauge audience interest. Deploying web analytics software like Google Analytics and Chartbeat, they will track web activity and then turn the most popular posts into segments on air.[63] (Producers will also use this analytics tracking approach to determine which segments should be turned into "an A+ post" after the broadcast, meaning the ones that will receive extra attention in terms of adding visuals, text, and other contextual elements.)[64]

Nowadays the most important online work, the *Lehrer* producers explained to me, actually happens in advance of a broadcast.[65] Especially when they are looking to tackle a subject incorporating nonregular public radio listeners, such as life in New York's ethnic immigrant communities, being active online ahead of the broadcast can help to draw in fresh voices. Lehrer regularly tackles social justice issues and topics that address marginalized or underrepresented populations, yet the NPR public radio audience remains relatively stratified (predominantly white, upper-middle class, educated, and older—although WNYC's audience is more culturally diverse than most public radio audiences).[66] Reaching out to other communities in advance through social media can bring new voices and perspectives to the mediated conversation.

Additional strategies for using the internet to strengthen the radio program include creating a "segment buildout" a few hours beforehand, basing the preliminary digital content around the producers' notes and other prep work.[67] This gives the online audience an early preview of the show and establishes a space for listeners to leave comments once the live broadcast begins. The post can then be quickly updated afterward to reflect what actually happened during the broadcast, including adding newsworthy quotes or photos to accompany the archived audio. In addition, if a producer live-tweeted a segment, those tweets can be embedded into the website to create a "fake transcript." These supplemental materials tend to be more shareable than audio and thus can attract attention online beyond the show's listener base. In any event, this environment of sharedness adds new layers of audience participation and extends the lifespan of a daily talk show like *Lehrer*, while maintaining a sense of liveness—immediacy and presentness—to its output.

Building Engaged Radio Audiences Online

The Brian Lehrer Show was an early adopter, though by the late 2000s, nearly all radio stations and individual programs had established a presence online. It may have taken more than a decade for usage to become widespread, but today it would be hard to find a station or show without a fairly robust website, a Facebook page, Twitter and Instagram accounts, and so on.[68] Nevertheless, these sites and platforms are often little more than marketing tools for publicizing schedules and events, promoting on-air personalities, running contests and giveaways, and generating traffic for advertising purposes. Adopting digital media marketing logics, the hope is that this audience "engagement" will secure brand loyalty.[69] (As to my earlier point about the origins of WNYC's website, these web platforms are often managed by public relations staffers or dedicated social media personnel, operating separate from the editorial staff.) The listeners may be active in the sense that they are connecting with the broadcaster via social media—"liking" or sharing posts and occasionally adding bits of commentary—but this engagement is mostly promotionally driven, and the exchanges it generates tend to be generic and superficial. In other words, throughout much of the traditional radio industry, the web is merely being used to measure and monetize the audience. There remains a vast distance—if not a disconnect—between the broadcaster and the listener. Rarely is there any of the consequential debate and community building that Lehrer evoked in his comment about facilitating "a politically diverse public forum." This is especially true of commercial radio; however, even public radio has recognized the internet as a significant revenue stream. Public radio websites and podcasts are exempt from FCC rules governing noncommercial content and therefore can be freely "monetized."[70] Thus, public radio's web presence is open to the same exploitation critique typically reserved for commercial media entities.[71]

The difference resides in the ends being sought: Are radio producers pursuing streaming listeners, social media activity, and the like as a means to bolster ratings and attract advertisers or donors? Or are they doing so with the goal of diversifying content and audiences in the context of public sphere principles? To put it another way, is "audience engagement" merely a euphemism for converting listeners to customers (or in the case of public radio, donors), or does it mean building an online community in pursuit of the public good? The fact of the matter is that it is not necessarily an either-or situation. Particularly in the case of US public radio institutions in the neoliberal era, stations and networks like WNYC are divided by a commitment to a public service mission and a need to attract private funding for survival, which in recent decades has meant adopting market-based strategies and competing directly with commercial media for the attention and dollars of "listener-members."[72] On the one hand, WNYC professes to produce "groundbreaking news, content, and cultural programming that invites ongoing dialogue" and to "explore ways to be an essential resource for New York City's diverse communities promoting inclusion, awareness and intercultural engagement."[73] On the other hand, its executives employ commercial media marketing discourses and logics, speaking of their programs and websites as "products" and of finding ways to "monetize" them, including inserting "brand messages" (audio ads) into all of its podcasts and online streams.[74]

WNYC has an entire "Digital Content" team of a half-dozen employees who manage its main social media accounts (e.g., @WNYC on Twitter and Instagram, WNYC Radio on Facebook). The Digital Content staff also produces video and other multimedia content exclusively for the web. Yet each of WNYC's original programs has its own website and social media accounts that its producers control. Although they are all appearing under the WNYC banner, they are in fact separate channels operating under different motives. While the digital staff focuses on interacting with social media influencers and generating web traffic through tabloid-style "clickbait" practices, the producers concentrate on using social media as a newsgathering tool and a means to connect with the audience for editorial purposes.[75] This results in a bifurcation in public radio's approach to the internet over the past decade-plus, where producers for talk and news radio programs like *Lehrer* see digital media as an organizing tool for community building, while executive management sees digital media as an opportunity for revenue generation. There is a symbiotic relationship between them. The two need not cancel each other out, but there is obviously a risk that the public service motives may erode over time, especially during periods of budgetary crisis.[76]

Conclusion

"Talk radio was the web before it existed. Everyone calling in was a reporter, sharing information," observes Brian Lehrer.[77] In many ways, Lehrer is spot on: talk radio truly was the first interactive form of mass media. The "new media" phenomena of participatory media, citizen journalism, and crowdsourcing all have clear parallels to the "old media" of terrestrial talk radio. There are, of course, differences between them—most notably that broadcast radio remains top-down in organization. The airwaves are controlled by a relatively small number of institutions, and professional media producers like Lehrer ultimately control who and what gets heard on the air. But at least in principle, the incorporation of social media and other forms of internet-based "interactivity" into traditional talk and news radio programs creates opportunities for increased inclusion and diversity.

The internet helps talk radio be an even better version of itself. Particularly in the case of public service-oriented programs like *The Brian Lehrer Show*, networked digital media present opportunities for more voices—and more diverse voices—to be heard. They enable producers to connect more directly with their audience, and thus be more effective in assisting the formation of a wider public sphere. These developments are all a boon for radio's sociability, amplifying both the direct participation in the program itself and also the social interaction that develops elsewhere around the radio programming through its extended life online.

Moreover, the internet has revived talk radio and spread the talk radio logic to other popular media. Talk and news radio formats are especially accommodating of convergent journalism production practices. They are also well suited to the "anytime, anywhere" modes of mobile media consumption. Audiences can listen, as well as participate from, practically anyplace with smartphones and social media—strengthening the sociability of radio and the special relationship the medium has with quotidian activities and personal expression. The expansion of personal media communicative practices throughout the networked digital media landscape (as elaborated on in chapter 3) means that the intimacy and reflexivity that were once distinctive of talk radio have become everywhere commonplace. Local, community-oriented call-in talk radio shows like *Lehrer* operate as much as platforms for the sharing of personal *experiences* as they do for opinions (though the two, of course, are often tightly integrated). Broadcast talk radio's emphasis on individual self-expression and first-person witnessing is now customary throughout contemporary media.

CHAPTER FIVE

Hang the DJ?

Music Radio and Sound Curation in the Algorithmic Age

Standing onstage at Apple Inc.'s annual Worldwide Developers Conference (WWDC) in June 2015, Eddy Cue, the information technology company's head of software and internet services, declared, "The truth is, internet radio isn't really radio. It's just a playlist of songs."[1] Apple's splashy product launches are key to the corporation's public image, momentous in design and overflowing with sweeping, idealistic rhetoric.[2] Cue was introducing the digital music-streaming service Apple Music, specifically its Beats 1 feature: Apple's "24/7 worldwide radio station" that would broadcast live to over one hundred countries exclusively on the internet. Although he did not name names, Cue's playlist remark was unequivocally a swipe at popular internet radio services like Pandora that use computer algorithms to autogenerate personalized playlists. A few minutes earlier in the presentation, record producer and Apple Music executive Jimmy Iovine explicitly called out algorithms for failing to curate playlists with the same savvy and dynamism as a human being. "Algorithms alone can't do that emotional task," he asserted, admonishing the audience, "You need a human touch."[3]

To compete in a crowded online music-streaming market that included Pandora, Spotify, Google Play Music, Rdio, Deezer, and Tidal, Apple was seeking to differentiate itself through a commitment to "human curation"— the 2010s Silicon Valley buzzword for the involvement of old-fashioned editors, publishers, programmers, radio hosts, and disc jockeys (DJs) in the packaging, re-presenting, and promoting of media content.[4] It was adapting strategies from Beats Music—the music-streaming service Iovine had cofounded with musician Dr. Dre and sold to Apple in 2014 as part of a $3.2 billion deal for their audio product company Beats Electronics—and MOG—a social network-themed music-streaming service that Beats acquired in 2012. Apple Music's "human touch" was provided through a proliferation of playlists crafted by "the most talented music experts"—mainly music journalists and celebrity guests. These preselected sets of songs were made available to subscribers for anytime, on-demand streaming. In addition to the curated playlists was the personality-driven internet radio station Beats 1, which was led by BBC Radio 1 DJ Zane Lowe along with the young tastemaker DJs Ebro Darden (of New York hip-hop radio station Hot 97) and Julie Adenuga (of London dance music station Rinse FM). Seeing as Apple Music was billing itself as "a revolutionary streaming music service," it is more than a little ironic that its two biggest innovations were conspicuous remediations of old media forms and practices: handpicked playlists harkening back to carefully composed FM radio DJ playlists and a radio station staunchly committed to live broadcasting.[5]

The 2015 launch of Apple Music coincided with a shift in the recorded music industry from downloading to streaming. That year, streaming—through paid subscription and free ad-supported services—surpassed digital downloads for the first time as the industry's largest source of revenue.[6] Throughout the 2000s, the story of the recording industry had been of declining physical sales (compact discs) and increasing digital downloads (legal MP3s), spurred by Apple's iTunes Music Store and iPod portable media players. This shift ran concurrent to, and was overshadowed by, a steep decline in overall recorded music sales, widely attributed to "illegal" file-sharing through peer-to-peer (P2P) networks like Napster. Nevertheless, attention in the recording industry remained focused on an ownership model, only transitioning sales from the purchase of physical CDs to digital files instead.[7] The widespread adoption of smartphones like Apple's iPhone, plus the spread of high-speed mobile broadband connections, kickstarted a new wave of online music-streaming services around 2010: Spotify, Pandora, Rdio, Slacker, Google Play Music, Beats Music. Music listening steadily shifted from downloading to streaming during the 2010s, and all predictions suggest this streaming trend will continue onward.[8] (As a point of clarification, *online music-streaming services* like Apple Music are not internet radio since they allow on-demand access to any artist or song in their catalog, although they do offer radio-like features as part of a suite of utilities, including Beats 1. In contrast, Pandora is a streaming radio "pureplay" service.)[9]

This chapter explores the rise of streaming music in the 2000s and 2010s, focusing on tensions surrounding internet music radio and the use of computer algorithms to personalize the listening experience. It centers on the case of Pandora Internet Radio, which has been the most listened to online radio platform in the United States for most of that timespan—currently claiming seventy-two million active users.[10] Pandora lets listeners create customized radio "stations" by picking a favorite artist or song, after which the service's algorithms do the work of selecting the music they hear. Since the mid-2000s, internet music radio has mostly done away with the DJ-hosted program format, favoring instead continuous streams of unadulterated music tailored to the specific tastes of individual listeners. While widely used by audiences, these personalization services have provoked laments about the loss of human connection from the likes of Apple's Iovine and Cue, as well as more grounded critiques about a rise of "algorithmic culture" that reduces publicness and perpetuates elitism.[11] Yet radio's sociability derives from much more than just the presence of a human voice; it is equally about listeners turning on the apparatus to connect with something beyond themselves. A DJ-less stream of music can provide that sought-after companionship, creating a score for the rhythm of the listener's daily life. Radio's liveness is little changed, too, since the listener is still tapping into an endlessly copresent stream of audio—the act of listening is still live because listeners are experiencing it in the now, in their present time. Services like Pandora can also offer a form of social exchange, in that the audio stream remains largely outside the listener's power, that relinquishing of control being a key part of radio's sociability.

The presence of algorithms obscures a significant amount of human labor and content curation that is occurring behind the scenes. Pandora's algorithm is powered by the Music Genome Project: a database of over 1.5 million songs, each painstakingly evaluated and hand-coded by skilled music analysts. Thus, while the cultural work of DJs has been offloaded to computers—or at least it *sounds* that way to listeners who rarely hear human voices on the stream—there are still human beings actively involved in the content curation. The Pandora case study demonstrates how these invisible workers are making decisions about the music's cultural meaning and value that, in turn, shape the listener's experience. This is the work of cultural intermediaries, and there is an intentionality and sociability in these processes that retains a distinct radioness. These "human curators" are everywhere involved in music-streaming services nowadays, signaling that there is a reemergence of the radio worker under way.

Music Radio Online before the Rise of Music-Streaming Services

Eddy Cue's claim that internet radio was "just a playlist of songs" was both an exaggeration and entirely ahistorical. For starters, in 2015, there were a multitude of music format radio stations simulcasting their over-the-air feeds online. Also, many internet-only stations/networks with broadcast-style DJ-hosted music radio programs had existed since the 1990s. Net-Radio started all the way back in 1995, explicitly copying the programming and business models of a terrestrial radio commercial music station, presenting webcasts with an eclectic mix of vintage and modern rock and roll.[12] The Minneapolis-based online radio network, after being acquired by software company Navarre, expanded to more than a hundred channels of mostly music programming, though it eventually shut down in 2001 due to a lack of advertising and new investment following the dot-com bubble burst.[13] Legacy media institutions like *Rolling Stone* magazine and MTV dabbled in music radio webcasting during the late 1990s. Other more narrowly focused music webcasters included the recording industry mogul Quincy Jones, who created Qradio in 1998, an online radio network and multimedia web resource dedicated to African music.[14] Over the years, numerous web start-ups built niche stations focusing on a particular music genre or a freeform iconoclasm, very much in the spirit of college radio and celebrated community radio stations like New Jersey's WFMU. With deep ties to the local independent music scene, webcasters such as Los Angeles's Little Radio and New York City's East Village Radio hosted live music performances, featured tastemaker DJs and drop-in celebrity guests, and generally functioned as de facto cultural centers for a community of artists, broadcasting their local scenes to the world.[15] Other small broadcasters, such as Seattle's KEXP and Ohio's WOXY, transitioned to online-only webcasts either temporarily (in KEXP's case) or permanently (in WOXY's case) following financial troubles. Once online, they found much larger audiences for their innovative alternative/indie music programming.[16] Beginning in 1999, the website Live365 was a long-running distribution platform for broadcasters to simulcast their terrestrial radio streams, and it also allowed listeners to create personalized music radio stations for webcast. Not unlike Apple Music, it played up the sociability angle with its slogan "Internet radio programmed by real people, not an algorithm." Live365 shut down in January 2016 following the implementation of increased sound recording performance royalty rates by the US Library of Congress's Copyright Royalty Board (CRB), though it relaunched in January 2017.[17]

These are but a mere handful of examples: thousands of webcasters, big and small, produced music radio exclusively for the internet from the 1990s through the 2010s. Though some adopted the automated playlist model critiqued by Cue, many more reproduced the traditional live broadcast radio model of DJs spinning records. Most failed to survive to the present day, either starved of funding by reticent investors and advertisers or falling victim to a volatile regulatory landscape. In the late 1990s and early 2000s, a tremendous amount of energy and enthusiasm surrounded music radio on the internet, especially as a vehicle for discovering independent artists and "alternative" music genres not typically featured on mainstream broadcast radio's dominant "hot" and "hit" formats. In a *Rolling Stone* essay in 2000, music journalist Rob Sheffield mused about "the addictive power of radio" and the pleasures of getting lost online for hours amid unpredictable streams of music, declaring that "the lure of Web radio is intense, an insomniac's wet dream, the golden age of wireless." He added, "You can hear a spontaneity and intimacy that American radio has lost at a time when a handful of media oligarchs impose their uniform playlists on a nation of DJs who wouldn't know how to pick their own songs even if they were allowed to."[18] Many music fans shared visions of a vastly expanded music radio landscape, filled with new internet-only stations tapping into the deep reservoirs of unknown or underappreciated records from around the globe and across the decades.

The problem with this vision of expanded music radio choice was that, under the Digital Millennium Copyright Act (DMCA) legislation passed in 1998, *webcasters*—defined as broadcasters transmitting only on the internet—were subjected to higher royalty rates than terrestrial or satellite radio broadcasters.[19] As an industry, radio is not regulated uniformly in the United States, the result of a patchwork legal and licensing framework developed piecemeal as each new transmission technology emerged (terrestrial broadcast, satellite, internet). Under the DMCA, terrestrial radio broadcasters only pay songwriter performance royalties (i.e., publishing fees to songwriters, but nothing to performers), and they are granted affordable blanket licenses from the major performing rights organizations (ASCAP, BMI, SESAC). Satellite radio broadcasters pay both songwriter and sound recording performance royalties (i.e., fees to both songwriters and performers), currently set around 15 percent of revenue—a cost that is higher than terrestrial radio but still a "grandfathered" royalty rate.[20] Webcasters, in comparison, are required to pay performance royalties for both publishing and sound recordings, and these are assessed on a per-song, per-listener basis rather than capped through a blanket license or a revenue percentage fee.

The way these commercial webcasting regulations are structured, it is nearly impossible to achieve economies of scale: the more listeners webcasters attract, the more royalties they have to pay, and yet advertising revenues or subscriptions do not scale at the same rate. Thus, while webcasters need a large audience to attract advertisers, the incremental costs can rapidly outstrip revenue. Webcaster royalties increase exponentially regardless of earnings, creating a situation where popular webcasters (such as WOXY) essentially become victims of their own success, their royalty bills far exceeding their advertising profits.[21] There have been some loopholes created for "small webcasters," including educational and nonprofit services. However, these loopholes are temporary; the royalty rates are reset every four years, creating a perpetual state of uncertainty that causes many producers to steer clear of music webcasting entirely.

This issue of recorded music royalties for internet radio is dizzyingly complex and not the primary focus of this analysis. Royalties are easily the single most widely researched and reported topic related to online music radio, and this preoccupation with the legal and economic aspects of recorded music distracts from other significant cultural dimensions of music radio online.[22] (Truth be told, these issues have less to do with music *radio*, per se, than they have to do with the recorded music industry desperately trying to exploit whatever revenue sources it can in the wake of decreased physical record sales.) Nevertheless, an initial lack of regulation—even after the DMCA became law in 1998, webcaster royalties were not put in place until 2002—enabled a period of expansion and experimentation in online music radio that quickly declined in the mid-2000s, as webcasters were weakened by insurmountable licensing fees. An atmosphere of uncertainty and fear was created by the fluctuating royalty rates, which combined with aggressive legal threats made by the recording industry, specifically the RIAA. On a number of occasions over the past twenty years, large groups of webcasters ceased operations en masse over impending royalty rate increases that they feared would bankrupt them or put them in legal jeopardy. Today, the online music radio landscape is dominated by legacy broadcasters whose simulcasts are exempt from the digital performance right, such as iHeartMedia, as well as by a small number of very big players, such as Apple and Pandora, which are able to negotiate their own royalty rates with music publishers and record companies directly or can attract audiences massive enough that their advertising profits can offset their licensing costs.[23]

Returning again to Apple's Eddy Cue, as a distinctly promotional discourse his "just a playlist of songs" statement at the WWDC was not directed at the full breadth of internet-only music radio available to listeners in 2015. Despite the royalty burden and other challenges, there were still many hundreds, if not thousands, of boutique, micro-niche music streamers operating worldwide.[24] But these small webcasters posed little threat to the multinational behemoth Apple. Rather, Cue's comment was directed at the handful of music-streaming services with which Apple was directly competing for the ears of millions: Pandora, Spotify, and Google Play Music paramount among them. In this context, Cue's comment was a bit more accurate, as each of these streaming services ventured into radio, to one degree or another, yet they all relied heavily upon algorithmic processes to generate customizable "stations" or "channels." Beats 1 was designed, then, to bring the (hu)man back into the machine. Or, as the editorial team behind Apple Music's predecessor Beats Music put it in a promotional piece, "Gone are the days of robots telling us what to listen to."[25]

Man-Machine-Music:
Pandora and the Shift to Algorithmic Radio

The thing is, yes, this is a piece of software that you are using but it's not really technology that is doing this—it's musicians, really. Pandora is just a way to take the knowledge that these musicians have and make it available to lots of people at the same time. So, sure, it is technically a piece of software but the recommendations are really human recommendations, when you think about it. The whole genesis came about because I became pretty good at figuring out what people like in music; as a composer, in writing music that they would like. I became pretty good at recommending stuff and that's what Pandora is supposed to be: the small indie record store owner who knows you personally, knows what you like and has this encyclopedic knowledge of music. So, I wouldn't really consider what we have as a technology solution per se. It uses technology but it uses

musicians' ears more than anything. It's a musician who makes the recommendation to you—they just do it via a piece of software.
—Tim Westergren, Pandora Internet Radio cofounder, 2006[26]

We are living in the age of algorithms.[27] In the modern era of pervasive computing and big data, practically every bit of information in our daily lives—from the music we listen to, to the news we read, to the movies we watch, and even the personal messages we receive from family and friends via social media—is filtered through computational procedures known as algorithms. These software programs and protocols have been labeled "social algorithms" by social scientists, since they work to classify, sort, and prioritize what information we see and how we see it.[28] "Software has become a universal language, the interface to our imagination and the world. [It is] a layer that permeates contemporary societies," writes Lev Manovich.[29]

An *algorithm* is in essence a set of instructions that is fed into a computer in order to sort data for the purposes of solving a defined problem.[30] They are so abundant and amorphous today that algorithms have come to be colloquially understood as simply "the things computers do."[31] Precisely because of their ambiguity and omnipresence in modern life, algorithms are equally inscrutable and taken for granted. The fact that algorithms are automated computer functions—the products of the rigid quantitative logics of computer science and mathematics—means they are widely presumed to be neutral and objective, isolated from the human instincts of taste and social bias. To be sure, numerous media scholars have exposed the flaws in such thinking, revealing how algorithms manipulate information and can be used as instruments of social control.[32] Algorithms, plus the data they operate upon to produce song recommendations and the like, are in reality created by humans through laborious, manual processes. Without data and instructions supplied by people, "Algorithms are inert, meaningless machines."[33] Simply put, algorithms are but one small part of a broader network of people, processes, materials, and machines that we call "technology."

Tim Westergren's explanation of how Pandora Internet Radio produces its customized radio stations reveals the huge amount of human labor and creativity that is hidden behind the service's algorithmic music streams and its austere computer interface. There are no DJ voices to be heard on a Pandora "station," but that does not mean the music is unmediated or purely random. His statement gives the lie to Beats Music's assertion that Pandora and its ilk are just a bunch of "robots telling us what to listen to."

Yes, there are algorithms and databases and other computational elements mediating the listening experience; at the end of the day, though, there are people instructing those "robots" as to what music to choose and how to choose it. Or, as Westergren put it, "It's a musician who makes the recommendation to you—they just do it via a piece of software." The analysis in this chapter circumvents the objects of the algorithms themselves and instead looks at the larger internet radio production culture in which these machines are put to work. In particular, how exactly does Pandora operate and what are the interactions between the curators and analysts who program its algorithms? Who are these "musicians," as Westergren calls them, and what methods, knowledge, and expertise are structuring their approach to music radio on the internet?

An algorithm is a precisely controlled series of steps implemented to solve a problem. The problem Pandora was initially designed to solve was that of navigating the overwhelming quantity of music content available to audiences, which existed in massive volume and was also growing at breakneck speed. In the late 1990s and early 2000s there was no shortage of music radio options online, as Sheffield's earlier "golden age of wireless" quote suggests. The internet erased (most of) terrestrial broadcasting's technological and regulatory boundaries, meaning that listeners were no longer limited to whatever over-the-air signals their antennae could pull in—online they could tune in to droves of radio stations from across the state, the country, or the globe.[34] Moreover, the proliferation of peer-to-peer file sharing networks (e.g., Napster, LimeWire, Kazaa), legal music downloading platforms such as Apple's iTunes Music Store, emerging subscription music-streaming services like Rhapsody, as well as MP3 blogs and music sampling sites, all meant that *access* to recorded music was hardly a problem. The problem for listeners was *search*: how to find what they wanted, as well as how to uncover new sounds amid this abundance of choice. In such an environment, Jeremy Wade Morris and Devon Powers propose, streaming became "a metaphor for unlimited access to content."[35] This cornucopia of musical choice brings to mind the utopic "celestial jukebox," the vision of an online platform that instantaneously dispenses any music imaginable to any device.[36] The case of AudioNet/Broadcast.com (chapter 2) showed that at the turn of the century, the one-stop web portal began to emerge as the dominant model for media distribution. The internet of the 1990s was transforming from "an infrastructure of connection" (e.g., email, message boards, chronologically displayed news posts, relatively static websites) to one in which the everyday online experience was dynamic and personalized, each individual user's encounter with media content being generated by an algorithmic sort.[37]

In this landscape of abundant choice, simply aggregating hundreds of radio streams or millions of downloadable songs did not satisfy for long. New technologies and techniques began emerging around 2000 for matching musical artists, albums, or songs to listeners' individual preferences. These were alternately known as *music recommendation services* or *music discovery services*. Beginning in 1998, TheDJ.com (later acquired by Spinner, then purchased by AOL and folded into Netscape Radio) pioneered the automated, DJ-less streaming music model, offering more than a hundred radio stations divided up by micro-genres and specialized themes (e.g., "Fem Rockers," "Modern Love," "Brit Pop")—though listeners could not customize the stations and the music was programmed by staffers, not algorithms.[38]

It was in 1999 that Tim Westergren initiated the company that would eventually develop into Pandora. A pianist who performed with folk and alternative rock bands in the San Francisco Bay Area and who transitioned into composing music for independent films, Westergren partnered with computer engineer Will Glaser and Jon Kraft, a business executive and former Stanford University classmate, to form a company called Savage Beast Technologies.[39] With $1.5 million in venture capital seed funding, Savage Beast began as a business-to-business (B2B) technology company intent on solving the problem of music discovery by creating the Music Genome Project (MGP), a database that classified songs by specific musical attributes.[40] This database was then paired with algorithms that matched the sonic attributes "to an individual's unique musical tastes."[41] Savage Beast licensed this music recommendation data platform to consumer-facing businesses, including music retailers Tower Records, Best Buy, and Barnes & Noble, who used the technology to power CD listening kiosks in their brick-and-mortar stores and online retail sites. "The vision was to provide software to help people buy more music in music stores," said Pandora's former CEO, Joe Kennedy, adding, "Unfortunately, that was a dying business."[42]

Therefore, Pandora did not start out as an internet radio service. It was only after its initial business model faltered around 2004 that Savage Beast began looking for other applications of the Music Genome Project database. Westergren et al. eventually realized that "radio was the best use of what we do."[43] The impetus for the Pandora Internet Radio platform actually came not from Westergren or his music staff but instead from

Kennedy, a former senior manager for E-Loan financial services and the automobile manufacturer Saturn. He had been brought in as CEO in 2004 to steer Savage Beast through its financial struggles.[44] The company had an internal website called "Safari" that allowed employees to type in the name of any song and view a list of the MGP's closest matches. Sitting at their computers in Savage Beast's Oakland offices, they could set the site to continuously play the matching songs in order. "Some of us would spend the day running matches and listening to them," explained Kennedy. "Although our motivation for doing so was to refine and further develop the Music Genome Project, it gradually occurred to us that in some sense we were hearing a personalized radio experience."[45] When it became clear Savage Beast needed to pivot its business, Kennedy proposed turning the MGP music recommendation engine into a consumer-facing internet radio service. He even came up with the new company name: Pandora.[46] On August 29, 2005, Pandora launched as a commercial, subscription-based, internet-only radio service.

Pandora did not initially embrace the label "internet radio." Rather, following its roots as a service supporting the recorded music industry, it preferred a discourse of "music discovery" and "music recommendation." The company initially described itself as an "online music discovery service." The word "radio" was only mentioned once in the company's inaugural press release, deployed to explain the "tailored stream of songs" that listeners created, analogizing these streams to "online radio stations."[47] Although it long operated under the company name Pandora Internet Radio, the service actually started out in 2005 branding itself simply as "Pandora" with the tagline "created by the Music Genome Project." That tagline was changed to "radio from the Music Genome Project" from May 2007 until November 2008. Then, beginning in November 2008, the official name and logo were changed to "Pandora Internet Radio."[48] (The name subsequently reverted back to just "Pandora" in 2012.) Pandora representatives began using radio as a touchstone in interviews as early as 2006, though they typically did so reluctantly and cloaked in qualifiers. For instance, CTO Tom Conrad described Pandora as "sort of a one-click radio" to differentiate the service from other online music portals.[49] The management appears to have resorted to "radio" not out of an affinity for the radio medium but instead as a way to make legible to the public a service that did not have other close correlates. Critics and audiences began referring to Pandora as "internet radio" almost immediately, however, and reluctantly the company itself took up the identifier.

Still, Pandora representatives frequently cast aspersions on radio, sometimes subtly and other times bluntly. Beginning in 2007, the Pandora home page proclaimed, "It's a new kind of radio." Yes, it acknowledged, Pandora's music streams were like radio, but they were different too—the "new" implying a progression, a forward movement to a more advanced state. This is an established convention of remediation: to reference and build upon an existing medium—taking advantage of its cultural status and the public's prevailing expectations—while simultaneously denigrating it, positioning the new media form as somehow better than the previous version and corrective of its perceived flaws.[50] Throughout the late 2000s, Westergren and his colleagues routinely referred to terrestrial radio in their promotional discourse, yet they almost always did so by creating a caricature of contemporary music radio as nothing but corporate conglomerate-controlled stations with homogenized, Top 40 hits-driven playlists. "A radio station only gets to send out one station that everybody has to listen to. Consumers are so different it is very hard to program that one station to accommodate everyone, so it winds up being the best radio for the largest number of people. . . . If you are going for the largest listener base you will tend to play the predictable, well worn hits and well known artists," reasoned Westergren.[51] No mention was ever made of the many college radio stations and small local community radio stations that still programmed diverse music genres and styles, often with no regard for commercial markets or mass tastes. This specter of a post–Telecommunications Act of 1996 Clear Channel apocalypse was a convenient straw man for Pandora because it validated the company's personalization approach, which management claimed was driven by the quality of the music and how well it was suited to the listener's unique taste rather than its popularity or commercial appeal.

Nevertheless, the "radio" label was flexible, and Pandora benefited from it whenever possible. In the post-Napster era of the mid-2000s, internet music services like Pandora used their associations with radio's streaming distribution model to discursively relocate the discussion away from the troublesome subjects of music downloading and file sharing. Taking ownership of "radio" helped Pandora to legally and economically position itself outside of the music-downloading controversy, as well as set itself up to potentially lay claim to terrestrial radio's favorable licensing rates.[52]

The Pandora website and mobile apps operated (as they still do today) by allowing listeners to create customized music streams—or radio "stations"—by selecting a favorite artist or song. (Listeners could also tune into friends' stations as well as preset genre stations.) Beginning with this initial "seed," Pandora's algorithms chose additional songs the listener might enjoy, based upon similar musical qualities or attributes. The idea of "music discovery" implied that, unlike a system built purely on musical genres or sales charts, it would lead listeners from a familiar artist to new, unexpected ones. Although listeners could suggest the station's starting point, they could not select a particular song or artist to hear on demand, the way they could with an online music-streaming service like Spotify. Playback controls were limited; a station could be paused but a listener could not go backward to replay songs or otherwise make a particular song or artist play more frequently. Indeed, the Pandora music player interface was rather rudimentary—and it has changed remarkably little since 2005. Control of the station was limited: the listener could give each song a "thumbs up" or "thumbs down" rating plus skip songs or tell the system "I'm tired of this track," meaning it would not play that particular song again for a while. However, skips were limited to six per hour per station or a maximum of twenty-four per day across all stations.

All this meant that, although listeners were able to directly influence a small portion of the music heard on their stations, the vast majority of the songs were selected seemingly at random by Pandora's algorithms. Rather than give users ultimate control over the playlist, these preference features were designed to provide the system with feedback so that its algorithms could better tailor the listening experience to the individual's personal tastes. Such a setup met technologically focused definitions of radio, since the music stream was always "live" in the sense that it was received in the same moment as it was transmitted. In terms of the phenomenology of listening, the listener still experienced a continuous feeling of directness and presence. Arguably, the ability for the listener to make subtle changes to the radio station through thumbs-up and thumbs-down ratings made Pandora's channels seem even *more* alive than a standard broadcast station, since the content was dynamic and volatile.

Pandora did not broadcast a single channel or set of channels, which is a key way it distinguished itself from broadcast radio and most other internet radio stations. Essentially, it was microcasting millions of unique stations at any given moment. Since stations were not preprogrammed and literally existed only for the individual listener (each of whom could create up to a hundred stations), there were no DJs on Pandora. All a listener heard was the stream of music, plus occasional audio advertisements. This lack of DJ intervention was one of the reasons why its stations were alternately called *playlists*: Pandora itself slipped back and forth between a discourse of "stations" and "playlists." Nevertheless, Pandora was said to create a "lean back" or "hands off" listening experience akin to traditional radio, since after the initial seed was chosen the listener need not do anything and the music would automatically play in an uninterrupted, real-time stream.[53] This is the primary reason why Pandora was legally classified as radio for music-licensing purposes: it was designated a "noninteractive" service since listeners could only express general preferences about what they heard and when they heard it, compared to "interactive" music-streaming services like Spotify where listeners could search a vast catalog and listen on demand to whatever they wanted, whenever they wanted, and as often as they wanted.[54]

In the eyes of the law, radio is a "noninteractive" medium, defined primarily through degrees of user predictability and control. Even though Pandora listeners could influence the music the service provided, they could not predict exactly when or whether a specific song would play. In practice, this is a level of control that is not exceptionally different from traditional broadcast radio, where listeners can actively switch between stations to curate their personal listening experience. A listener may call up and request a song from a DJ, though if or when it gets played on air is unpredictable. The sociability of radio entails that the medium is a space for the audience and Pandora's algorithmic DJ to interact, but also that the audience is not able to control the music fully and must trust the algorithm to an extent.

The Music Genome Project and the Invisible Labor of Algorithmic Radio

The central resource that Pandora relies upon to make its personalized radio stations—and the thing that most sets Pandora apart from other internet radio services—is the Music Genome Project database. Predating Pandora itself, the MGP was envisioned by Westergren and developed with Stanford University musicologist Nolan Gasser. Its millions of songs have been analyzed according to a taxonomic system identifying particular musical attributes, such as a song's melody, harmony, rhythm, lyrics, instrumentation, and orchestration. Each of these musical attributes is given a numerical value (rated on a 1–5 scale), turning them into data points (metadata) that customized algorithms can then use to match any given song with other songs in the database sharing similar musical traits. This reliance on matching—a form of linking—roots Pandora firmly in networked digital culture, as the *link* is a fundamental component of the web. The MGP essentially aims to take music apart and reverse-engineer it, evaluating each song by its acoustic elements. Pandora seeks to "capture the essence of music at the most fundamental level," an objective achieved through formal musical analysis.[55]

Taking inspiration from the Human Genome Project and its goal of mapping all the genes of humans, the Music Genome Project treats major groups of musical characteristics as "chromosomes" that are full of multiple minor characteristics, or "genes."[56] For example, "vocal delivery" is a chromosome, and genes in that group include "light or breathy," "smooth or silky," "gritty or gravelly," and "nasal."[57] Each chromosome might have anywhere from a dozen to fifty genes.[58] There actually is not one music genome but a dozen: pop, classical, world, Celtic, Indian, African, jazz, reggae, electronic/dance, R & B, gospel, hip-hop/rap. The number of 450 genes gets bandied about in the press and accentuated in Pandora's own promotional discourse, although in reality the different genomes are assessed according to different sets of genes: there are 450 genes for classical music but only 200 genes for pop music—and songs from the pop genome actually make up the bulk of the MGP database.[59] (The pop genome covers rock, funk, blues, folk, and country—what Westergren has called "the mainstream genres.")[60] Nevertheless, each track in Pandora's streaming library has been examined and coded by a trained music analyst, a process that takes anywhere between ten minutes and a half hour per song. These professional analyses are the primary source data for Pandora's entire music discovery system. They are the sum of what the MGP knows about the music, forming the basis of practically everything done by the algorithms.

The stated mission behind Pandora and the MGP is to expose listeners to new music based "on what the music actually sounds like," as opposed to more commonly referenced markers such as genre, popularity, or other users' previous consumption behavior.[61] Part of Pandora's oft-repeated origin story is that Westergren was a struggling touring musician who observed firsthand that the majority of musical artists were forced to live gig to gig without any record label support or marketing. Additionally, he felt most popular music was rarely the best available music. If only musicians could get around the bottleneck created by cultural gatekeepers like A&R reps, music critics, and corporate radio playlist makers and put their music directly in front of audiences, he believed, music fans would eagerly take to these lesser-known artists. "I was passionate about helping working musicians to find the audience they deserved."[62] Westergren's method to achieve this goal was to renounce basically all the social and cultural aspects of music that, in his mind, were preventing people from actually hearing the music for what it was. Thus, the basic idea of the MGP was to "approach music from almost a scientific perspective," according to chief architect Nolan Gasser.[63] This is why the MGP focuses on distinct and measurable musical characteristics that are ostensibly objective and observable, such as the presence of an acoustic guitar or the use of a triple-meter time signature.

While an average listener might be able to pick out a song's melody or its instrumentation, far fewer are able to identify more complex musical attributes such as the use of modality or syncopation, yet alone analyze these traits uniformly across a multitude of recordings. As a result, the MGP requires analysts with considerable experience in musical performance and composition. Pandora refers to these workers as "music analysts," though they are sometimes also called "musicologists." In point of fact, most analysts do not have degrees in musicology, nor does their work resemble musicological research. Instead, they are "musicians with day jobs," as a *New York Times* writer observed.[64] Pandora goes out of its way to hire working musicians—as analysts and in other staff positions, too.[65] Although accounts vary, Pandora employs between twenty-five and one hundred of these music analysts. All combined, they add as many as ten thousand songs to the MGP each month.[66]

Yet not just any musician can get the music analyst job. "We employ rigorous hiring and training standards for selecting our music analysts, who typically have four-year degrees in music theory, composition or performance," Pandora's executives assured potential investors in 2011, "and we provide them with intensive training in the Music Genome Project's precise methodology."[67] In addition to being academically trained in music, as part of the application process prospective analysts must take a test "that is, in essence, a music theory exam."[68] If they are hired, they are subsequently put through around forty hours of training per genome, and they must "internalize" "a fairly extensive manual for each genome."[69] Thus, as musicologist Jason Kirby has observed, Pandora requires a combination of "both 'street' and 'book' smarts" from its analysts that is quite rarified. This selectivity results in the imposition of inherent taste and knowledge hierarchies.[70]

Pandora's method is known in data management circles as a *content-based recommender system*, since it constructs its recommendations based upon descriptive information about the song recordings themselves.[71] This contrasts with more widely used automated recommendation techniques like collaborative filtering, which typically do not require platforms to create any original data, since they draw upon user-generated behavioral data and ratings and metadata like product descriptions provided by manufacturers. In contrast, Pandora's music analysts are laboring extensively to create new cultural paratexts. Every song in the MGP database makes up an entry containing far more content than just the artist's original sound recording: the audio is surrounded by hundreds of points of descriptive metadata created by the analysts, which place the music into a broader cultural context and relate it to other music.

Pandora's focus on measuring and manually annotating the qualitative features of musical recordings is an act of curation. The MGP staff members are selecting and organizing a collection of music for public consumption, much the same way an art curator handpicks paintings for a museum, a collection development librarian acquires books for a library, or a radio DJ chooses playlists for airplay. Moreover, these cultural workers are directly shaping not just what music makes it into Pandora's song catalog but also how that music should be heard. The metadata created from their musical analyses, combined with the playlist algorithms built by Pandora's programmers, determines what music supposedly naturally belongs together (i.e., "a good match"). These matches in turn influence what listeners hear, and when and where they hear it.[72] Therefore, Pandora is staffed by skilled cultural intermediaries who are doing work that is fundamentally similar to that of a DJ or radio station music director. It is simply that their labor is a degree removed, shrouded behind computer algorithms and a scientific discourse that purports to objectively and definitively classify culture.

Pierre Bourdieu's concept of *cultural intermediaries* has found renewed purchase in a digitally networked media environment in which new forms of "creative" work proliferate.[73] Cultural intermediaries are those workers in the information economy who have "occupations involving presentation and representation" and who produce symbolic value for goods and services—normally individuals working in marketing, fashion, design, counseling, and education.[74] Media professionals such as journalists, newspaper and magazine editors, and broadcasters are cultural intermediaries, since it is their job to select what content is presented to audiences and to frame the meaning and cultural value of the content. They are cultural workers whose primary function is as gatekeepers and tastemakers, mediating between media producers and audiences. Media scholars have extended this cultural intermediary concept to include the technologies of recommender systems that organize and frame online media use. Jeremy Wade

Morris argues that streaming music algorithms are "infomediaries" that possess increased power and control over audiences.[75] For Morris, cultural intermediation today is a hybrid process that involves "a combination of human actors and technical devices."[76] It is difficult to separate Pandora's human analysts and playlist engineers from the algorithms and user interfaces to which they delegate much of their intermediary work. Still, the human workers' contributions to the MGP are unmistakably a curatorial practice and a form of cultural intermediation in the most classic sense of Bourdieu's concept.

DJs are prime examples of cultural intermediaries in the Bourdieuian sense, as it is their main job to present and comment on musical recordings. Not coincidentally, another common name for a radio DJ is a *presenter*. Historically, DJs have held positions of considerable cultural power, playing instrumental roles in promoting recording artists, establishing new musical genres, creating cultural identities for local music scenes, and generating shared moods and emotions among an audience.[77] There is much more to the work of a DJ than announcing the records being played on air. "The DJ defines her/himself through music; it is a vital part of her/his life," argues professional DJ and scholar Steve Taylor, who further explains, "This enthusiasm for the form is the basis of an ongoing campaign to make others aware of music's potential. The DJ is constantly discovering new music and contextualizing older music within an ever developing canon."[78] The term "disc jockey" implies a skillful maneuvering of the records: it is a position that requires charisma and an impeccable sense of timing. A DJ must understand the emotional qualities of the music in combination with the feelings of the audience, choosing the right song for the moment.

The great pleasure of listening to music on the radio is found in moments of *serendipity*, when the DJ brings together a pair or set of songs in a way that is unexpected or idiosyncratic but wholly unique.[79] This ability requires substantial embedded knowledge on behalf of the DJ, in terms of understanding music history, possessing command over a breadth and depth of recordings, and also having an awareness of the audience and a capacity to anticipate listener desires. Undoubtedly, many radio stations today have music directors who manage DJ playlists and set a "rotation" of songs that must be regularly broadcast. Since the 1950s, format radio has been populated by consultants with central programming control.[80] Nevertheless, many DJs still have autonomy over their playlists, and the romanticized image of the charismatic individual still dominates our popular culture discourse of the DJ.[81]

With automated internet radio like Pandora, there are no longer identifiable personalities punctuating the stream of music, as the algorithms
have taken over the physical occupation of "spinning" the records. These
are not insignificant changes. Yet there are human laborers who are still
adding context and symbolic value to the music and crafting the sequence
of songs, which is among the most important cultural work of a radio DJ.
Despite the lack of human DJs at the forefront, Westergren has said that
"we want Pandora to feel like it's talking to you"—and that effect is primarily achieved through the moments of serendipity and surprise that the
MGP's unconventional matches create.[82] The purpose of the MGP is to
"link songs to one another in pleasing and surprising ways."[83] The connecting tissues between songs are the music analysts' descriptions. Even
though analysts are not able to deliberately pair together individual songs
the way a DJ might, how they classify the individual traits of a song directly
determines how, and if, it will be matched with other recordings. If an
analyst rates one pop song's vocals high on the "nasal" scale and another
pop song's vocals high on the "gravelly" scale, the likelihood is low that the
algorithms will ever pair such disparate sounding tracks. Pandora and its
algorithms have not eliminated the participation and influence of cultural
intermediaries in radio and popular music, they have simply changed the
intermediaries from DJs to analysts and made their intermediation all but
invisible to the audience.

Taste and Cultural Intermediation in Pandora Internet Radio

There is one glaring problem: the foundational assumption structuring
the MGP—the idea that music can be objectively analyzed—is highly sus-
pect. Strip away the algorithms and all the dazzling technology, and the
main thing differentiating Pandora's song selection method from that of
traditional broadcast radio's is that its choices are based on the supposedly
neutral components of musical form, as opposed to marketing reports and
advertising budgets and critic judgments and all the other economic and
sociocultural factors that allegedly get in the way of appreciating music for
music's sake. Pandora puts forward a distinctly egalitarian rhetoric. "The
Music Genome Project is completely blind to popularity. So it's really a
level playing field," says Westergren. "The only data used to build playlists
are the musicological details, and the thumb up/thumb down feedback.
That was a central part of Pandora's foundation years ago—creating equal
opportunity for every talented musician."[84] Moreover, Westergren believes

that song selection should "not [rely on] the opinion of an editor, and not [be] based on what other people like,"[85] adding, "Our intention is not to be the arbiters of what's good or bad."[86] "Our mission is to play only music you'll love"—the "you" here being individual listeners, who get to influence the music stream through their taste expressions.[87] It is Westergren's assertion that the music analysts can be neutral, unbiased judges and that the only person whose taste influences Pandora's music recommendations is the listener. While Westergren readily admits that Pandora is "mapping taste," he somehow believes this can be done with the company's cultural cartographers never imposing their own taste in the process.[88]

Such a proposition falsely presumes that formal musical analysis is an objective science. Yet the discipline of musicology (of which music theory is a subfield) is not neutral. It arose at a specific time and in a specific context (nineteenth-century Europe), and as such it codified particular musical practices and notions of "good music." Residing at the core of musicology are a set of critical assumptions: works are autonomous; art possesses transcendent qualities; the individual artist should be the focus of critical and historical examination; listening should be detached and contemplative, and thus analysis should be text centered.[89] The entire premise of the Music Genome Project rests on these ideological principles, including the notion that recordings can be analyzed as discrete aesthetic objects completely independent from the cultural sphere (i.e., socio-musical practice, the social class structure, the lived experience of the audience). While the MGP does break from historical musicology in some ways, such as an attempt to incorporate affective elements like mood and aspects of performance (not just composition), it does so through an equally problematic positivist Enlightenment ideology of scientific classification and comprehensiveness. The methodological and ideological rigidity of the MGP is entirely out of touch with the "New Musicology" that has challenged the cultural authority of academic musicology since the 1970s and prioritized studying the ways in which music functions in social settings.[90] The MGP continues a preoccupation with structure and analytical formalism severed entirely from cultural context. Moreover, it privileges a musical vocabulary and body of advanced musical knowledge that conflicts with Pandora's egalitarian rhetoric, as the formal training required for employment means its musical analysts are members of an elite social group. This is a high-skill job requiring a specialized college degree; Pandora's analysts are part of what Bourdieu identifies as the *new petite bourgeoisie*.[91] Put simply, the music analysts bring their ingrained social class and educational assumptions to the task.

No matter how many "genes" the MGP includes, there are always going to be features of a musical piece that are either purposefully or inadvertently ignored. Every taxonomy is a classification scheme that excludes at the same time it includes. By its very nature, a musical taxonomy is built on subjective, value-based decisions about what aspects of the music are important or unimportant. This is not a criticism of Pandora alone; any cultural recommendation system inevitably makes such judgments. Pandora's algorithms are designed to weigh certain "genes" and "chromosomes" more heavily than others when they select songs. Which factors are given greater or lesser weight is not publicly known; however, the mere fact that they are unevenly balanced conveys that Pandora's programmers have made decisions about their cultural importance. Perhaps most significantly, the MGP's reductionist emphasis on formal traits privileges the elite body of knowledge that is academic music theory, composition, and performance. Pandora's conscious decision to exclude sociocultural factors like historical context, popularity, critic opinions, or community practices not only implies that certain types of cultural knowledge and experience are inconsequential, it also rejects many of the reasons why people might enjoy and want to listen to music in the first place. (Though as detailed below, various sociocultural factors do manage to creep into the MGP—a fact that further emphasizes the error in trying to arbitrarily separate "the music" from the culture through which music attains its meaning and value.) Westergren and company falsely suppose that music can exist in a cultural or temporal vacuum.

Contrary to the discourse of objectivity and rational efficiency, the subjective decisions and biases of these cultural workers get encoded into Pandora's internet radio platform. While Westergren calls the MGP taxonomy "ridiculously academic," a closer inspection of the "genes" reveals an evaluation process that is strongly editorial and resembles popular music criticism as much as it does music theory.[92] While some of the characteristics measured in the MGP can be identified and defined rather precisely, such as chordal patterning, a good many more of them—and the terminology Pandora uses to define them—are hardly as exacting. For instance, drums are described as either "tight" or "booming," which are ambiguous classifications and not ones music theorists use normally. Other genes measure clearly subjective characteristics that have nothing to do with musical form in itself: aspects of mood and emotion, such as how "joyful" lyrics are or whether a song is "emotionally intense" or "motion-inducing"; level of virtuosity and how "busy" a solo instrumental performance is; whether the track has a discernable "influence."[93] With jazz recordings, for example, analysts identify whether there is an "R&B" or a "smooth jazz" influence.[94] Such a judgment relies on genre distinctions that Pandora repeatedly claims it ignores. Westergren has stated that "genre is useless" in the context of the MGP, since the system is designed to focus only on formal aspects of the musical composition and performance.[95] Yet the notion of genre is baked into the MGP's very structure: the different genomes are clearly based on traditional genre distinctions (e.g., pop, hip-hop, classical). A Pandora blog post even plainly describes the genomes as "categorized by genre."[96] Elsewhere, Westergren explains that "genre is really just a collection of genes to us."[97]

The notion that the MGP is immune to cultural factors such as workers' value judgments or mainstream popularity erodes even further when the curatorial process of how sound recordings reach the music analysts is taken into account. The analysts themselves do not choose the music that ends up in the company's music library. Pandora has a "music curation team" whose job it is to select and screen the songs to be added to the streaming catalog.[98] (Pandora does not add entire albums to its catalog, only handpicked songs.) Contrary to the democratizing notion of the celestial jukebox, Pandora is operating within a decidedly selective tradition of music and culture—an elitist, authoritative, "apostolic vision for culture" that Ted Striphas argues is becoming more normative with the spread of algorithms.[99] Unlike other companies that employ machine-listening techniques that can analyze thousands of songs in a day, such as the Spotify-owned Echo Nest, the time- and labor-intensive nature of its analytical process means Pandora can only add a few hundred songs per day to its catalog. However, this constraint dovetails with the company's quality over quantity ideology: their stated aim is to catalog only "good music." "Ironically, I found over the years that the fact that we couldn't go fast was a big advantage," states Westergren. "The problem that needs solving for music is not giving people access to [millions and millions of] songs."[100] Forthrightly, he told the *New York Times*, "We struggle more with making sure we're adding really good stuff."[101] No matter how you approach it, any notion of "good" is ultimately going to be a subjective, taste-based cultural judgment.

The curation process begins with Pandora's music buyers, who source new music from various sites. Steve Hogan, Pandora's music operations manager, explains that back in the Savage Beast days, he mainly turned to the music sales and radio airplay charts to determine what records to add to the MGP. "In a typical week, I would come in, research the *Billboard* charts and the *CMJ* charts, and make a list of albums we needed to get," Hogan explains. "Then I would drive over to Berkeley to the big record stores, and I would go on a shopping spree."[102] Westergren describes how the first thing Savage Beast did to build up the MGP in the early 2000s was to go through more than fifty years of popular music charts and acquire all the hit songs. "We went through the *Billboard* charts back [to] the '50s and got everything since then for the database. Various independent charts, went back to their beginnings as well."[103] This reliance on charts indicates that popularity influences the MGP after all. Pandora's equivalent to a library acquisitions team also turns to prominent music publications, retailers and distributors, and the record companies and recording artists themselves. Pandora's music buyer from 2003 to 2012, Michael Zapruder, explains that he would learn about new acquisitions by looking at popular music news websites like Pitchfork and the catalogs of tastemaker indie music retailers like Insound, Beatport, Forced Exposure, and Other Music.[104] Pandora is also in direct contact with the recording industry's major distributors and labels. As Westergren put it, "We go through independent labels, anyone with a reputation for good music."[105] They receive hundreds of submissions per week directly from record labels, publicists, and artists.[106] Pandora's music managers are adopting a distinctly editorial voice and acting as traditional gatekeeping cultural intermediaries. These activities are quite similar to the ways radio stations and individual DJs curate their music libraries and source their playlists.

While Pandora as a company avows that it is out to support independent musicians—Zapruder claims more than 90 percent of the music he bought for Pandora's library was released on independent record labels, for example—it nevertheless has a commitment to ensuring that all the most popular artists and songs are included in the streaming catalog.[107] Like any commercial music radio station, Pandora knows it needs the mainstream favorites on hand, as the audience will be expecting to hear those heavy hitters with regularity. Zapruder has stated that the music collection is built first and foremost to satisfy Pandora's listeners, which, in practice, means carrying all the popular favorites—and if artists are missing, they will get chastising emails and calls from listeners.[108] Westergren adds, "We need to have what people know because that is often what people use to start a station."[109] If listeners type in an artist or song name and cannot find it, this "missed" search might result in their abandoning Pandora. While the music analysts may be working from pseudo-objective criteria, the initial selection process is highly subjective. Moreover, popularity, sales figures, music critic opinions, and the like are all very much influencing what music gets included in Pandora's streaming catalog. The chart-topping hit artists and songs actually form the backbone of the MGP—they are Pandora's equivalent of a format radio station's rotation playlist, the established core around which the rest of the programming is built.

The Remediation of Radio Logics in Pandora Internet Radio

In terms of programming, Pandora substantially resembles traditional music radio, especially through its use of repetition. Although billed as a "music discovery" service, in reality Pandora uses its algorithms and accumulated listener data to mix up new music with a steady diet of familiar artists and songs. Eric Bieschke, the software engineer who developed Pandora's playlist algorithms, explains the algorithms are working properly when they determine "when this and only this listener is ripe for a burst of discovery, or a familiarly beloved song from their childhood, or their favorite song from last summer. It's all about hitting people with the perfect song in the perfect moment."[110] Pandora's "playlist team" regularly runs experiments on its audience to determine the right mix of familiar and unfamiliar music. "Some of [these experiments] are about increasing exposure to new music. Some of them are about cycling songs in and out of a rotation so you're actually hearing the same concentration of really good music, but it's spaced out in such a way that people don't perceive it as a repetitious experience," says Bieschke.[111] This careful balance of novelty and repetition is hardly a new strategy: it is precisely what good broadcast radio DJs have aspired to achieve with their music playlists for more than a half-century.

Despite Pandora representatives' critiques of the homogenization of chart-driven mainstream music radio, many observers have lamented the frequency of popular middle-of-the-road hit songs on Pandora's stations. One former employee observes that this type of repetition became especially pronounced after Pandora's 2011 IPO: "It used to be that you'd put in Modest Mouse and then hear all these crazy college indie bands. That was how it was created. It was great. But people in the Midwest hated it. Now, you put in Modest Mouse and you hear Maroon 5. It's much more like [terrestrial] radio. Some people got angry, but the majority like the changes."[112] As Pandora's audience broadened, it became increasingly beholden to the same commercial logics of the iHeartMedia and Cumulus

Media chains. It has resorted to the tried-and-true broadcasting strategy of repeating the biggest hits and most popular artists, again and again.[113]

Repetition of hit songs has become increasingly prevalent on US commercial radio during the past decade, and Pandora has not coincidentally followed that lead.[114] Pandora's analytics team has studied terrestrial radio programming strategies and verified those findings with their own internet radio-listening experiments. "The reason terrestrial radio repeats so much of the same darn music is because that is the thing getting people coming back to the radio station," Bieschke reveals. "It's the most annoying thing about terrestrial radio, but it's absolutely by design."[115] Pandora has concretely found that its oft-touted mission of "surprisal" is in direct conflict with the average listeners' preferences, which prioritize familiarity and repetition.[116] It is thus inaccurate to regard the serendipity of radio as being about playing entirely new music: the surprise factor comes from a DJ playlist that mixes songs we already know and love in new and unexpected ways. The music is familiar, only the sequencing is new.

It is an elitist presumption that adventurous "music discovery" is even a desirable activity for most listeners. Few people turn on the radio or open a music-streaming app to be challenged: they want to hear the music they like. Pandora's extensive data has shown that listeners are particularly intolerant of "discovery" (i.e., new music) while at work, since they are presumably putting Pandora on as background music. As a result, Pandora's algorithms generate more repetitive playlists during the daytime hours.[117] In addition, numerous critics have looked at Pandora's recommendations and concluded that the service actually deepens listeners' tastes rather than broadens them. It is more likely to connect listeners to similar acts in the same genre as the original artist or to unfamiliar "deep tracks" in a favorite artist's catalog than it is to make entirely unknown and musically diverse matches.[118] The takeaway here is that Pandora's playlist algorithms are purposefully designed to appropriate programming conventions—most of all a reliance on repetition—that have been prominent in traditional broadcast radio for generations.

Observers remark that Pandora "ignores the crowd," in that it does not use collaborative filtering techniques or other recommendation strategies that take into account social data and the listening habits of other users.[119] That is false: Pandora uses collaborative filtering, and, in recent years, it has increasingly come to rely on data analytics to program listeners' radio stations. (It has also added machine listening to supplement the music analysts' manual process.)[120] Each listener's every action is tracked on Pandora, creating billions of points of "contextual feedback." Circa 2014, Pandora had collected more than 35 billion thumbs-up/down data points alone.[121] Pandora's "playlist team" of data scientists and software engineers constantly tinker with factors like song repetition, as well as whether time of day or geographic location impacts music preferences, if listeners prefer more local artists or national acts, whether they will tolerate alternate takes like live recordings and acoustic versions, or if more weight should be given to the thumbs-up or thumbs-down button.[122] (The answer: thumbs-downs are given greater weight.)[123] These various tests are run platform-wide and the collective results are used to tweak Pandora's "master algorithm." This means that other listeners' preferences are impacting the music individuals hear in their personal stations. Pandora is quietly using listeners' activity for more than just understanding who they are personally and what music they enjoy: they are also using it to shape the experience of the entire audience.

Importantly, the engineering staff are cultural intermediaries too. And with Pandora relying increasingly heavily on analytics, these workers' labor and cultural values are shaping the music programming on Pandora as much as those of the MGP's music buyers and analysts. There are competing ideologies at play within Pandora nowadays: the analysts are focused on pseudo-objectively analyzing songs to enhance music discovery and appreciation; the buyers are keenly aware of keeping the audience happy by populating the catalog with the most in-demand songs/artists; and the data analysts are interested in identifying large-scale patterns among the entire audience, particularly for the purpose of maximizing the session length and "return rate." That is, the data analysts are supporting the subscriptions and ad sales staffs by keeping people listening and keeping them coming back—and, in practice, this means emphasizing repetition and de-emphasizing discovery.[124] Economic interests are pushing Pandora to operate more conservatively, skewing the platform away from its stated mission of music discovery. The result is that it has developed to function institutionally and sound aesthetically more and more like the commercial music radio Westergren and his colleagues have discursively defined themselves against.

Whereas Pandora started with a predominantly human curatorial approach to internet radio, it has increasingly grown to depend on algorithms and computational techniques such as machine listening, plus the data scientists and software engineers who manage these machines. (Pandora operates more than seventy-five different algorithms, or "recommenders," that all feed into the "master algorithm.")[125] "We're never purely doing the musician-musicologist-expert technique and we're never doing the pure data scientist-machine learning-engineering technique," former chief scientist Bieschke explains. "We pull all of these people together—engineers, data scientists, musicians, musicologists, curators—and put them all in the room and have them come at the problem from different directions. Sometimes the insight comes from people with music expertise and sometimes the insights come from the people looking at the pure data. Oftentimes, the true breakthroughs come when we're crossing those two worlds."[126] The key point here is that there are people—cultural intermediaries— involved at every stage. Perhaps counterintuitively, as Pandora's music recommendation system has become more infomediated, there are actually more people involved in getting the music to the audience, each and every one of them adding symbolic value to the music and molding the radio listening experience along the way.

Conclusion

Internet music radio services like Pandora are still distinctly *radio*, despite the fact that they are automated and personalized. Specifically, I am challenging the conclusions of radio scholars who have contended that these automated streaming services are, in fact, *not* radio because of their absence of human communication.[127] Chris Priestman contends that music-only services lack intentionality in their production, which in turn removes the element of sociability. However, there is a significant amount of human labor and intention behind the Music Genome Project and the production of Pandora's customized radio "stations." Although this algorithmic music radio may not conform precisely to traditional conventions of the radio DJ/presenter guiding listeners and their radio listening experience with a storylike narrative, the construction of the music stream is far from random or lifeless. Theories that define radio through a limited conception of human conversation—literally "real presenters talking from real studios in real time"—preclude not only large swathes of internet radio but also a good many terrestrial radio stations from being considered "radio."[128] For decades now, FM commercial music radio in the United States has embraced a "more music / less talk" programming strategy that favors long blocks of uninterrupted music and "announcers" who take up a restrained presence on air.[129] Since the mid-2000s the fastest-growing music format in FM radio has been the DJ-less Jack FM, which one critic describes as "essentially function[ing] as a 1,000-song iPod set on 'shuffle.'"[130] Radio definitions that romanticize a certain style and era of imaginatively programmed music playlists delivered by magnetic on-air personalities end up privileging a very particular notion of radio that does not actually match with the reality of how the medium has existed in reality for some time.

Apple Music's Beats 1 also draws upon a romanticized discourse of exuberant DJs versus soulless algorithmic robots. Eddy Cue promises, "And it isn't just algorithms. It's recommendations made by real people who love music, and they're our team of experts."[131] But of course, Pandora is similarly backed by a team of experts. Algorithms are only abstractions: there are always people working behind them, feeding them data and programming them to do this and that. Online music radio platforms are built by people, and algorithms follow the instructions set by those individuals. There is hardly a music radio webcaster or streaming music service today that does not employ human curators, editors, or content analysts in active roles. There is a convergence of technology happening: everyone from Spotify to Pandora employs a hybrid methodology of collaborative filtering, machine listening, and multiple types of content-based recommendation. But there is also a convergence of man and machine, of computational and human sorting processes.

Many of the platforms that once prided themselves on *not* using human music editors or curators (such as Spotify and Google) are now putting humans out front, in the form of radio or radio-like features. Spotify now delivers automated song/artist-based personalized radio stations. It also compiles weekly two-hour-long personalized playlists, or "mixtapes," for its users through its "Discover Weekly" feature.[132] These streams and playlists are automatically generated by Echo Nest's algorithms, which rely on a mix of collaborative and content-based recommendation. However, Spotify also offers "In Residence" radio shows, which are prerecorded, "podcast-like" programs featuring famous artists playing and speaking about their favorite music.[133] Tidal also features artist-curated playlists. Amazon provides curated playlists to its Prime Music customers. Before shuttering in late 2015, Rdio offered a Pandora-like personalized radio feature called "You FM," a series of "expertly curated stations" (preprogrammed playlists, really), and simulcast feeds for nearly five hundred Cumulus Media live broadcast stations.[134] In the recent years, various other smaller online music start-ups have focused on handpicked playlists or hosted live webcasts, including Shuffler.fm (relaunched as mobile app Pause.fm), This Is My Jam (now shut down), Turntable.fm (also now shut down), and Dash Radio.

Google employs a team of approximately fifty curators for its Google Play Music All Access platform. Google acquired the staff in 2015 along with the curated playlist service Songza, which lives on within Google Play Music through the "Listen Now" thematic playlists feature.[135] Even Pandora has launched "handpicked" New Music Stations that provide "an elevated listening experience" with a preprogrammed weekly mix of new releases across about a dozen different genres.[136] Pandora has also offered up "personalized soundtracks" in the vein of Spotify's "Discover Weekly" playlists, as well some 250 hand-curated "Featured Playlists" focused around various genres, artists, moods, and activities.[137]

As online music radio vastly expanded during the decade from 2005 to 2015, it primarily developed along two separate trajectories. On the one path, there were the algorithmically driven personalized streaming radio platforms like Pandora. On the other path, there were the traditional live broadcast-style webcasters and carefully curated playlist services like Songza. In the past few years, these models have converged. All the major online music-streaming services now offer some mix of personalized algorithmic radio streams, broadcast radio-like hosted stations or programs, and human-curated playlists. Of course, Apple Music, Spotify, Google, Tidal, and the like then pair these radio features with on-demand streaming music catalogs containing forty million-plus songs. (Pandora launched its own Spotify-esque on-demand music service, Pandora Premium, in 2017, building on the assets of the defunct Rdio, which it acquired in November 2015.)[138] It is notable that "radio" for these massive online music platforms is only one option in a portfolio of services. Apple Music launched with the slogan "All the ways you love music. All in one place." Moreover, the company claims that Apple Music is "three things": a "revolutionary music service," "24/7 global radio," and "connecting fans with artists," in addition to the vast iTunes Music Store. Thus, radio is positioned as one piece in a larger digital music puzzle. But in the late 2010s, radio has nonetheless become an increasingly important piece.

While algorithms are hardly going away, online music radio and streaming services are realizing they need human curators to add layers of context. In the rush to provide audiences with personalized services and massive libraries of content, many of the social and cultural aspects of music listening were lost, including the shared communal experience of tuning into the same channel as your neighbors and loved ones or the simple joy of relinquishing control for a while and waiting to hear what song the DJ plays next. As DJ Ebro Darden of Beats 1 muses, "In the simplest terms, people like people. Social is the world we live in. Human curation is in and around someone that you trust or someone you just met. It's like walking up to a bar to have a drink or sitting next to somebody listening to something. That's what we're trying to create: a gathering moment, sitting around discovering music together."[139] This is an illuminating assessment, but Darden nevertheless gives a rather one-dimensional perspective on sociality. Again, a social exchange takes place in the mere act of a Pandora listener choosing to let someone else select the music—or something else, in the case of an algorithm. Furthermore, whatever sociability might be lost by not having a DJ identify and explain the playlist can be recaptured through the sharedness of social media: Pandora listeners can share a favorite station or track with friends via email or social media. A new displaced communal listening is occurring through social media tools that are built into internet radio platforms and apps like Pandora, and it is leading to a new kind of sociability that replaces the real-time simultaneity that Darden and Beats 1 privilege with a more fully participatory, audience-centered sharedness.

Internet music radio comes in many shapes and sizes: the live broadcast model is not the be-all, end-all of radio. Personalized music streams and curated playlists still serve many of the same functions of serendipitous music discovery and communicative exchange that are key features of radio's sociability. Also, Darden himself is a veteran radio personality, and many of the new "editors" and "curators" at places like Apple Music and Google Play are former radio producers, as well as music journalists and A&R representatives. They are essentially doing the work of radio station music directors and DJs. The reestablishment of these cultural intermediaries from the radio and popular music industries means a restoration of elitist gatekeeping. Cultural intermediaries never left, truth be told, as this chapter's examination of the invisible labor of Pandora's music curators and analysts has shown.

In the realm of online music, there has been a remediation of radio as both a cultural form and a social function. Chris Becherer, a senior vice president for the online music-streaming service Rdio, astutely observed, "There's a group of people that have been doing curation in this country for over 100 years quite effectively, and that's local broadcast radio DJs."[140] This is yet another example of the story coming full circle, of "new media" start-ups like Pandora supposedly breaking away from the "old media" industries and practices only to end up returning to radio's tried-and-true ways in the end.

CHAPTER SIX

Touch at a Distance

The Remediation of Radio Drama in Modern Fiction and Nonfiction Audio Storytelling

"We're used to thinking of sounds as being about something—speech is always about something. But it feels to me more like touch. Sound is kind of touch at a distance." Stanford University psychologist Anne Fernald makes this claim to cohost Robert Krulwich in a 2006 *Radiolab* episode on the musicality of language.[1] The quotation hints at the exceptionally affective power of sound—specifically, Fernald is referring to speech between mothers and their infant children. Over the past decade, her statement has become a mantra of sorts for radio producers and fans seeking to demonstrate the "intimacy" of radio, podcasting in particular. *Radiolab*'s hosts, Krulwich and Jad Abumrad, even used Fernald's "sound is like touch at a distance" line as the centerpiece for a 2006 talk on how they use sound in their radio series.[2] Fernald's statement and its subsequent use by *Radiolab* nicely sum up two significant developments in US radio since 2005. First, it encapsulates the increasing emphasis on emotion and affect in long-form radio, from both the producers creating the content and the critics and audiences receiving it. And second, its inclusion in an episode about the boundary between language and music, and the suggestion that sound is something more than speech (by which Fernald means more than mere content or information), declares that there is a poetics of sound. To put it another way, sound—and specifically, the intricately composed sound of radio shows and podcasts like *Radiolab*—should be regarded as an art.

There has been a flourishing discourse of radio as art and as a medium for creative experimentation since the mid-2000s. The greatest advancements in radio as an art form have occurred in the area of *audio storytelling*, a broad category of fiction and nonfiction programming united by the use of narrative and other dramatic techniques, as well as a composed sonic aesthetic. And these developments have principally happened through the emerging practice of podcasting, albeit often in a symbiotic relationship with traditional broadcasting institutions and conventions.[3] An all but defunct form on US broadcast airwaves, the *radio drama* (or radio play) has found increasingly receptive audiences online in the past decade. The *radio feature-documentary* has been even more successful, serving as the primary space for the exploration of "creative audio" and "sound-rich audio stories."[4] Indeed, much of the industry growth and audience enthusiasm surrounding podcasting both before and after the breakout success of *Serial* in the fall of 2014 has occurred around the broad genre of "storytelling podcasts"—which is little more than a new name for the old form of the narrative radio feature-documentary.

This chapter explores some of the more prominent aesthetic as well as cultural and industrial trends in the audio storytelling category during the decade from 2005 to 2015, which have been especially impactful on the current third wave of podcasting. It does so through case studies of the podcast *Welcome to Night Vale* and the radio/podcast series *Radiolab*. As this chapter's title suggests, the remediation of radio drama is presented as a framework for better understanding this creative wave in both fiction and nonfiction soundwork. This approach no doubt makes sense for audio drama podcasts like *Night Vale*, though it may be less immediately clear how it applies to reality-based programming like *Radiolab*. However, as the analysis here shows, contemporary journalistic and documentary audio storytelling is heavily indebted to the themes, tropes, and techniques of radio drama as well.

Defining Audio Storytelling

The term *audio storytelling* has sprung up in the past decade to describe narrative-driven fiction and nonfiction programs that focus on individuals and their personal experiences. Scholars like Siobhan McHugh have loosely referred to this as a "crafted audio" approach, implying a commitment to narrative structure and sound design not commonly found in the majority of live radio broadcasting or "chatcast"-style podcasts.[5] Elsewhere, McHugh distinguishes between "American" and "European" audio storytelling styles: the "American" style more documentary and journalistic, described as "hand-held" and "host-driven" with reporters placing themselves at the heart of the program and personalizing the narrative; the "European" style more poetic and choreographed, heavily aestheticized and rooted in older methods of dramaturgy.[6] In both traditions, audio storytelling involves an impressionistic presentation of lived experience.

Audio storytelling programs are prerecorded and usually heavily edited, their soundtrack comprising a mix of "tracks," "actualities," "ambience," and music and sound effects.[7] In radio production discourse, *tracks* are the host's, narrator's, or reporter's voices; these are typically recorded in the controlled environment of a studio and provide the editorial message, giving the story its direction. *Actualities*, or "acts," are the other voices and action sound that create conversation with the host/narrator/reporter—the most common "acts" being subject interviews, plus archival clips of speeches and foregrounded natural sounds. *Ambience*, or "ambi," are sound beds recorded on location (i.e., outside a studio) and added to the mix to evoke scenes and places for listeners. Scored music and canned sound effects create additional sonic variation and dynamism. Combining all of these discrete sonic elements is labor intensive, requiring careful editing and composition, which is among the reasons why this type of "crafted audio" is distinguished from other programming within the radio and podcast industries.

The audio storytelling approach is rooted in drama and theater as well as literary journalism and narrative nonfiction. The notion of a "story" implies the presence of narrative and plot. A story is not merely information about a topic or a recounting of events. It must involve characters, scenes, and a sequence of actions that creates conflicts—the *drama* emerging from problems or complications—and typically leads to a resolution or moment of revelation.[8] Although these techniques are fundamental for dramatic art and literature, they are equally essential to nonfiction storytelling. As a matter of fact, Neil Verma observes that the majority of today's radio and podcast producers are coming to audio storytelling through a journalistic framework, influenced by contemporary documentary-style programs like *This American Life* rather than historical anthology or serial radio dramas.[9]

Notably, the concept of storytelling also has contemporary impact beyond just radio and podcasting: *digital storytelling* has been an exceedingly popular form and practice within the fields of digital humanities and media education since the mid-1990s. It has subsequently found commercial value in transmedia and other multiplatform/cross-media franchising strategies. Typically expressed through digital video and multimedia web projects, digital storytelling "is a workshop-based practice in which people are taught to use digital media to create short audio-video stories, usually about their own lives."[10] This idea of "ordinary" people being taught how to produce digital media themselves is rooted in traditions of community arts practice and social justice oriented notions of empowerment through self-expression. The goal of digital storytelling being, according to media scholars John Hartley and Kelly McWilliam, "to give a voice to the myriad tales of everyday life as experienced by ordinary people in their own terms."[11] (These digital storytelling objectives will be explored in depth during chapter 7's analysis of the nonnarrated audio storytelling mode and the podcast *Love + Radio*.) The essence of digital (and audio) storytelling is based around real-life memories and personal commentaries, and increasingly these perspectives are narrated in the first person.

I observe six main types of programs that receive the audio storytelling label, or what we might call the *six modes of audio storytelling*: live storytelling; personal documentary; historical narrative; the nonnarrated story; the feature-documentary; and audio drama. Quite simply, the *audio drama* mode comprises *Welcome to Night Vale* and other fiction programs for radio and podcast, in serial and anthology formats. The *feature-documentary* mode consists of shows like *Radiolab*: host-driven narratives that emphasize characters and personal stories but nonetheless adhere to journalistic forms and practices (e.g., investigative reporting, assembling evidence and experts, commentator analysis). The *nonnarrated* mode will be the main focus of chapter 7 and the *Love + Radio* case study: it is most simply defined as a monologue narrated by the subject of the story rather than a host/reporter.

The *live storytelling* mode mainly consists of anthology-style programs in which individuals tell short personal narratives on stage in front of a live audience. These shows are an extension of the oral storytelling clubs and festivals that grew out of the 1960s folk counterculture and that have been popular in the United States since the 1970s. Updated for a younger generation of urban professionals, representative live storytelling radio/podcast series include *Mortified, Risk!, True Story*, and, most famously, *The Moth*. Rather than stories emerging out of journalist interviews or personal audio diaries, these stories tend to be highly performative and scripted (or at least well rehearsed) to elicit laughter and other audience reactions. Also known as *platform storytelling*, these events most closely resemble stand-up comedy or slam poetry performances.[12] In fact, the focal point of The Moth nonprofit organization's activities is a series of "open-mic storytelling competitions" it calls "StorySLAMs."[13] Here the storytelling is self-conscious and taken out of the everyday in order to be admired as a type of performance art. Notably, too, series like *The Moth* regularly feature prominent literary and cultural personalities—that is, not "ordinary" people. For these reasons, the live storytelling mode might also be thought of as *personal narrative performance*.

The *personal documentary* is essentially an audio autobiography or memoir. With parallels in the *radio essay* format, it is a serialized first-person narrative in which the narrator gives a subjective account of some aspect of his or her life. Modern personal documentary podcasts are also often *metanarratives*, in that they reflexively call attention to themselves and the process of production. The progenitor of this mode is *StartUp*, the podcast series "about what it's really like to get a business off the ground." For its first season in 2014, former *This American Life* and NPR *Planet Money* producer Alex Blumberg provided "a transparent account" of his journey building the for-profit podcast network Gimlet Media.[14] *Millennial*, which launched in January 2015 and ended in August 2017, adopted a similar entrepreneurial story, following the twentysomething Megan Tan as she navigated the postgraduate job market and transitioned from a planned career in photojournalism to one in radio journalism. Other personal documentary series are even more starkly private, such as *First Day Back*, which documents the filmmaker Tally Abecassis's attempt to return to her career after having kids and her struggle to maintain a work-life balance. *How to Be a Girl* is a series about gender and identity, guided by a single mother, Marlo Mack, as she tells the story of her six-year-old transgender daughter.

The *historical narrative* mode is a rather singular style, at least compared to these other first-person audio storytelling modes. It typically features a third-person narrator offering an interpretation or perspective on events of the past, dramatizing the proceedings using music, sound effects, archival recordings, and even staged re-creations or clips from film and television. The structure is decidedly literary: each episode has a plot and a thesis, a cast of characters, moments of action and conflict, and a climax. Unlike the rotating "talking head" expository style of a Ken Burns documentary film or television series, however, podcast series like *Dan Carlin's Hardcore History*, *You Must Remember This*, *Cocaine & Rhinestones*, and *The Memory Palace* filter historical events through a single narrator's perspective, which can be very clearly biased or impressionistic. The stories, then, most closely resemble historical essays or books, the host functioning as a guide. Nevertheless, the often very poetic style of a podcast like *The Memory Palace*—

with its emphasis on overlooked moments and people in history, and its attention to individual's emotional lives and humanity—personalizes it and aligns it with the form and politics of the other audio storytelling modes. The popular cycle of true crime podcasts that followed the breakout success of *Serial*—including series like *S-Town*, *In the Dark*, *Up and Vanished*, *Dirty John*, *Mogul*, and *Crimetown*—frequently combine elements of the feature-documentary and historical narrative modes.

Welcome to Night Vale and the Audio Drama Mode of Audio Storytelling

Something unusual happened in the podcasting world in July 2013: a relatively unknown, independently produced serial drama about the strange goings-on in a fictionalized American Southwest town suddenly leapt to the top of the Apple iTunes Podcasts charts, surpassing chart mainstays like *WTF with Marc Maron*, *The Nerdist*, *TEDTalks*, and US public radio powerhouses *Radiolab*, *Freakonomics*, and *This American Life*.[15] Debuting a year prior in June 2012, the twice-monthly podcast series *Welcome to Night Vale* had received approximately 150,000 downloads across its first twenty-five episodes. With the help of an outpouring of fan support on social networking website Tumblr, the series abruptly gained 150,000 downloads in a single week. In total, *Night Vale* had 2.5 million downloads in July 2013 alone and another 5.8 million downloads that August.[16] In the ensuing half-decade, the podcast regularly placed in the US iTunes charts, and it charted internationally too.[17] *Night Vale* emerged as a genuine cult phenomenon, spawning countless sold-out live shows, multiple novels and script books, and a planned television series.[18]

The surprise popularity of *Night Vale* was remarkable for two reasons. First, it was an independent podcast succeeding in a space increasingly dominated by professional media producers, especially parties affiliated with the NPR public radio network or celebrity entertainers, such as Maron and fellow comedians Adam Carolla, Joe Rogan, and Chris Hardwick. Second, *Night Vale* was an audio drama—a format that was central to radio's Golden Age from the 1930s to the 1950s, but which had been almost entirely absent from US airwaves in recent decades.[19] A small but rapidly growing niche, some of the more widely known fiction series to appear alongside *Night Vale* during podcasting's third wave include *Homecoming*, *The Truth*, *The Message*, *lif-e.af/ter*, *Limetown*, *Ars Paradoxica*, *The Bright Sessions*, *The Phenomenon*, *Fruit*, *Bronzeville*, *Black List Table Reads*, *The*

Cleansed, Our Fair City, The Black Tapes, The Magnus Archives, Knifepoint Horror, Getting On with James Urbaniak, Serendipity, We're Alive, The Leviathan Chronicles, Archive 81, The Amelia Project, The Unexplainable Disappearance of Mars Patel (targeted at a young adult audience), *Secrets, Crimes & Audiotape* (a musical), and *The Thrilling Adventure Hour* (a live staged production with which *Night Vale* has collaborated). *Night Vale*'s creators have also issued other audio dramas under the "Night Vale Presents" banner, including *Alice Isn't Dead, Within the Wires, It Makes a Sound*, and *The Orbiting Human Circus*.

This case study examines *Night Vale* as representative of this emerging radio drama revival, and especially of the ways podcasting has opened up spaces for a new wave of creative audio production. *Night Vale* and its post-2010 fiction podcast confederates are part of the third wave of podcasting, yet this newfound creativity is arriving through a remediation of older radio forms, techniques, and styles—most notably radio drama, along with local community radio and fringe categories of talk radio. The supposed newness of podcasting is significantly tempered when the full range of radio's history and forms are taken into account.

Welcome to Night Vale tells the story of Night Vale, a small desert town somewhere in the Southwestern United States, where strange, supernatural events are commonplace: a hovering Glow Cloud rains dead animals and controls the local school board; the Sheriff's Secret Police are run by an unknown leader and surveil the town's citizens with Gestapo-like tactics; The Dog Park is surrounded by an electrified fence, hosts meetings of mysterious Hooded Figures, and is off limits to dogs and people; unidentified lights and sounds regularly emerge from nearby Radon Canyon but citizens are prohibited from speaking about them; a subway system spontaneously appears; a shady private corporation, StrexCorp, runs the neighboring town of Desert Bluffs and steadily attempts to infiltrate Night Vale; and a tiny underground city is located underneath the local bowling alley. What is more, these unusual entities and occurrences are treated as mostly normal by the series' narrator, Cecil Palmer (voice acted by Cecil Baldwin), a radio host at Night Vale Community Radio (NVCR). A hybrid of the science fiction, horror, and comedy genres, the podcast utilizes a serialized narrative form, weaving together a story that features more than fifty recurring characters and an expansive geographic universe with over sixty locations. Yet, for much of the series' run, the only voice the audience hears is that of Cecil. The podcast is narratively framed as a radio broadcast with the podcast listener positioned as a listener to Cecil's news and talk radio program on NVCR. While Cecil is occasionally joined by a guest or plays prerecorded material on air, the majority of *Night Vale*'s 150 episodes (and counting) consist of Cecil, the radio professional, reporting on the actions of the townspeople and quoting other characters in what is effectively an ongoing monologue.

Night Vale's creators, Joseph Fink and Jeffrey Cranor, are experimental theater writers and performers, not trained radio professionals. They were members of the New York Neo-Futurists, a performance art-oriented experimental theater collective, when they started the podcast. *Night Vale* was developed independently and has no affiliation with any radio station, podcast network, or other established institution. It is distributed free of charge, and episodes remained commercial-free for years (preroll and postroll ads were eventually added in 2017). Prior to accepting ads, the creators supported themselves and the production through crowdfunding, merchandising, live performances, and select premium or bonus content, namely paid downloads of live show recordings.[20]

The way Fink and Cranor tell it, *Night Vale* is a profoundly DIY production. The pair collaboratively write the scripts. Baldwin, a friend from the New York Neo-Futurists, records all the audio alone in his apartment on a $60 USB computer microphone. Original music is mostly provided by another friend, Disparition (aka Jon Bernstein). And Fink edits the show in his apartment on Audacity, a free digital audio editing software application.[21] Audio files are uploaded to Libsyn, an inexpensive podcast hosting service, and then aggregated free of charge to iTunes, Spotify, and various other podcast apps and internet radio platforms. Apart from social media accounts through Twitter, YouTube, and the like, the podcast has never paid for any advertising or promotional campaigns. Indeed, marketers have turned *Night Vale* into a case study for "viral marketing" done right.[22] Its success has been almost entirely driven by online fandom.

In many ways, then, *Night Vale* can uncritically be viewed as a textbook case of podcasting as a type of disruptive user-generated content. Indeed, in the mid-2000s the journalistic discourse on podcasting frequently framed it as a disruptive practice that operated outside the existing media content industries, and in turn was capable of threatening those traditional institutions. Podcasting is "the latest form of jailbreak media that has plain old citizens pulling up the microphone and mainstream media running scared," David Carr aggrandized in the *New York Times* in 2005.[23] That discourse of disruption continues today, albeit slightly tempered, with podcasting frequently positioned as a direct competitor to traditional media. For instance, podcasting's audience growth is often correlated with declines in broadcast radio and television consumption.[24] Much of the existing scholarly discourse on podcasting has similarly focused on its prevalence among independent, nonprofessional producers working outside the traditional media industries (i.e., amateurs).[25] Podcasting is typically categorized as a type of deprofessionalized, deinstitutionalized "personal media" alongside other Web 2.0 forms of user-generated content, such as blogging and YouTube videos.[26] As such, it is strongly connected to discourses of produsage, participatory culture, and broader theories about the democratizing effects of the internet—all of which assert, to varying degrees and ends, that networked digital media like podcasting are opening up the media environment for an increasingly active audience, empowering more diverse cultural production and eroding traditional hierarchies between media producers and audiences.[27]

There is a romantic privileging of amateur and independent podcasters in the growing scholarly research on podcast production cultures. Moreover, there is a tendency in this research to focus on radio/podcast genres, styles, and forms that are presumably new, or at least socially or culturally distinct from the content of mainstream broadcast radio.[28] However, the fact of the matter is that the podcasting field circa 2020 has become highly professionalized and increasingly consolidated. Almost every major American media outlet participates in podcasting: television networks like CBS, ESPN, and MSNBC; the new media companies Vox, Vice, Boing Boing, and HuffingtonPost; old-guard print media outlets like the *New Yorker* and the *New York Times*. As discussed in the conclusion, Radiotopia (PRX), Gimlet Media, and a growing number of podcast networks have developed, further consolidating podcast production and distribution.[29] Accompanying this consolidation has been the proliferation of a few distinct styles borrowed from traditional radio production, including the public radio feature-documentary approach examined later in this chapter. There is certainly still space in podcasting for innovative, independent, and nonhegemonic voices and perspectives. Yet as the industry becomes more crowded and dominated by a handful of major players, the risk is that those voices will become harder to hear, and it will be more difficult for new things to happen.

All of this is to say that *Night Vale* represents the potential for podcasting to foster a new wave of creative audio production, and yet it is also something of an outlier in the nascent US podcasting industry. It is an independent production that is made on a shoestring budget by a small creative team with no ties to the traditional radio industry or any other major media institution. Still, one should avoid calling them amateurs or nonprofessionals, since they are, in fact, trained theater writers and performers. In other words, they are very much media professionals, just not radio professionals—and this is the case with many newcomers to podcasting during the practice's ongoing third wave. Artists from outside the radio industry, such as Fink and Cranor, are seeing in podcasting an opportunity to explore their work (theater, in this case) through a different medium. Nevertheless, despite the discourses of disruption, deprofessionalization, and deinstitutionalization mentioned earlier, a truly independent production is a rarity among the ranks of today's most consistently popular US podcasts.

Night Vale is an audio drama, an iteration of the radio drama form that has all but disappeared from US airwaves since the early 1960s. As Neil Verma has shown, radio drama was a staple of US network radio programming during the medium's Golden Age from the 1930s through the 1950s. It was a bastion for innovative dramatic storytelling and performance that greatly influenced both television and film conventions.[30] With the cultural ascendance of television in the midcentury, though, the US radio networks ceased to regularly produce drama on radio, due to the loss of advertiser sponsorship, programming shifts, and the breakup of the network model.[31] Radio drama has continued to enjoy considerable popularity and institutional support in some foreign countries, in particular the United Kingdom. There, the BBC with its public service monopoly has maintained a commitment to high production quality radio drama, broadcasting hundreds of hours of radio plays each year, often featuring the talent of the nation's biggest stage and screen stars.[32] In the United States, there have been occasional radio drama productions aired since the 1960s, such as *CBS Radio Mystery Theater* (1974–82), *The National Radio Theater of Chicago* (1973–86), the National Public Radio anthologies *EarPlay* (1971–81) and *Playhouse* (1981–2002), and the various productions of Joe Frank (late 1970s–2017). In recent years, a few prominent public radio stations have mixed fiction radio pieces into arts anthology programs, such as *The Organist* on KCRW (2013–present), *The Acousmatic Theater Hour* on WFMU (2008–9), *The Next Big Thing* on New York's WNYC (1999–2006), and the CPB- and NEA-funded and independently syndicated series *Hearing Voices* (2001–12).

Still, for nearly a half-century there has been extremely little in the way of new radio drama commissioned in the United States, and most of what has been broadcast since the 1990s consisted of independently produced one-offs. There was practically no original internet-only audio drama made during the 1990s and 2000s, apart from Pseudo's couple short-lived experimentations and the few programs syndicated by the Internet Multicasting Service (discussed in chapters 1 and 2). If remembered at all today, radio drama is mostly regarded as a historical phenomenon, vintage programs nostalgically revisited through occasional "old-time radio" (OTR) broadcasts on local public radio stations and by fan communities on websites like Internet Archive.[33] A few public radio networks, such as Wisconsin Public Radio, stage live radio dramas for broadcast every so often; however, these are almost always adaptations of classic plays, movies, or novels, such as *The Time Machine* and *Forbidden Planet*, and not original works.[34] In this light, *Night Vale* is quite unique, reviving a long-fallow format and, in the process, expanding the universe of soundwork available to today's listeners. For many younger audiences, *Night Vale* is their very first introduction to audio drama. Without podcasting—combined with affordable digital production tools and inexpensive distribution platforms like iTunes—there would likely be no contemporary audio drama made in the United States, as there is no institutional support for fiction programming within the nation's broadcast radio industry.

Night Vale's distinctiveness arrives through a remediation of these older radio forms, techniques, and styles—even though its producers and fans often distance the podcast from the medium of radio. The series is not quite as new or unprecedented as it first seems when placed into proper historical and cultural context. This is not meant to discount the podcast's creativity: it is a skillfully written and acted series, and its particular mixture of genres and themes is highly original. Befuddled critics often resort to recombinant phrases to describe the series: "where David Lynch meets *The Twilight Zone*"; "one part David Lynch, one part New Weird, and one part Bizarro fiction"; "like *A Prairie Home Companion* set in an arid version of the author Stephen King's community of Castle Rock."[35] Other reviewers regularly compare the podcast to the work of authors H. P. Lovecraft, Lemony Snicket, and Douglas Adams. These descriptions can be read simply as attempts to make legible a text that is unfamiliar. Yet, as these appraisals suggest, much of the discourse surrounding *Night Vale* privileges literary, theatrical, or even filmic sources.

Despite these similarities to other media, *Night Vale*'s core elements are all deeply rooted in radio. The series is actively remediating an array of radio genres, forms, and practices—not only radio drama but also local community radio and fringe forms of talk radio. Through Jay David Bolter and Richard Grusin's theory of remediation, it is possible to see how *Night Vale* revives classic radio while also reconstructing it for audiences in the hypermediated world of the twenty-first century.[36]

The first radio form that *Night Vale* remediates is the radio drama. The podcast is a scripted, dramatized serial narrative that is written, performed, and produced to be heard. This fundamental structure and approach are hallmarks of radio drama.[37] In interviews, *Night Vale*'s creators, while usually regarding radio with a necessary deference, nevertheless discursively locate its roots in the theater and oral traditions. "The inspiration came less from radio and more from the much older tradition of storytellers and monologists, people standing in front of you and telling you a story," claims Fink. "That's really where *Welcome to Night Vale* came from for me, finding a format where you could find a single voice and have that single voice tell you a story. I think it was a lot more to do with the storytelling tradition for me than for radio."[38] It is perhaps not surprising that Fink, trained in the theater rather than radio, would position *Night Vale* and himself in this way. Nonetheless, plays and speeches tend to be written and performed as single, nonserialized pieces; the serialized, multiepisode audio drama is the domain of radio.[39] In terms of medium specificity, too, the podcast's recorded, nonlive status lends itself more to radio than the stage.

Night Vale's use of genre is also highly reminiscent of mid-twentieth-century radio drama.[40] The mystery, science fiction, horror, and fantasy genres that the podcast mixes certainly all have rich histories within twentieth-century theater and literature. However, these were among the most popular types of programming during radio's Golden Age, with series like *Suspense* (1942–62), *Inner Sanctum Mystery* (1941–52), *The Whistler* (1942–55), and *X Minus One* (1955–58). Even prestige sustaining series, such as *The Mercury Theatre on the Air* (1938) and *Columbia Workshop* (1936–43; 1946–47), featured mysteries and thrillers with supernatural themes not dissimilar to *Night Vale*. Moreover, a number of classic radio's most acclaimed psychological thrillers, including "Sorry, Wrong Number" (first broadcast on *Suspense*, 1943) and "The Hitch-Hiker" (first broadcast on the *Orson Welles Show*, 1941), are essentially one-person shows. Even "The War of the Worlds" (first broadcast on *The Mercury Theatre on the Air*, 1938)—perhaps the most famous of all American radio dramas, alongside "Sorry, Wrong Number"—features a lengthy twenty-minute closing act that consists almost entirely of voiceover narration from Orson Welles as Professor Richard Pierson.[41] Thus, there is considerable precedent in radio drama for the monologue format that Fink cites. Moreover, American radio and theater were intertwined during this era, with many of the New York theater community's writers, directors, and performers working simultaneously in both industries. Plays that were produced to be listened to rather than seen—designed to take advantage of the intimacy of the voice and the listener's own imagination (as is the case with *Night Vale*)—would have been culturally and aesthetically recognized as radio, not live theater.[42]

The second radio form that *Night Vale* remediates is community radio. Most obviously, Cecil's broadcasts from NVCR are the narrative frame of the podcast. Community radio is more than just a useful storytelling device, however: it is deeply embedded in nearly every aspect of the series. While the topics Cecil discusses may be unusual—the Children's Science Museum exhibit "The Moon Is a Lie," the commercial airliner that suddenly appears and disappears inside local homes and schools, the cat that is stuck hovering in the NVCR men's bathroom, the faceless old woman who secretly lives in everyone's homes—the presentation of this bizarre information is actually quite normal. Each episode/broadcast is divided up into segments typical of local news and talk radio: news reports, community calendar notices, public service announcements, live-read advertisements, traffic and weather, as well as "breaking news"-style interruptions with updates on unfolding events. A running gag in the series is that "the weather" is, inexplicably, a musical performance. There are various other segments, such as the "Emergency Dream Broadcast System" and horoscopes, which draw upon standard radio conventions. Despite exploiting their meanings for often humorous effect, it is nonetheless these radio practices that structure the series and provide the context in which the narrative operates as entertainment. Moreover, diegetically the series is quite faithful to the concept of broadcasting. For instance, when Cecil's love interest Carlos the scientist is attacked by the residents of the tiny underground city, Cecil is unable to join him because he must continue to broadcast live.[43] He cannot break from the strictures of the broadcast program. Breaking news is regularly delivered to Cecil on air via phone calls and notes from station staff. In these ways, the verisimilitude of community radio is central to the podcast.

Notably, Cecil's delivery of the strange and alarming goings-on in Night Vale is mostly deadpan. He *sounds* like a serene, reassuring local radio host. This ironic juxtaposition creates much of *Night Vale*'s distinctiveness and charm. The series' basic use of speech, music, noise, and silence—"the primary codes of radio"—corresponds with standard radio production.[44] Cecil's DJ chatter creates continuity between the different elements (e.g., the news, traffic, weather), giving the podcast its basic narrative structure and also its distinct identity: it is a mode of speech that is intrinsically rooted in radio broadcasting. Cecil's words are tightly scripted yet presented in such a way that sounds natural and spontaneous, so as to give the appearance of informality, creating a sense of intimacy between the broadcaster and the audience. This scripted but impromptu sounding mode of address is another definitive characteristic of radio.[45] Paddy Scannell points out that "the central fact of broadcasting is that it speaks from one place and is heard in another," the result being that radio talk "attempts to bridge the gap by simulating co-presence with its listeners."[46] Cecil creates this copresence by addressing the audience directly, using words and phrases such as "dear listeners." He also often breaks from script at times, discussing his affection for Carlos or expressing his distaste for certain townspeople, such as the Apache Tracker ("What an asshole that guy is!") or Steve Carlsberg ("I don't even want to read an email from that jerk").[47] These moments serve to endear Cecil to the listener, reducing broadcasting's separation between addresser and addressee.

Indeed, podcast listeners might take Cecil's quirks and digressions, where he gets emotional and rants about his personal life or opinions, as evidence of how *Night Vale* deviates from traditional radio. Such an argument, though, would be based on a rather narrow view of radio programming and styles. Much of the journalistic and fan discourse surrounding *Night Vale* compares it to NPR public radio-style programming, where hosts present with a civil, pseudo-objective vocal tone. The fact of the matter is that news and talk radio is much more diverse than just the "socially conscious" and "serious, carefully modulated, genially authoritative" "NPR Voice"—or, for that matter, the bombastic, snark-filled style of Howard Stern or Rush Limbaugh.[48] There is a wide range of local radio in the United States where the mode of delivery can be much more informal and personal, including public radio stations operating outside the NPR network, grassroots low-power FM (LPFM) community stations, and college and other educational stations. It is not so uncommon to hear hosts on these broadcast radio channels mix news with opinion, and even divulge impassioned confessions. The same goes for CB and ham radio, where the informal conversational mode is the status quo.[49] Compared to these sources, Cecil's more idiosyncratic moments are not so unusual.

The third radio form that *Night Vale* remediates is fringe talk radio. The aesthetic of individuals talking into a microphone and narrating all that they see, think, or feel—part of the "intimate" style for which podcasting is so often heralded—is hardly a new technique, even if it does seem novel compared to most of the professionalized mainstream radio heard in the past few decades. There are many parallels between *Night Vale* and fringe forms of talk radio, such as late-night paranormal and conspiracy theory programs like Art Bell's *Coast to Coast AM* (1984–present; now hosted by George Noory), religious and right-wing conspiracy shows like *The Alex Jones Show* (1996–present; previously *The Final Edition*), and love, sex, and relationship shows like *Loveline* (1983–present; most famously hosted by the team of "Dr. Drew" Pinsky and Adam Carolla). On these programs, the frank, offbeat, or supernatural subject matter of *Night Vale* is the order of the day. To his credit, Cranor confirms the influence of some of these programs on his work.[50] The point here, though, is there is a long and rich history of more marginal forms of talk radio that *Night Vale* is drawing upon but which receives little acknowledgment in popular discourses of radio and podcasting—or in academic studies of radio, for that matter.

The *Night Vale* podcast remediates Golden Age radio drama along with more contemporary local community radio and other fringe categories of talk radio. The new practice of podcasting is not divorced from earlier media like radio. Instead, it is refashioning radio and even paying homage to it. Remediation involves a replication and simultaneous reworking of the old media. Indeed, much of the pleasure of listening to *Night Vale* comes from the ways in which it plays with the well-established conventions of radio broadcasting—satirizing them but also saluting them at the same time. It is remarkable how many contemporary fiction podcasts self-reflexively use the medium of radio and other audio-recording technologies and practices as key narrative devices, much as *Night Vale* does. For instance, *The Black Tapes* follows a radio journalist reporting on a paranormal investigator's unsolved cases. *The Bright Sessions* and *The Magnus Archives* are built around cassette tape recordings. *The Message* centers on a podcaster covering the decoding of an alien message by a team of cryptologists.

While *Night Vale* is in certain ways an outlier in today's podcasting field, some of its lessons can be extended to podcasting as a whole. When put into historical and cultural perspective, there is little that is truly "new" about podcasting, at least at a *textual* level, in terms of audio form, technique, and style. In recent years, Amazon.com-owned Audible has invested substantially in elaborate audio productions of popular novels (complete with ensemble casts, music, sound effects, etc.—not only professional narrators), in addition to original works created exclusively for Audible's digital audio platform.[51] These pieces get labeled "audiobooks" because of their industrial point of origin, and thus get discursively positioned outside of radio and podcasting. Yet they feature all the same sonic and textual features as radio-centric audio dramas, and it is prudent to analyze these and other fiction podcasts together as part of the same phenomenon.

Radiolab and the Feature-Documentary Mode of Audio Storytelling

Mixed in among a list of names of physicists, economists, computer scientists, architects, historians, geneticists, classical and jazz music composers, neurologists, civil rights attorneys, and poets was the name of a then still relatively little-known radio personality: Jad Abumrad.[52] It was September 2011, and the thirty-eight-year-old Abumrad—the creator, producer, editor, and cohost of *Radiolab*—was one of twenty-two Americans being awarded a MacArthur Fellowship by the John D. and Catherine T. MacArthur Foundation. The five-year grant, informally known as a "Genius Grant," is given "to individuals who show exceptional creativity in their work and the prospect for still more in the future."[53] In addition to the prestige of the fellowship, Abumrad received $500,000 of "seed money" for future projects with no strings attached. He joined Bill Siemering (Class of 1993), Hugo Morales (Class of 1994), and David Isay (Class of 2000) as one of only four radio producers out of more than nine hundred total recipients to be awarded the fellowship since its inception in 1981.[54] "Abumrad is inspiring boundless curiosity within a new generation of listeners and experimenting with sound to find ever more effective and entertaining ways to explain ideas and tell a story," the MacArthur selection committee proclaimed in the announcement naming him a Fellow.[55]

In the decade since, *Radiolab* has grown to become one of the most perennially popular feature-documentary series in the third wave of podcasting, as well as one of the most aesthetically influential programs in the audio storytelling field. Especially in terms of its use of music and sound design, *Radiolab* has set the gold standard in audio storytelling production, establishing a benchmark against which producers and critics measure the sonic creativity of other soundwork. The MacArthur Fellows committee's assertion that Abumrad's program "is inspiring boundless curiosity within a new generation of listeners" was hardly hyperbole. The only other program more influential in modern audio storytelling is *This American Life*, and perhaps the *This American Life* spin-off *Serial*. (Even Abumrad admits that *This American Life* and its auteur-like host-producer Ira Glass were an early influence on him when he was developing the *Radiolab* format in the early 2000s.)[56] Nevertheless, I have chosen to center my exploration of feature-documentary form and practice around *Radiolab* because *This American Life* (and *Serial*, more recently) has received more than its share of journalistic and scholarly analysis. Furthermore, *Radiolab* represents the next step in "sound-rich" radio production: a progression of *This American Life*'s character-driven narrative journalism approach (first established back in the late 1990s), upgrading it with a dazzling mix of music and sound effects and a looser, more reflexive and engrossing host narration style. The *New Yorker* has called Abumrad a "sound wizard."[57] Even Ira Glass agrees: "I marvel at *Radiolab* when I hear it. I feel jealous. Its co-creators Jad Abumrad and Robert Krulwich have digested all the storytelling and production tricks of everyone in public radio before them, invented some slick moves of their own, and ended up creating the rarest thing you can create in any medium: a new aesthetic."[58] It is that aesthetic, more so than *This American Life*'s even, that has distinguished the feature-documentary mode of audio storytelling during the third wave of podcasting, attracting new audiences and inciting other audio producers to push the boundaries of their craft.

Radiolab has become something of a gateway podcast for young audiences and aspiring creators: it is a program that introduces audiences—especially people who would not consider themselves broadcast radio listeners—to podcasting and gets them hooked on what the medium has to offer. Never mind that *Radiolab* was born, and continues to be, a broadcast radio program; it is not a podcast-only production, as is also the case with *This American Life* and numerous other popular "podcasts." Still, even though it is syndicated across the United States on more than five hundred public radio stations, more of *Radiolab*'s listeners encounter it as a podcast than a broadcast—its podcasts receiving over seven million downloads per month.[59] Thus, although it is problematic that its broadcasting ties are frequently effaced, the fact of the matter is that *Radiolab* represents contemporary podcasting in the minds of many listeners. Furthermore, *Radiolab*'s producers have steadily come to conceive of the program more and more as a podcast experience.

The origins of *Radiolab* lie in a late-night documentary anthology show on New York City public radio network WNYC's AM channel called *The Radio Lab* that debuted in 2002.[60] While it shares the same name (stylized slightly differently), it was very different from the program that audiences have come to know since its first full "season" in 2005. Abumrad studied creative writing and music composition at Oberlin College, graduating in 1995 with the intention of pursuing a career writing film music. While he searched for film soundtrack work, he took a dot-com web design job in New York. Before too long, Abumrad realized that he hated the day job and his composing ambitions were not going to pan out. He claims to never have had much of an interest in radio growing up but, around 1999, recognized "that the two things I liked to do"—writing and composing— "kind of met in the middle at radio."[61] That is when Abumrad began volunteering at public radio stations around New York City, first at Pacifica's WBAI and then at New York Public Radio's NPR affiliate WNYC. He learned local news reporting at WBAI, and at WNYC he started contributing more ambitious freelance pieces to nationally syndicated programs like *On the Media* and *Studio 360*, as well as producing for *The Next Big Thing*.[62] Following a 2001 restructuring at WNYC that opened up slots for more locally produced original programming, Abumrad was tapped by program director Mikel Ellcessor to produce and host a show that aired on Sunday nights from 8:00 to 11:00 p.m. on the AM frequency—which, Abumrad claims to have eventually found out, reached "no one, like, probably *zero* listeners" because the station lowers the power on its AM signal at night.[63] *The Radio Lab* went on the air in April 2002 with Abumrad working alone, the original concept being that it was a freeform "documentary showcase" in which he functioned as "a DJ spinning documentaries."[64] As the "DJ of documentary," Abumrad explains that "the idea of it back then was to just take all of these documentaries from around the world and just smash them together."[65]

Abumrad established himself as more than a mere curator. Using techniques that resembled sampling in hip-hop and avant-garde electronic music more than anything conventional within radio production, Abumrad would mix together archival material, clips he begged off public radio friends like Jay Allison, and sound art pieces he himself created, connecting it all together with elaborately mixed sound effects and original music compositions. "We'd get all these really weird stories, like a BBC story from Zimbabwe or some sound-art piece about dogs, and you would have to somehow make a connection between them," Abumrad says.[66] "I would try to mash avant-garde classical and electronica with classic news documentary with radio dramas with first-person storytelling and interviews."[67] The project was decidedly experimental; due to the small late-night audience and miniscule budget, the environment was low risk enough for

WNYC that Abumrad could treat *The Radio Lab* as "this anything-goes play space." He adds, "The 'lab' was because we didn't know what the fuck was going to happen. It was an experiment. I had no experience, so I didn't even know what the status quo was."[68] Even though many people have come to assume that the "lab" in *Radiolab*'s name refers to a scientific laboratory (since the show is famous for its science reporting), it in fact refers to *experimenting with radio* and its different forms, genres, and techniques, pushing the sonic limits of the medium.

While it is usually only referred to as an amusing anecdote (if it is referred to at all), this developmental phase of *Radiolab*, which lasted from 2002 to 2004, is incredibly significant for understanding the series' current aesthetic and its remediation of other radio programs and forms. First, the now-signature *Radiolab* style of intricately layered words, music, and noises with constantly shifting perspectives grew out of these experiments. Second, the *Radiolab* aesthetic, as innovative as it undoubtedly is, owes a great deal to the sounds of broadcasting's past—radio features and documentaries as well as radio drama and radio art.

Abumrad's music background tends to get cited by journalists as the source of *Radiolab*'s style and approach. Moreover, his frequent claims in interviews that "I had no idea what broadcasting *should* sound like" because he did not grow up a fan of radio tends to displace the focus away from radio form and practice entirely.[69] The show also routinely gets called "cinematic." Without question, Abumrad's editing and sound design techniques benefit considerably from his modern classical musical training, and he composes much of the music to custom-fit the show in a manner akin to film scoring. An in-house composer is a rare opportunity for an audio storytelling program, this original soundtrack in turn lending *Radiolab* much of its unique identity. Yet when Abumrad and his colleagues do occasionally open up about their radio influences, it is evident that the program's aesthetic indeed takes considerable inspiration from the medium that lends the show its name. *Radiolab*'s early developmental phase was a crash course in radio history for Abumrad. "I was just getting into public radio via the usual suspects, the news magazines and Ira [Glass]," he shares, adding, "I had this educational period where I basically listened to nothing but the history of radio, from Walter Cronkite to Jean Shepherd to Ira to now."[70] Executive producer Ellen Horne, who in September 2003 was the first staffer to join *Radiolab*, explains that Abumrad was especially interested in foreign radio documentaries that had "really immersive sound" but which did not have a broadcast home in the United States.[71] Abumrad has also pointed to the satirist and monologist Joe Frank and the renowned "audio artist" Gregory Whitehead as early influences.[72] Cult heroes in creative radio circles, both Frank and Whitehead have worked extensively in experimental audio drama—Frank from the 1970s up until his 2018 death and Whitehead from the 1980s through today. Horne also points out that even though it never gets cited in WNYC's promotional materials or discussed in the press, Golden Age radio drama was an important touchstone in *Radiolab*'s vision for itself and its sound early on, especially once Robert Krulwich began entering the picture in 2003. "There was a big theatrical opportunity in how you could use sound and how you make immersive sound," says Horne, who regards *Radiolab* as a modern-day form of "radio theater." She elaborates, "There's the experience of sitting around the wireless that we're reinventing through these theatrical experiences. From day one, that was our mission: to reinvent that in a way that was relevant."[73]

Robert Krulwich was a veteran broadcast journalist in his late fifties when he first met Abumrad and Horne in 2003. At the time, he was working in television as a science reporter for ABC News. The story goes that Abumrad was sent by WNYC to record Krulwich for a promotional spot. "So here he comes. He looks like a messenger kid on a bike," Krulwich says. After rewriting a prepared script that he deemed "ridiculous," Krulwich looked at a nervous Abumrad and asked, "Who are you?" It turned out that the two men, despite being some twenty-five years apart in age, shared a lot in common, including both being Oberlin graduates and both having worked at WBAI and WNYC. "It was like looking at someone living my life," says Krulwich.[74] The pair started meeting regularly for breakfast, and Krulwich began exploring his new friend's radio pieces. An episode Abumrad had made about *The Mercury Theatre on the Air*'s 1938 "War of the Worlds" broadcast particularly caught Krulwich's attention. "It wasn't just good. It was spectacular," Krulwich says. "What I particularly noticed was that it had a rhythm, a sense of music, that was different from what I expected or had ever heard."[75] "I said, 'Oh, my God, you are really good. You know stuff. Can I play with you?'" Krulwich adds, joking, "I kind of sat in his lap and wouldn't leave."[76] The pair started collaborating together in November 2003—or "playing together," as they both like to say—beginning with "a five-minute thing" that, Abumrad explains, immediately "had all the elements" for which the Abumrad-Krulwich version of *Radiolab* has become well known: "a strange skit"; "funny sounds and a surrealistic sort of thing"; "two guys bickering or talking and being confessional."[77] Then Krulwich began appearing on *Radiolab* as a regular guest, starting with the

December 2003 episode "Flight (Species Envy)." In February 2005 the series relaunched with Abumrad and Krulwich as cohosts. This is the *Radiolab* that listeners know today.

Abumrad receives the lion's share of the credit for *Radiolab*'s sound and style, and indeed he is the visionary behind the series' sonic aesthetic. He performs the editing and mixing and has final cut on every show.[78] Nevertheless, Krulwich's influence on the post-2004 *Radiolab* format is unmistakable, beyond just his contributions to the cohost banter. Although Abumrad explored topics related to memory, spirituality, medicine, biology, and the like, his earlier radio pieces tended to be more broadly focused on the humanities and the relationship of art and culture to the natural world.[79] It was not until Krulwich got involved that science topics took center stage and that the journalistic style of gathering a wealth of details, sources, and evidence took hold (even if their presentation of that evidence-based reporting is quite peculiar and singular). After all, Krulwich had been a broadcast journalist for thirty years, covering stories from Watergate to the Gulf War. He was a business correspondent for both NPR and CBS, as well as a science beat reporter. Among his many positions, Krulwich hosted the technology and culture miniseries *Brave New World* for *ABC News Nightline in Primetime* (1998) and, right at the same time that he was participating in the *Radiolab* relaunch, he signed on as host and managing editor of the PBS science program *NOVA scienceNOW* (2005–6, at which point he passed the show over to Neil deGrasse Tyson).

Krulwich not only brought his reporting chops to Abumrad's *Radiolab*, he also brought the theatrical sensibility that Horne described. Krulwich was well known at NPR in the 1980s for his "inviting singsong" style of reporting, the *Washington Post* calling him "the reporter whose hyper-conversational manner could make any complex topic seem accessible."[80] Similarly, *New York Magazine* praised him for his "freewheeling style," calling him "the man who simplifies without being simple."[81] Krulwich was renowned for his sense of drama. In the early 1980s, he famously created an eleven-minute radio opera about interest rates for NPR ("Rato Interesso"), loosely based on Mozart's *Don Giovanni*.[82] And he appeared on the *CBS This Morning* television program as the Singing Economist, performing arias about current economic trends.[83] This technique of making big, complicated ideas accessible and entertaining is perhaps *Radiolab*'s signature quality, alongside its experimental sound design.

At its core, *Radiolab* is a science show—also venturing into related fields of philosophy, technology, math, psychology, and the social sciences—that deals with "big questions": the universe, life, death, knowledge, language, war, and so on.[84] Much like *This American Life*, it does so through broadly themed episodes tackling very general topics: some episode titles include "Time," "Space," "Sleep," "Words," "Colors," "Laughter," "Sperm," "Cities," and "Who Am I?" It is a feature-documentary in that it is a (mostly) non-fiction program that mixes traditional elements of journalism and news reporting with more artistic and dramatic elements. If non-music radio programming can be placed on a continuum with straight news talk placed at one extreme and the fictionalized radio drama at the other, then the feature-documentary (or "documentary radio," as John Biewen and other contemporary radio producers refer to it) is situated somewhere in the middle, a mix of words, sounds, and music that merges the informational content of journalism with the form and emotion of art.[85]

Just as standard NPR news programs like *All Things Considered* are often called *news magazines*, in that they consist of numerous stories reported more in-depth than standard headline news-style broadcasts (and usually a mix of "hard" and "soft" news), the term *radio feature* insinuates an even more extended, highly focused examination of a story or topic, similar to a thoroughly researched and contextualized feature story in a print magazine. (As I suggest in chapter 7, there are parallels between the rise in popularity of audio storytelling and the trend for long-form journalism and literary nonfiction in print media.) Often, as is the case with *Radiolab* and also *This American Life*, multiple stories may be covered within an hour-long episode, but they nevertheless all connect to an overarching theme or narrative. This is not breaking news, even though it is journalistic in the sense that it is informative, educational, and occasionally timely and topical (at least in the sense that episode themes will be timed to match with particular holidays, seasons, or other cultural events).

Emphasis is placed on *characters*—people—whom the audience is invited to identify with through fleshed-out, exceptionally visualized scenes. This is where the artistic and dramatic elements come in: the focus of a feature-documentary like *Radiolab* is on storytelling rather than mere reporting of events and facts, and that is often achieved through vertically structured and intimate, slice-of-life narratives. NPR's news magazines may craft similar shorter segments (known as "radio packages"), but what really sets an audio storytelling feature-documentary apart is its meticulous attention to form. Biewen, the audio program director at Duke University's Center for Documentary Studies, calls radio documentarians "journalists/artists" who "use sound to tell *true* stories *artfully*."[86] This emphasis on

form can range from simply playing with voice and basic narrative structure to experimenting wildly with actuality sound and score music in a way
that verges on sound art. It need not always be "complex" or affectedly
avant-garde. Still, feature-documentaries are able to sidestep conventions
and engage in a level of experimentation that average broadcast radio or
podcasting rarely, if ever, does.

Radiolab tends more toward the "wild experimentation" end of that
spectrum, even though the core of its aesthetic is what Abumrad describes
as "the pleasant illusion of 'two guys chatting.'"[87] He is referring here to
the back-and-forth dialogue that occurs between himself and his cohost
Krulwich—a loose, conversational style that is also extended to the discussions between the hosts and their interviewees. Indeed, there is an emphasis placed on voice and narration—the voices of the hosts and interviewees
stitched together to recount experience. Still, the show is, at its most basic,
"about curiosity and discovery," to quote Abumrad again, and this inquisi-
tive, innovative spirit is extended from the show's focus on "big ideas" to
the way it explores those ideas creatively through sound.[88] Most notably,
the producers quickly and often abruptly butt voices up against each another, as well as layer voices on top of one other and then layer atmospheric
noises, sound effects, and music on top of (or underneath) it all.

An example: in the *Radiolab* episode "Patient Zero," the hosts examine
the concept of a "patient zero," meaning the case that can be identified as
the starting point of an outbreak.[89] They begin with the story of Typhoid
Mary, the woman commonly understood as the source of the first typhoid
fever outbreak in the United States, in the early 1900s in the New York
City area. *Radiolab* begins the episode with a cold open that features a pair
of producers, Lynn Levy and Sean Cole, randomly speaking to one another
("So have I said where we are? Am I on tape yet?"). They are clearly in an
outdoor environment, a brisk wind creating loud distortion in the microphone. Abumrad quickly identifies the producers but does not otherwise
introduce the story or the episode. Returning to the actuality sound, one of
the producers explains from the field that they are on an abandoned island
where a woman with an infectious disease was at one time quarantined, but
the exact location and identity of the woman remain unnamed. Then, Abumrad and Krulwich begin their host track narration, which takes the form
of an improvisational-sounding dialogue. Krulwich asks, "This is a story
that begins when?" To which Abumrad responds, "Well, actually, it starts
in 1906." The narration continues in this conversational back-and-forth
mode for a while, Krulwich playing the inquisitor to Abumrad's more au-

thoritative storyteller. Soon a third voice joins the conversation, that of an expert, University of Wisconsin–Madison medical historian Judith Leavitt.

This is clearly a storytelling mode compared to news radio's standard descriptive style of a host intro and hook followed by a reporter opening. In an episode of *Radiolab*, information is revealed at an almost rapid-fire pace, and yet the major plot points are only delivered incrementally. Instead of laying out all the most important aspects of the story in the lede, much attention is paid to context and creating a visual image for the audience. Rather than the thesis, characters, and scene all being set immediately, we are two minutes into the episode before it is clearly established that they are talking about a typhoid outbreak, more than three and a half minutes before it is announced that this is the story of Typhoid Mary, and it is not until after the four-minute mark that Abumrad and Krulwich announce the theme of the episode.

Radiolab is about exploring ideas—lofty, difficult, abstract ideas—and more than anything it achieves that through *experience*. Here experience is meant in a double sense: creating a fun, adventurous listening experience for the listener, as well as connecting, through descriptive personal stories, to universal thoughts and feelings that the audience will be acquainted with intimately. For instance, the vocal tone is loose, accessible, even jovial, with digressions and moments of humor interjected. The dual-narrator device functions to bring the audience into the story, Abumrad and Krulwich expressing amazement and asking each other questions in a way that often reflects what the listening audience is likely to be thinking. The banter also underlines the sense of discovery. When a startling point is revealed, for instance, the narrators stop and spontaneously declare, "Really?!" Moreover, the back-and-forth dialogue also functions as a kind of theater, a style more akin to a radio drama than a news report. This intimate, first-person narration builds tension and draws the listener in, akin to a group of friends sharing an amazing story at a bar.

Music is particularly integral to *Radiolab*'s aesthetic. Again in the "Patient Zero" episode, almost as soon as Abumrad and Krulwich's introduction starts, musical stings begin to creep into the piece. At first, these are curious sounding, modern classical style piano and string arrangements that quietly stay beneath the voices, mostly solitary notes that sound as though they are searching for something. As the Typhoid Mary story begins to build with Abumrad, Krulwich, and Leavitt describing the typhoid outbreak of 1906, the music perks up, horn bursts and tense strings serving to underline the impending danger. The voices and music continue this way, emphasizing and building upon one another in a montage fashion. Pauses and silences are interspersed to highlight moments of confusion or revelation. Indeed, these elements all work together to make *Radiolab* sound like the process of intellectual discovery—it is the research and problem-solving process manifested audibly. And it is an audio storytelling style that hybridizes a range of historical radio genres and formats from feature and documentary to news magazines, experimental radio art, and Golden Age radio drama.

Conclusion

Remediation reminds us of the importance of taking into account the full range of radio's forms, not only those that are the most recent, popular, or culturally dominant. This is true of audio drama as well as most of the feature-documentary podcasts that are exceptionally popular with modern audiences. Today's audio storytelling either has direct roots in broadcasting or, if closely scrutinized, is comparable to radio predecessors—and those unaware producers and audiences could learn a lot from revisiting radio's vast storytelling archive. As Josh Richmond contends, podcasting to date has mostly lacked an experimental or avant-garde "bleeding edge."[90] Most of the currently available content falls into a handful of predictable genres or categories—interviews, news, game shows, comedy, drama—each with distinct roots in broadcast radio. Too much of the scholarly and journalistic discourse about the internet and podcasting divorces these "new media" from the past and depicts them as disruptive technologies. Instead, through the examples of *Night Vale* and *Radiolab*, the continuities between the "old" and the "new" are mapped out. These audio storytelling programs are bringing much-welcomed diversity and ingenuity to the current podcasting environment, and doing so through a creative refashioning of radio's rich cultural heritage.

Regardless of these artistic debts, there are significant innovations at play in podcasting's revival and reworking of radio drama. US broadcasters have either undersupported or abandoned radio drama and other forms of long-form audio storytelling in recent decades. Radio drama, in the few instances when it even finds a home on today's broadcast dial, is usually relegated to a small local station and aired during a deserted corner of the schedule, such as Saturday late night, where it would be difficult to reach an audience of any notable size. Thus, podcasting is opening up much-needed spaces for these neglected radio forms and practices to exist—and to thrive.

Radio-internet convergence has also helped open up the radio archive to curious listeners and young producers who might wish to explore and innovate. This means access to a wider range of contemporary programs, including international productions of audio drama and radio art from Europe, Australia, and beyond that are not broadcast in the United States. Our listening is no longer confined to the programming choices of our local radio stations and networks. It also means access to a vast assortment of historical radio programs from the early and mid-twentieth century, many of which have only become widely available to the public in the past two decades. These programs are often saved and distributed via informal fan- and collector-run web archives, such as the Internet Archive's Old Time Radio Researcher's Group.[91] As a result, listeners can seek out programming beyond the major broadcast and podcast networks, while producers now have a much larger supply of material from which to draw ideas and inspiration. For all these very reasons, it is even more crucial to place contemporary audio storytelling in historical perspective, instead of treating audio drama or storytelling podcasts as new or exceptional phenomena.

Lastly, audio storytelling programs in this third wave of podcasting strongly uphold the sociability that is a central characteristic of the radio medium. In particular, these long-form creative audio programs emphasize a conversational narrative style and pay close attention to individual experience, everyday life, and subjectivity and affect. This focus reproduces the presence, immediacy, actuality, and liveness that are essential to radio broadcasting. Even recent radio dramas like *Night Vale*, despite the sci-fi and supernatural story elements, adhere to a presentation style that is mostly linear and direct, the primary content consisting of people's daily lives and delivered in a colloquial speech style. Indeed, a large part of what makes *Night Vale* so compelling—and the same goes for a nonfiction feature-documentary series like *Radiolab*—is the way it immerses listeners in the inner thoughts and the minutiae of the everyday life of an individual or a small group of characters. These quotidian intricacies are the stuff of radio.

CHAPTER SEVEN

Make Them Feel

Nonnarrated Audio Storytelling
and Affective Engagement

"It's not enough to make people think, you have to make them feel," states *Love + Radio* podcast producer Nick van der Kolk.[1] During the third wave of podcasting (2010–present), there has been a heavy push into introspective first-person audio storytelling that explores individuals' lives in great emotional depth and detail. This is especially true of productions originating within the podcasting sphere. Although digital storytelling and storytelling podcasts are gaining increasing purchase within the professional media industries (both public and commercial), the form is nonetheless rooted in more than a decade-old Web 2.0 principles of user-generated content, audience-led creativity, and utopic ideas about the democratization of media culture. Unmistakably, this focus on individuals and their personal experiences has meant an overwhelming tendency in audio storytelling toward confessional disclosure and themes such as family, friendship, romance, community, identity, youth, health, environment, education, and work.[2] This chapter extends the analysis of contemporary *audio storytelling* presented in chapter 6, shifting the focus to the centrality of affect and people's lived experiences found in podcasts made in the first-person *nonnarrated* mode.

The podcast *Love + Radio* is offered as a case study of the nonnarrated mode of audio storytelling that uses personal, conversational narratives to uniquely capture the contemporary structure of feeling. The analysis in this chapter highlights historical trends in public radio production cultures. It draws together theories and concepts from cultural studies and oral history to show how nonnarrated audio storytelling's foregrounding of the voices and lives of ordinary people functions politically by listening to others and expanding the range of public discourse. Although not without their shortcomings, nonnarrated podcasts like *Love + Radio* serve as a platform for sounding the voices of marginalized people, revitalizing radio as a tool for participatory democracy.

The Nonnarrated Audio Storytelling Mode

If interpreted literally, to hear that a story is "nonnarrated" may initially suggest that it is a pure presentation of dialogue, lacking any semblance of narrative form and structure. To be sure, a nonnarrated audio story displaces the authority of the radio/podcast host or *narrator*—though it does not remove radio producers from the equation—and presents *actuality* material in a relatively uninterrupted flow. Though the nonnarrated mode is far from being a freeform or unmediated affair. As a point of comparison, in film studies the term "nonnarrated" is used in contradistinction to narrated, thereby simply referring to fiction films without voice-over or off-screen narration.[3] In a nonnarrated film, all dialogue occurs onscreen, in situ. In contemporary radio and podcast production, the designation "nonnarrated" has come to refer specifically to a documentary audio piece where there is no narration from a host/reporter.[4] The segment consists entirely of a subject's voice—and typically that someone is an "ordinary" person (i.e., not a celebrity, politician, media professional, social elite, or other newsworthy figure).

This approach is commonly found in long-form feature-documentaries, where individuals tell their life story or recount a lengthy and involved event. The piece is recorded as a question-and-answer style interview, though it is edited to present only the subject's responses to the interview questions. If the reporter/interviewer is heard at all in the piece, these interjections are kept to a minimum. The recording is sometimes made in the field, meaning in the subject's home or a public place—and not in a studio environment—which lends it a further sense of "authenticity." The effect of editing the piece so that the interviewer is excised entirely is that the "story" presents as a monologue where the subject's voice stands alone. Referring to the main sonic components of an audio story—tracks, actualities, ambience, and music/sound effects—a nonnarrated story is all acts, no tracks.

The term "nonnarrated" is used here because it is widely adopted in the radio/podcast production culture. However, to call these radio pieces "nonnarrated" is in point of fact a misnomer, since the interviewee is both the subject of the discourse (the main character, in effect) and the narrator (the speaking subject). In first-person audio storytelling, it is incorrect to say there is no narrator merely because the producers have removed themselves from the final cut. The act of "let me tell you about this thing that happened to me" firmly situates the storyteller as the narrator. Indeed, due to radio's linear nature and its lack of visual or textual cues, practically all radio texts feature a narrator of some kind, with the rare exception of experimental radio art pieces. There needs to be some*one* to guide listeners through the events of an audio story.

This nonnarrated mode of audio storytelling may be small compared to the *This American Life*– and *Radiolab*-inspired feature-documentary mode, though it has seen an uptick in the past decade during the third wave of podcasting. *Love + Radio*, which began in 2005, has been joined more recently by podcasts such as *Everything Is Stories*, *ARRVLS*, *Long Haul*, and *This Is Actually Happening*. This is not to suggest that the nonnarrated mode is entirely new. The style of foregrounding the first-person accounts of ordinary people dates back to some of the earliest radio documentaries produced in the United States. Before and during World War II, for example, folklorist Alan Lomax and others in the Library of Congress's Radio Research Project, the Office of War Information, and the Armed Forces Radio Service produced radio pieces like *People Speak to the President* (1942) that centered around "vox pop" or "man on the street" interviews with ordinary citizens.[5] The nonnarrated mode has also lived a long, fruitful life on NPR since the 1970s, even if it is mostly confined to infrequent feature-length documentaries (produced through dedicated documentary units like American Public Media's American RadioWorks) or, more commonly, short radio pieces inserted into news magazine or arts-and-culture programs like *All Things Considered*, *Morning Edition*, *The World*, *Studio 360*, and *Hearing Voices*. Nonnarrated stories are one of a few formats that serve as the building blocks for NPR's signature programs, and thus are a key element of "the NPR sound." Indeed, the nonnarrated mode is much admired by radio makers invested in the art and craft of audio production, so much so that it is often the first assignment given to aspiring public radio producers at training programs such as Transom Story Workshops.[6]

Thus, much of the recent wave of creativity found in feature-documentary style podcasts is arriving through a remediation of older radio forms, techniques, and styles. The most prominent practitioners of non-narrated audio storytelling over the past few decades have been Dave Isay, Joe Richman, and the Kitchen Sisters, the production duo of Davia Nelson and Nikki Silva. Though much of their careers have been spent working as independent radio producers, all four have deep roots within the public radio industry: Isay with *All Things Considered* and *Weekend Edition*; Richman with *All Things Considered*, *Weekend Edition*, and *Car Talk*; and Nelson and Silva with *All Things Considered* and *Morning Edition*. They have also each held teaching positions training young public radio producers.[7] Isay's early career, which culminated with the much-heralded 1993 radio documentary *Ghetto Life 101*, was focused on creating "audio profiles of men and women surviving in the margins," or what he calls "sound portraits."[8] Then, in 2003, Isay founded StoryCorps, a nonprofit organization designed "to instruct and inspire citizens to record high-quality audio interviews with family, friends, and community members."[9] These oral history interviews, modeled on the practices of the Works Progress Administration in the 1930s, are archived at the American Folklife Center at the Library of Congress. Segments air regularly on NPR's *Morning Edition*, as well as through a *StoryCorps* podcast. Much like Isay, Richman's approach to nonnarrated audio storytelling centers on giving his subjects audio recorders and training them to record themselves, so they can "report on their own lives and histories."[10] The results are what he calls "radio diaries" that tell "the extraordinary stories of ordinary life."[11] Richman originated this approach in 1996 with a series called *Teenage Diaries*—the segments for which aired on NPR's *All Things Considered*—and he has since expanded it to include audio diaries from prisoners, prison guards, retirement home residents, among many others.[12] He likens it to a form of citizen journalism—although Richman and his team, who operate through a New York–based nonprofit organization called Radio Diaries Inc., ultimately select and edit the material that actually gets heard by the public.[13] The WNYC public radio series *Radio Rookies* adopts a similar approach, teaching New York City teenagers to document and tell their own stories. The Kitchen Sisters' signature approach to nonnarrated audio storytelling has been to create historical documentaries out of archival sound recordings, random "found sounds," and home recordings, many of which are submitted by public radio listeners.[14] One of Nelson and Silva's most acclaimed series was *Lost & Found Sound*, which produced segments that aired regularly on NPR's *All Things Considered* between 1999 and 2005. Richman has also crafted similar nonnarrated histories using archival interviews and historical recordings. Both Richman and the Kitchen Sisters today have podcast series distributed through the Radiotopia podcast network: *Radio Diaries* and *The Kitchen Sisters Present* (formerly *Fugitive Waves*), respectively. In each case, the podcast is primarily a platform for crafting new narratives out of historical recordings, quite often repurposed content they originally produced years and even decades ago for public radio broadcasts. The efforts of Isay, Richman, Nelson, and Silva to get subjects to record themselves and to build stories out of historical home recordings shares much in common with the user-led democratic media philosophy of digital storytelling.

Love + Radio and the Nonnarrated Mode

The *Love + Radio* podcast series mainly consists of *nonnarrated* stories.[15] While it does not invite its subjects to record themselves à la *StoryCorps* or *Radio Diaries*, its producers embrace a mostly hands-off interview style that more closely resembles anthropological participant observation than a traditional one-on-one radio interview (known as a "two-way" in radio jargon). *Love + Radio* was chosen for this case study of the nonnarrated mode of audio storytelling because it was one of the first podcast-only series to adopt the nonnarrated approach, and currently it is among the mode's most popular practitioners. Its production team has also experimented with the conventions of the nonnarrated form more than any other program during the third wave of podcasting.

The story of *Love + Radio* begins in college radio in 2004, when producers Nick van der Kolk and Adrianne Mathiowetz were undergraduates at the Bard College student-run station WXBC, cohosting a show called *This One Time*.[16] It was there, in early 2005, they started experimenting with what would eventually become the first incarnation of *Love + Radio*.[17] After graduating in spring 2005, van der Kolk and Mathiowetz both took internships at Public Radio Exchange (PRX) in Cambridge, Massachusetts, which before long turned into full-time editorial positions for the public radio content distributor.[18] Once there, they began producing episodes of *Love + Radio* on the side. The series debuted in October 2005. It was initially distributed as part of NPR's short-lived podcasting initiative Alt.NPR, which was developed to carry edgy, "experimental" original-to-podcast content targeted at younger audiences.[19] Van der Kolk and Mathiowetz produced about a dozen episodes together between 2005 and 2007, until Mathiowetz departed for a web manager position at *This American Life* and a second career in photojournalism.[20] Nick van der Kolk—who himself has cycled through a number of jobs in the public radio industry, including with Chicago Public Media's Vocalo and *Snap Judgment*—has remained the one constant behind *Love + Radio* over the past decade. He has been joined by a handful of regular collaborators, most notably the producer and sound designer Brendan Baker.[21]

There are a number of distinct phases in *Love + Radio*'s development, marked by changes in aesthetics, personnel, and institutional support. The producers have retroactively divided these phases into "seasons" on the podcast's website, even though the earliest seasons span multiple years and uneven episode runs.[22] The first phase lasted from 2005 to 2008, during which time van der Kolk and Mathiowetz worked together and episodes were distributed through Alt.NPR. These earliest episodes—which the NPR directory dubbed "stories of post-college confusion"—were mostly anthology pieces organized around themes (sex, guns, violence, secrets, urination, ghost stories, dating), in what might be described as a Generation Y version of *This American Life*.[23] While they eschewed host narration, they also had a tendency to haphazardly mix multiple perspectives, and, moreover, some segments were clearly scripted performances. There were even a few staged readings of fictional short stories. It was toward the end of this first phase that van der Kolk, after Mathiowetz's exit, began moving more fully into the nonnarrated mode with the in-depth profiles of individual subjects that has since come to define the series. *Love + Radio* went on hiatus from fall 2008 to fall 2010, returning in November 2010 with support from Vocalo and Chicago Public Media. That brief second phase ran from 2010 to 2011. From 2012 to early 2014, the series had a short stint of a handful of episodes for WBEZ, making up the third phase.[24] Then, in February 2014, *Love + Radio* entered its fourth phase when it became a founding member of PRX's Radiotopia podcast network.[25] The podcast has consistently operated as a full-time production since 2014, though it shifted distribution from Radiotopia to the subscription-based podcast network Luminary in 2019.[26]

The nonnarrated style and sound of *Love + Radio*—the subjects speaking their mind in a mostly uninterrupted manner while carefully composed editing, music, and sound effects subtly structure the story—coalesced midway through the second phase, when van der Kolk first collaborated with Brendan Baker. It was then that their April 2011 episode "The Wisdom of Jay Thunderbolt," about a Detroit man who runs a strip club out of his house, won the Third Coast International Audio Festival's Gold Award for Best Documentary. It was the first-ever podcast to win the festival's top honor.[27]

An episode that exemplifies the effectiveness of *Love + Radio*'s non-narrated approach would be 2013's "Jack and Ellen."[28] It is the story of a young woman, Ellen, who escapes her job as a Subway "sandwich art-ist" by becoming "a professional blackmailer." Specifically, she pretends to be a fifteen-year-old boy on Craigslist in order to lure in pedophiles and blackmail them—a practice known as pedobaiting. Through her alter ego "Jack" (and a few other identities), she estimates she has extorted at least one hundred people for a total of $30,000 to $40,000. To begin, the subject matter of the episode is sexual solicitation and pedophilia: topics that are discussed on public radio from time to time, albeit in a rather narrowly defined and conservative manner. *Love + Radio*, in contrast, spends a full thirty-one minutes on Ellen's story, discussing the topics with much more candor than broadcast radio ever could, leaving the more explicit and unsa-vory details intact rather than tiptoeing around them. While pedophilia is nearly universally abhorred, Ellen's catfishing and blackmail are also widely detested (and illegal) activities. Her story, therefore, is rife with moral am-biguity: the actions of the pedophiles are unethical and illegal, yet Ellen's exploitation of them is too. Since van der Kolk and his coproducers (Bren-dan Baker and Mooj Zadie, for this episode) refrain from narrating Ellen's story and attaching a fixed moral lesson to it, listeners are openly invited to draw their own conclusions about the ethical dilemmas presented in the audio story.

The fact that the audience is presented with Ellen and her point of view (and not that of the victims of her pedobaiting, for instance) means listeners are encouraged to empathize with her, at least initially. This is one area where the benefits of radio as a medium for storytelling emerge: unlike print, which only gives us access to *what* was said, the aurality of radio also gives us access to *how* it was said. Radio captures the tonality and pacing of the subjects' speech along with their subtle dynamics and narra-tive rhythms, which carry a whole subtext of affect and meaning.[29] In the beginning of her story, Ellen is very matter-of-fact about her decision to impersonate a gay man online and then blackmail other men. Hearing her voice, she sounds unfeeling and remorseless, even laughing flippantly over her choice to make the Jack character very skinny, or what gay men call a "twig." "Pedophiles love twigs," she chuckles smugly. At this point, it is dif-ficult to empathize completely with someone who *sounds* so brusque, cold-hearted, and self-serving, even if we feel no sympathy toward the people of whom she is taking advantage. In fact, as she explains the inner workings of her fraudulent scheme in greater and greater detail, her voice still sounding detached and callous, it is hard not to begin feeling a small amount of pity for these presumed pedophiles, whose vulnerability Ellen depicts rather with glee.

It is not until the seven-minute mark that the audience actually gets formally introduced to Ellen, and she really begins to explain her motivations for the pedobaiting. Discussing her personal life outside the fraud, Ellen's tone becomes lighter and more affable. Her pace gets less terse, which makes her sound more open and expressive. We begin to understand how she came to be where she is, even if she is breaking the law at every step along the way—stealing DVDs and other small resalable items rather than working a proper job. There is no doubt: Ellen is a criminal. Yet we learn that she was always "a creative person" who used to write fan fiction, and that the catfishing may in some small way be an extension of those artistic propensities. We also learn that her mother knows what she does and approves "because she thinks pedophiles are pigs." And we discover that she herself sees the error of her ways, although similar to drug dealers (an analogy she herself draws) the easy money of pedobaiting is simply too good to pass up given her otherwise modest career options. Partway into the episode, we learn that Ellen herself is a lesbian, which she admits puts her in a morally complicated position. "Being a lesbian myself, I was a bit put off by the idea of threatening to out people because that's really awful," she admits, adding, "I felt like a traitor because I, myself, was in the closet at one point. . . . So I didn't really want to exploit that fear, and that made it hard to not feel guilty." These small details, while they do not justify her actions entirely, nonetheless add layers of moral complexity to the situation.

Over the course of the podcast, Ellen develops from a rather one-dimensional, unmerciful caricature into a dynamic character who is conflicted by being a blackmailer but who manages to find ways to rationalize her actions anyway. "They can be a dad, they can be a doctor, they can be gay. But they all still want to have sex with children. That's what I had to keep in mind, to allow myself not to feel too shitty about what I was doing. At that point, I put aside our common bond," Ellen concludes. It would be nearly impossible to capture this dynamism in a four-minute *All Things Considered* radio piece: only the long-form structure of a nonnarrated feature can reveal such emotional textures and intricacies. And it is not just through what she says but how she says it—the aurality—that this dynamism is achieved. Moreover, the fact that the producers do not try to frame Ellen's behavior, as most broadcast journalists would be inclined to do, shows respect for the subject's voice and perspectives—even if her views do not conform to the more socially acceptable morals presumably held by the producers themselves.

Nonnarrated audio storytelling programs like *Love + Radio* are not unmediated, nor are they made by pseudo-objective broadcast journalists or documentarians. Their producers recognize the value of using editing and nonverbal sound (music, sound effects, ambience, and natural sound) to enhance the affective power of a story. This is where nonnarrated programs overlap with feature-documentaries in the *Radiolab* vein, and where *Love + Radio*'s experimental tendencies are revealed. "Jack and Ellen" is a nearly straight nonnarrated story, in that Ellen's is the primary voice heard in the radio piece (apart from a few places around the twenty-minute mark where the interviewer, Mooj Zadie, interjects momentarily; van der Kolk also offers a brief introduction, which only states the title of the episode along with sponsorship information and an explicit content warning). Yet Ellen's personal narrative is very dialogic, containing multiple characters or "voices." In addition to her point of view, there are times when she speaks in the first person as "Jack" and her different online personas, plus there are times when she voices other people, including correspondence received from men she blackmailed. To bring these multiple voices to life and give the story a more conversational quality, sound designer Brendan Baker modified Ellen's voice using digital pitch-shift plugin effects.[30] In fact, the radio piece begins with a heavily pitch-shifted voice that has obviously been distorted in postproduction. The audience is dropped into the narrative mid-conversation. It is not yet evident who is speaking or why the voice is distorted, and thus listeners could very well presume that the vocal effect is in place simply to disguise the identity of the speaker. The pitch of the voice is fairly baritone, and thus reads as a masculine voice. Then, around three minutes, the distorted "male" voice is discussing naming his online alias when an undistorted female voice interrupts. "It was Jack," the woman says, her voice faintly breaking through layers of music and atmospheric digital noise, akin to a distant radio station signal intruding into a broadcast feed. "Hi, my name's Jack," the distorted male voice repeats, almost in unison. Another fifteen seconds later, the female voice returns, explaining, "That was the story." The male and female voices oscillate back and forth like this for another half-minute, at which point the female voice begins to steadily take over and it becomes clear that they are the same person. The woman, who we eventually learn is named Ellen, is the story's narrator, and the distorted male voice is her alter ego Jack.

It is the sound design that tells listeners these different voices are the same person—it is never explicitly stated in the narration. Elsewhere in the episode, we hear various pitch-shifted voices representing Ellen's pedobaiting victims, her blackmailer persona (who is not Jack but rather someone who accesses Jack's accounts without his permission and threatens to expose the pedophile), and her roommate and her best friend, both of whom chastise Ellen for her actions. These are not dramatized re-creations (with one exception); the recordings are mostly all of Ellen speaking the different roles in the first person, then manipulated by Baker in postproduction to sound like a cast of characters. Beginning around twenty-four minutes, Ellen recounts a story where she nearly gets caught for her crimes, at which point her undistorted voice and the pitch-shifted voice of the mark, then followed by a police detective, are layered directly on top of one another, resulting in them speaking the same lines together. This artificial double-tracking makes the producers' manipulation of the audio conspicuous and completely transparent, if it was not already obvious to the listener by that point in the piece.

In addition to the vocal processing and other sound effects, *Love + Radio* uses music in unconventional ways. The convention for most contemporary nonfiction radio is to use music primarily to make a point. If there is rising action in a story and a speaker is building up to an important plot point, for instance, then the music will enter in the middle and help escalate the momentum. Elsewhere, music will emerge unexpectedly to underline the drama of a moment. Or alternately, the music will be pulled back, the sudden silence highlighting a key moment in the narration. After a central idea or emotional peak in a story, editors will also often insert a "music post," where the music will play in the clear for a few seconds without any talking. In all of these cases, music is being employed as a narrative hand-holding device, directing audience attention and establishing pacing as well as mood. Music is "a little flashlight that helps us get our ideas across," says *This American Life* editor Jonathan Menjivar.[31]

Love + Radio utilizes these conventions in places, though the general approach is to push back against writing and production techniques that are overly instructive and didactic. Baker claims that they want to avoid "when the music or sound design is a bit too 'on the nose'" and instead "push toward something more intuitive and avant-garde." He explains, "We start with the voices from our radio stories and then add disparate sounds to create a kind of music."[32] In "Jack and Ellen," for example, the same music recurs as a leitmotif throughout the episode every time Ellen delves into her blackmail scheme—a propulsive, skittering electro-jazz instrumental track. Thus, the content of the subject's speech is driving the soundtrack rather than the soundtrack propping up the content. Instead of sampling prerecorded music, Baker also has a propensity for taking sound effects or other nonmusical sounds and looping them until they become small musical gestures. (These include ambient noises serendipitously caught on the interview tape that he then repurposes as music samples.)[33] These noise-like digital distortions resemble the electronic music genre known as *glitch*, and give episodes like "Jack and Ellen" an anxiousness that underscores the edginess and moral ambiguity of the content.[34] On occasion, words and phrases will be layered and syllables will be slowed down or repeated, either to connote a montage-like compression of time or to suggest a break in the interview. Or sometimes these effects will be imposed simply to create "a musicality to the dialogue," says Baker, adding, "Our interviews 'sing' along with the score."[35] Despite these assorted alterations and artistic flourishes, the editing and sound design in *Love + Radio* only strengthens emotions that are already in the subject's narrative. It is intricate but understated; it does not exploit the narrative or maneuver the audience in a heavy-handed manner.

The Participatory Mode of Feature-Documentary Audio Storytelling

Love + Radio's producers themselves point out that the level of production in episodes like "Jack and Ellen" is not feasible or even appropriate for all audio stories.[36] The series deviates from the nonnarrated mode for select episodes, most often taking up a *participatory mode* in which the producers interact directly with their subjects.[37] This participatory mode is best classified as a subset of the *feature-documentary* mode of audio storytelling described in chapter 6, though it also retains key elements of the nonnarrated approach. Typically, these episodes consist of edited two-way interviews between a producer and a subject with little or no music and sound effects added. Nevertheless, the approach and the feel of the *Love + Radio* interviews differ considerably from most interviews heard on commercial talk radio or public radio news magazines, such as *Fresh Air*. For starters, these are still ordinary people (i.e., not celebrities, politicians, media professionals, social elites, or other newsworthy figures), and the topic of the discussion is the interviewee's life story, instead of news and current events. In addition, the interviews are routinely recorded outside of a professional studio, either on location or over the phone/computer, the rough audio quality contrasting with the polish of studio broadcasts. This raw quality discursively functions as a signifier of intimacy and authenticity. To that end, the producers also leave in (or even accentuate) phone greetings, call distortion, and other "mistakes" as a way of simultaneously creating an un-edited feel and reflexively prodding the listener to hear the documentary as a construct or representation.

Unlike a pseudo-objective observational documentary where the inter-viewer/producer is unheard and discursively removed, this participatory mode embraces the radio producers as participants and emphasizes the ac-tual, lived encounter between the interviewer and interviewee.[38] After all, an interview is not a realist representation of the subject: it is an interaction that exists specifically for the broadcast or podcast—and it is often a highly composed and performed interaction at that. This is perhaps why the *Love + Radio* producers place at least a few small interviewer interjections into even their more straightforward nonnarrated radio pieces like "Jack and Ellen." These interjections could otherwise be edited out, but instead they linger as a reminder that there is an intermediary present in-between the speaker and the listener.[39] The *Love + Radio* interviewers also embrace an occasionally confrontational but mostly hands-off style that allows their subjects to speak at length, which in the spirit of the nonnarrated mode foregrounds the subjects' life story above all else.

"An Old Lion, or a Lover's Lute" is an example of a *Love + Radio* episode done in this participatory mode.[40] The 2014 episode consists of an inter-view between producer Ana Adlerstein, a twentysomething white woman, and Jerome, a fifty-one-year-old African American man who catcalled her on an Oakland, California, street. Running just shy of thirty minutes total, the episode's first segment consists of Ana stopping on the sidewalk, turn-ing her audio recorder on, and politely but suspiciously questioning Je-rome about his rude behavior toward women. As they are talking, Jerome's ex-girlfriend, the mother of his teenage son, happens to walk by and joins in the conversation, describing the couple's complicated relationship (she is gay and now married to a woman) and telling the story of how he picked her up at a McDonald's years ago. It evolves into a boisterous exchange that captures all the energy and humor you might expect from three com-plete strangers who only just met by happenstance on a noisy city street. The episode's second segment consists of a follow-up interview between Ana and Jerome that, though it starts out jovial, quickly turns argumenta-tive and then, somewhat unexpectedly, heartfelt. Jerome is a self-professed womanizer whose attitudes toward women are plainly sexist and ignorant.

He is a walking, talking stereotype of a chauvinist, and it is easy to imagine that if he were featured in another radio piece he would be represented as a one-dimensional, superficial "player" and made an object of scorn. On *Love + Radio*, though, we are actually allowed to hear him try to rationalize his opinions and explain why he behaves the way he does. What Jerome says is often disturbing: "If you walk down the street looking like that, any man in his right mind gonna stop, and they gonna look, and say, 'Hey, how you doing?' It's because of the way she look. . . . She looked like she was easy because of her outfit she had on"; "Any man see that, first thing come to his mind, 'I'm gonna need that.' It ain't nothin' personal"; "I don't want no woman that been all wore out and everything. . . . I prefer somebody with less miles"; "If your face ain't that pretty, your body has to at least be petite and nice. I do not like fat womens." And Ana is not afraid to confront him on his discrimination: "Wait. Hold on. I don't believe you"; "Wait. Wait. Blue jeans? You're wearing blue jeans right now!"; "Are you talking about a woman like a car right now, Jerome?"; "Why does it matter how many men [a woman has] had sex with?"; "You're being a total hypocrite"; "Jerome, this is gross." Yet even when Ana challenges Jerome, she does not shut the conversation down or change the topic, as might be the case if this interview were happening in another forum. She allows Jerome to speak his mind freely. Moreover, toward the end of the interview, the conversation gets more earnest as Ana asks him about his children, his experiences growing up, whether he has ever had his heart broken, and so on. The result is that a man who many people would write off as a close-minded bigot is still given respect and portrayed with humanity. The listener is left to judge Jerome based on his words—and *Love + Radio* allows him to say plenty.

Love + Radio, Structure of Feeling, and the Politics of Audio Storytelling

The affordances of podcasting—unrestricted length, low production and distribution costs, absence of FCC regulation, less need for broad audience appeal—allow producers like van der Kolk to tell longer, often very layered stories that range in premise from the mundane to the sensational, placing special emphasis on individual experience and affect. Audio storytelling podcasts like *Love + Radio* tend to feature subcultural or marginalized subjects who are either entirely overlooked by most popular media or, if they are represented, who are framed as deviants, social pariahs, or one-dimensional stereotypes: sex workers, cult members, stroke survivors, con artists, hitchhikers, prisoners, radical political activists, drug dealers, and the like. Nonnarrated podcasts, in particular, give these individuals a platform on which to voice feelings that are transgressive and frequently challenge the ideological status quo. Moreover, while the nonnarrated mode is ultimately an aesthetic choice for the producer, it is well suited to the podcast format because there is less need for host narration in a podcast, since listeners have presumably sought out the program and they will consume the story linearly from beginning to end. That is, listeners will not be randomly tuning in to the middle of the story, as is common with broadcast radio, and thus they do not need the constant hand-holding of a host with periodic reminders about the topic of the program, the names of the speakers, and so on. "We just throw people in the deep end and let them flounder," says van der Kolk.[41] The result is that the audience needs to listen attentively in order to follow along. Furthermore, listeners are immersed in the narrative, hearing the motivations behind people's actions. These questions of *why* are typically left out of headline-news style reports, which focus only on the *what*, the consequences. In the least, audiences are encouraged to reflect on the subject-narrator's stories and draw their own conclusions. The prolonged, intensive focus on these individuals' feelings, emotions, and concrete human experiences uniquely captures the contemporary structure of feeling.

At its most basic, storytelling is, in the words of Walter Benjamin, "experience that goes from mouth to mouth."[42] It is an oral speech tradition in which the narrators are often "nameless," ordinary people conveying useful information—practical advice, a moral, a proverb or maxim—as it is "woven into the fabric of real life."[43] Radio, more than any modern mass medium, is uniquely equipped to capture and elevate the art of storytelling. Text can transcribe verbatim the content of a speech act, but it cannot relate crucial communicative aspects of the voice such as tone, stress, volume, speed, pitch, accent, dialect, and other elements of the emotional content of speech and the ways in which the words are spoken. Again, this acoustic landscape of sound and hearing is its *aurality*.[44] Admittedly, film and video provide the added benefit of displaying body language and contextual clues about the particular time and place of a speech act—although audio recordings can capture many traces of these embodied and spatial attributes too.[45] As Jonathan Sterne has warned about "the audiovisual litany," I do not wish to idealize or essentialize sound in a way that suggests that it draws audiences into the world while vision separates audiences from it.[46] However, film and video do tend to privilege the visual over the auditory, and certain topics and ideas are much easier to convey via sound than visuals. Moreover, at a production level, shooting video is more obtrusive than gathering audio, and speakers tend to be more self-conscious and restrained when being recorded on camera compared to audio.[47] For these reasons, radio is particularly well suited to the everyday, experiential type of storytelling Benjamin described.

The *everyday* is a foundational concept in cultural studies and the basis for the field's investigations into popular culture and lived experiences, especially those of disenfranchised or marginalized social groups. Raymond Williams's idea of *structure of feeling* describes the shared values of a particular group or society in a specific historical moment, defined through the actual affective social character of a period rather than the selective and idealized notions of many cultural analyses. "It is that we are concerned with meanings and values as they are actively lived and felt," writes Williams, to which he adds, "We are talking about characteristic elements of impulse, restraint, and tone; specifically affective elements of consciousness and relationships: not feeling against thought, but thought as felt and feeling as thought: practical consciousness of a present kind, in a living and interrelating continuity."[48] Rather than "feeling" in a purely emotional or sensorial context, Williams is referring to a nexus of social relations and experiences: people's daily interactions with the objects and practices of everyday life, and how those interactions yield reactions that structure how they live their lives. It is not quite a cultural unconsciousness but also not entirely an ideology or holistic worldview. For Williams (and with him the whole project of cultural studies), it is essential to study popular culture texts and, especially, the culture of the working and lower classes in order to develop a truly democratic account of culture.[49] Thus, the attention to subcultures and marginalized groups found in audio storytelling podcasts like *Love + Radio* is in step with the larger aim of cultural studies to capture the entire felt social character of a period, and not only selective, typically middle- and upper-class ways of thinking and being.

It is important to acknowledge that the structure of feeling is a historical configuration that can only truly be observed retroactively. What is remarkable about podcasting—from the huge amount of insipid discussion-based "chatcasts" to this more recent explosion of crafted audio storytelling—is that it is resulting in a vast digital archive of audio recordings of predominantly ordinary people talking about their experiences, their identities, their memories, and their opinions. These are narratives and points of view that are rarely represented in other mass media, at least not with this level of attention and care. "What we are looking for, always, is the actual life that the whole organization is there to express," Williams writes, adding, "The significance of documentary culture is that, more clearly than anything else, it expresses that life to us in direct terms, when the living witnesses are silent."[50] In the future, these nonfiction podcast recordings will be an invaluable resource for historians and cultural analysts seeking to understand the lived experiences and values of particular groups or classes of people in the early twenty-first century.[51] The stories people tell about themselves and others, while it may occasionally seem merely amusing or even frivolous and self-indulgent, do have great social and political significance—and could potentially be indispensable for understanding the lives of those who lived the culture.

In this way, the audio storytelling examined in this chapter can be seen as a form of oral history, particularly that which is made in a first-person nonnarrated mode. Oral history is defined as the "collect[ion] of memories and personal commentaries of historical significance through recorded interviews."[52] Integral to the very idea of oral history is the importance of individual life experience—so much so that the term "oral history" is interchangeable with other phrases such as "life history," "life story," "oral biography," and "personal narrative."[53] Oral historians work from the premise that all sources are subjective and that this fact should be taken as an advantage rather than a disadvantage. The collective memory or collective consciousness of a society, in other words, can only be truly understood through the careful consideration of a plurality of subjective voices.[54] The focus, then, is on many people's lives and how they think about their experiences, as opposed to only the narratives of social elites, abstract cultural theories, or purely structural analyses. The goal is to produce what anthropologist Clifford Geertz called *thick description*: providing a detailed account of a phenomenon and its context, especially through a large number of testimonies rather than a single view of the experience.[55] As an academic field of study, oral history initially grew out of post–World War II developments in social history in the United Kingdom and socialist concerns for recording "people's history" through the words and experiences of the working classes. In the United States, oral history took hold in the 1970s with the rise of multiculturalism and historians' newfound commitment to writing history "from the bottom up."[56] Area studies researchers took up oral history and ethnography as ways to make many voices heard and to challenge the historical and philosophical biases of Western research methods. The escalation of audio storytelling through podcasting at least has the potential to spread oral historians' political commitments to listening to others and to expanding the range of discourse.

Oral history is a trained discipline with strict standards and established principles, and many oral historians would take umbrage at the suggestion that audio storytelling pieces produced by professional audio journalists are a form of oral history. Nevertheless, the nonnarrated mode, especially, shares a dedication to the recorded in-depth interview (the centerpiece of oral history research) and to the interviewer suppressing the urge to talk and *listen* instead.[57] Unlike a traditional one-to-one radio interview, which functions more like a dialogue between the interviewer and the interviewee (think: Terry Gross on *Fresh Air*), the role of interviewers in audio storytelling pieces is to ask meaningful questions and then step back. They are functioning more as a witness than a participant.

Oral historians might view any manipulation of the audio interview as unethical. Storytelling, including audio storytelling, differs most strongly from formal oral history by also being highly attentive to style and structure, principally through the fundamentals of character, plot, and dialogue.[58] All talk is *not* storytelling—at least not formal storytelling, as it is conceived within literary and folklore studies. A story is a narrative (not a chronicle), which by definition requires selectivity and careful consideration of form and technique. While many oral history subjects may already be gifted storytellers, a story that is going to translate on radio/podcast and make listeners stay tuned typically needs at least some editing and slight restructuring. In the least, it will be edited to have a beginning, middle, and end, as well as a plot and point of view. This typically means that recorded conversation will be spliced up and presented out of its original order. Similarly, segments of the interview may be regrouped by theme or tone, or to create conflict and suspense. The interview will also be trimmed of the digressions and repetitions that are so commonplace in ordinary speech. The end product should not be a distortion of the subject's presentation of the self, rather a distillation of it instead.

What is more, *entertainment* is an essential element of all successful storytelling.[59] Importantly, audio storytelling in this way contrasts with investigative audio journalism and more expository modes of radio documentary, which offer an interpretation of the events they recount and/or make a case advocating for a cause or a specific point of view. The aim of storytelling, on the other hand, is to entertain audiences and motivate them to draw their own conclusions. Storytelling is a performance, albeit a performance that is close-up rather than distant or isolated. "The storyteller takes what he tells from experience—his own or that reported by others," writes Walter Benjamin, "And he in turn makes it the experience of those who are listening to his tale."[60] As an oral tradition, the act of storytelling situates the narrator in a relationship with a larger audience. "Storytelling is a form of giving," writes Amy Spaulding, by which she means, "You can preach with a story or sell with a story or teach with a story, but true storytelling should be a gift, with no demands that the story be interpreted in a particular way."[61] This is not to say that audio storytelling is apolitical, rather that the audience be invited to interact with the story and interpret its meaning individually.

Ultimately, the focus on the voices and lives of ordinary people—and, in particular, marginalized people whose points of view are too often unheard or underrecognized in other mass media—is what most distinguishes nonnarrated programs. Ostensibly, this is the mission of many feature-documentary-style public radio programs, too. The premise of *This American Life*—the archetype of the modern US feature-documentary mode—is to cover "everyday lives, personal lives."[62] Jay Allison, the founder of Public Radio Exchange (PRX) and one of the most influential figures in US public radio, has called *TAL* host Ira Glass "a champion for the Many Voices"—a reference to Bill Siemering's founding vision that NPR should "speak with many voices and many dialects."[63] Yet *TAL* has been critiqued for actually representing a rather narrow range of people and perspectives that resemble the demographics and tastes of the show's predominantly white, upper-middle-class, educated audience. As such, it is not truly pluralistic.[64] Furthermore, *TAL* is by its own admission didactic: Glass and his reporters are always "mak[ing] some really big statement," extracting from the stories of others a lesson that is then instilled on the audience. The everyday voices—the subjects—merely provide "anecdote[s]"; the radio producers are the arbiters of meaning.[65] In contrast, nonnarrated audio stories give the subjects the first and last word, for the most part. The message of a nonnarrated story is either set by its subject or left ambiguous, to be interpreted by the individual listener. Thus, knowledge and power are much more balanced and open in the nonnarrated mode.

Where podcast series like *Love + Radio* break away from *Teenage Diaries*, *StoryCorps*, and the many iterations of the nonnarrated mode that have appeared on broadcast radio over the years is in the prominence and the frankness of the content. Nonnarrated stories are not relegated to a short segment on *All Things Considered* every Friday afternoon, or the occasional (and increasingly rare) special feature-length documentary. In these podcasts, the nonnarrated stories are the entire focus of attention. As a result, they allow their subjects to speak uninterrupted for lengthy periods (fifteen, thirty, even sixty minutes) with minimal editing and little interjection from the producers. This approach serves to immerse the audience in the narrator's worldview. They also frequently delve into frank, explicit, or controversial content that would not be suitable for US broadcast radio. This means lots of curse words and graphic depictions of sex and violence. More consequentially, though, it means representations of people and perspectives that broadcast radio program directors would be apprehensive to put on the air. *Love + Radio* has created portraits of pimps, convicted murderers, phone sex operators, bank robbers, fetish models, and pedophiles. These are highly sensitive and controversial issues that, no matter how delicately they were discussed, would run a high risk of offending average public radio listeners, and also potentially running afoul of FCC regulations. Moreover, due to the ambiguity of the nonnarrated mode, these representations could be read as sympathetic to, if not supportive of, opinions and behaviors that even many liberal-minded listeners might find unethical or morally abhorrent.

This emphasis on subcultures and other diverse, marginalized perspectives in *Love + Radio* and the lion's share of other audio storytelling programs is a boon to classical progressive notions of liberal democracy, that is, the idea of a truly equal and morally pluralistic society where all individuals can pursue their interests fairly. Liberal democratic ideals such as these were, after all, central to the "participatory democratic philosophy" that guided National Public Radio's formation in the 1970s. And many of the public radio-trained producers behind today's audio storytelling programs still claim to hold these ideals dear, even if NPR as an institution has since become highly privatized and moved away from such social justice goals.[66] Nevertheless, this concentration on untraditional subjects is also not without its potential shortcomings. Too often, the "marginalized" voices in *Love + Radio* end up resembling a rogues' gallery of misfits and eccentrics. There is the avant-garde pianist turned boxing promoter turned gangster Charles Farrell in "Sesquipedalian." There is the crime scene cleanup specialist Tim Reifsteck in "Aftermath." There are the "rogue taxidermy" artists Nate Hill and Takeshi Yamada in "Animal Parts." There is the sex fetish "humiliatrix" Ceara Lynch in "Thank You, Princess." There is the illusionist and exotic animal performer Julia Christiie in "The Magical World of Eva Julia Christiie." There is the surrealist street performer Clyde Casey in "Another Planet." There is the balloon fetish model Amy

H. in "Dirty Balloons." Certainly, plenty of *Love + Radio* episodes focus on less ostentatious characters and premises, yet there is nevertheless a kitschy fascination with the weird, the outlandish, and the sensational throughout the podcast series' run.

A penchant for the quirky and the unusual is widespread throughout US nonfiction radio of the past two decades. *This American Life*'s stories typically revolve around a narrative surprise—a "twist" or a series of twists.[67] Meanwhile, *Radiolab*'s producers aim "to lead people to moments of wonder."[68] In storytelling, no matter the medium or genre, "surprise" is recognized as one of the core strategies for creating narrative pleasure.[69] Taking inspiration from these preeminent programs, a good many younger radio producers have replicated in their own work this storytelling technique of the pursuit of the unexpected. With *Love + Radio*, it is not that the surrealist street performer's or the balloon fetish model's life stories lack value, rather that the conventional wisdom that a "good story" needs narrative surprise leads radio producers to seek out a too-narrow group of subjects and narratives. As a result, the promise of audio storytelling to capture the full range of everyday life across all classes and cultures—and particularly the most disenfranchised voices of racial and ethnic minorities, immigrants, the underclass, and so on—is going at least partly unfulfilled.

In terms of inclusion and diversity, it is notable, too, that the majority of the subjects featured in *Love + Radio* are white or (seemingly average) middle-class people. Despite their lives taking unusual directions into the fringes of society, these are ultimately individuals with whom most podcast listeners—who on average are young (eighteen to thirty-four), white, affluent, and highly educated—are likely to identify.[70] The story of Tom Justice being a prototypical example (as told in the episode "Choir Boy"): a white, middle-class suburban Illinois high school class president, DePaul University graduate student, and potential US Olympic Team cyclist, Justice robbed dozens of banks simply for the thrill of it. As Justice himself shrewdly observes in the *Love + Radio* episode, he could have been the poster child for the all-American boy.[71] Jason Loviglio observes about *This American Life* and NPR on the whole that even though public radio in the 2000s–2010s still promotes an image of liberalism and utopian pluralism, the stories it tells reflect the audience it serves. The oft-cited "intimacy" of public radio (and public radio-inspired) programs, he argues, comes from the fact that the listeners, subjects, and producers form "a remarkably coherent and reflexive community."[72]

The emphasis in audio storytelling on individual experiences can also be seen as problematic. The neoliberal critique leveled against NPR over the last few decades is that if there is not a consumer demand for a type of programming or representation, then it will not exist, no matter its educational or social importance.[73] Loviglio, again, argues that this situation has not only resulted in programming that appeals to the interests of the service's predominantly affluent, educated, white, and English-speaking audience, it has also favored character-driven "slice of life" stories that privilege personal struggles over societal problems. These character-driven stories, says Loviglio, obscure the larger issues of politics, policy, and other deeply rooted institutional and historical factors, promoting the neoliberal moral that "the impetus for social change resides in the character of specific people, not in the intolerable social conditions that many people confront."[74] This line of critique could certainly be applied to *Love + Radio*. The 2014 episode "The Silver Dollar," for instance, features an African American musician, Daryl Davis, whose approach to overcoming white racism is to personally befriend Ku Klux Klan members. Davis claims former KKK imperial wizards and grand dragons among his close personal friends. Responding to African American people who accuse him of being an Uncle Tom, Davis says, "I pull out my robes and hoods and say look, this is what I've done to put a dent in racism. I've got robes and hoods hanging in my closet by people who've given up that belief because of my conversations of sitting down to dinner and they gave it up. How many robes and hoods have you collected? And then they shut up."[75] In fact, the episode ends on that statement. Although van der Kolk is heard asking Davis some questions during the radio piece, there is no host narration and no "really big statement" provided for the audience. Listeners could not be blamed for interpreting the lesson of Davis's thirty-five-minute-long life story as being that racism is an individual problem and not a societal problem, and that the best way to effect positive social change is through personal action rather than any political initiative or collaborative effort. This is a decidedly neoliberal message.

The "inwardness and historical amnesia of neoliberal subjectivity" is a legitimate risk of nonnarrated audio storytelling due to its stress on individual experience over sociohistorical context and cultural analysis.[76] Nick Couldry argues, however, that *voice*—the ability for people to give an account of the world and their place within it—is actually the best way to challenge neoliberal political power.[77] Couldry uses voice as a metaphor for agency and self-expression; he is not referring specifically (or at least not exclusively) to vocalized speech or sound media. Nevertheless, his concepts and the narrative strategies he outlines map precisely onto the audio storytelling practices described here. Critics of neoliberalism too often dispense altogether with the level of individual perspectives, either viewing affect as irrelevant to political economy or positioning experience only as determined or constrained by external discourses (false consciousness, essentially). For Couldry, though, the ability to subjectively give an account of oneself and one's experiences in the world (i.e., voice) is a basic feature of human life.[78] In fact, it is the one power that even the most marginalized and oppressed populations still possess.

Most people are already engaged in using their voices in this way, and thus simply calling for "more voices" in public, while important, is not enough. The real issue is valuing those voices and the conditions from which they emerge—that is, voices are embodied, and thus listening to a voice means recognizing speakers as well as the "landscape" in which they speak (i.e., the communities they represent, the time and place of the speech act).[79] That someone's life can become a *story*—a narrative that is heard by others and, in the process, recognized as having value—is one crucial way of showing that the person's experiences, and thereby the person's life, count. This is what audio storytelling is doing, especially non-narrated stories that make individual personal narratives the focal point. Podcast series like *Love + Radio* are creating spaces for (marginalized or at least stigmatized) people to speak on their own terms and tell us about their lives, and for us to listen to those self-narratives and give them meaning and purpose.

Audio storytelling is observing culture from the bottom up, generating the evidence of ordinary people on the ground that can be used to support a more expansive and fully open politics that can challenge neoliberalism. The trick is that we cannot simply take these voices or spaces as given. To paraphrase Couldry, spaces of voice are inherently spaces of power, yet in order to challenge the neoliberal status quo these spaces need to be connected to the political and economic areas of life.[80] Audio storytelling is helping a more diverse range of people to be made visible and recognized as having voice; however, real social and political change will come only if these new voices challenge the dominant norms and strategies. In the past few years, critiques have been raised from within the public radio industry itself about audio storytelling producers and subjects being overwhelmingly white, upper-middle class, and male.[81] Moreover, stories like Daryl Davis's, as honest and inspirational as they may be, ultimately do not call into question dominant racial ideologies such as colorblindness. Putting these personal narratives into a political context also means linking them together somehow to broader social movements. There is a certain incoherency to episodic series like *Love + Radio*, as well as *This American Life*, which switch up the subject and theme every episode. When the series as a whole is looked at together, the individual subjects' stories end up being just that: individual—isolated and alone. *Love + Radio*'s avoidance of didacticism is, on the one hand, its greatest strength in that it does not impose a reporter's lesson on the subject's life story. On the other hand, the series possesses no wider social or political message that unites the disparate stories together, outside of some vague sense of diversity or "the universality of individual experience"—which arguably distances the listener from an engagement with politics.[82]

In sum, podcasting has revitalized and strengthened audio storytelling as a tool for participatory democracy. In listening to the intricacies of other people's life stories, we are encouraged to develop empathy, and we may learn something about the complexities of society and of our own lives, too. At the same time, an overemphasis on sentimental, unselfconscious, or self-absorbed narratives risks mistaking self-expression for social action. Looking forward, there is a need for personal narratives that are more reflexive and critical, and particularly that focus on navigating uneven social structures. There is also a need for radio/podcast series or platforms that pull together specific communities of voices: people from one city or region; groups of workers within a single industry; members of a shared socioeconomic, racial, ethnic, gender, or sexuality group. Even in a nondidactic, nonnarrated series that maps individual experiences across an identifiable community, listeners are invited to hear the connections between the varied personal narratives and think more carefully about the everyday conditions from which the stories are emerging, rather than reflecting upon the stories as solitary or unusual.

Ear Hustle is one such model podcast series. Presented in the feature-documentary mode, it centers on the first-person accounts of prison life, as told entirely by inmates within a single institution, California's San Quentin State Prison. Prisoners are an underrepresented group in US society, and the approach here is to lend a sympathetic ear to a population that is normally maligned and ostracized. The Radiotopia-distributed podcast is cohosted by Earlonne Woods, a San Quentin prisoner, and Nigel Poor, an artist who volunteers at the prison. Other prisoners contribute to the podcast as reporters, musicians, editors, and sound designers. (Woods was actually released from prison in 2018; he had his sentence commuted by

California governor Jerry Brown—the governor citing Woods's contributions to the podcast as "a positive example.")[83] Compared to an investigative journalism podcast series like *In the Dark* or even *Serial*, *Ear Hustle* adopts a more "personal" style, focusing on individual stories and first-person perspectives. It is not explicitly "news"-y or political. No doubt, *Ear Hustle* is tightly constrained when it comes to making explicit critiques of the prison system: since it is produced within San Quentin, episodes must be approved by prison administrators before they can be released. Nevertheless, when the series is listened to episode after episode, many sociopolitical messages emerge: about race and gender relations in society; about class and the effects of unequal educational opportunities and structural unemployment; about the criminal justice system and the severity of "three-strikes" habitual offender laws.

Moreover, narrowly thematic podcast series—gathering, say, the lived experiences and disenfranchised voices of DACA immigrants, or migrant workers, or the inner-city poor—can best explore the systemic nature of racism, economic inequality, and other social problems by highlighting commonalities rather than isolated or exceptional cases. They could more easily facilitate online social interaction and community building that would bridge between the private selves of the individual radio storytellers and coordinated social actions in the public sphere. That is, the podcast could serve as a starting point for organized political action, or a public face for a political movement or civil rights organization. Lastly, if audio storytelling can be expanded through the workshop practices and teaching programs of digital storytelling, there is significant potential for it to further empower disenfranchised voices. More effort should be made to put audio storytelling skills in the hands of ordinary people (à la *StoryCorps* or *Radio Diaries*), not just professionals already operating within traditional radio institutions and systems.

Conclusion

This chapter has addressed the influx of audio storytelling, specifically in the area of nonnarrated podcasts, that emerged in the decade following podcasting's 2005 breakthrough. Many of the forms, practices, and institutions discussed here only fully took hold around 2013–14—and thus these processes are still unfolding and the history is still being lived. Even so, there are a number of historical trajectories in this audio storytelling field that we can map out and at least make predictions about for the near future.

Principal among them is the prevalence of remediation. Through podcast-ing, producers are greatly expanding the range of radio programming available to listeners, particularly in the area of feature-documentaries and nonnarrated stories. These are programs that are either too niche, too con-troversial, too avant-garde, or simply too lengthy or unusually structured to fit into the rather conservative and strictly formatted schedules of most US broadcast stations. Established radio producers, many of them frus-trated by the perceived lack of flexibility and innovation in the broadcast environment, are flocking to the relative openness of podcasting. This dis-course of creative freedom, for example, is persistent throughout a 2014 roundtable discussion with the Radiotopia members: "The podcast has been an incredible experience to rethink what we do and frame it differ-ently and play and do things that we couldn't do in the context of, say, *All Things Considered*," says *Radio Diaries*' Joe Richman.[84]

Meanwhile, podcasting has become an increasingly popular platform for media-makers as well as social activists from outside the radio industry. These outsiders venturing into podcasting include the investigative news-paper reporters behind true crime podcasts like *Dirty John*; the documen-tary filmmakers behind *Up and Vanished*; the fiction writers creating *Lore*; the musicians exploring their craft in *Song Exploder*. Seasoned radio makers and newcomers alike are all turning to old, preexisting radio forms, tech-niques, and styles to create their "new" programs. Podcasting is opening up much-needed spaces for these neglected radio forms and practices to exist. And, lo and behold, they are not just existing, they are prospering and finding exceptionally receptive and enthusiastic audiences.

While they owe a debt to historical radio, these contemporary podcast series are hardly retrograde or unoriginal. Remediation always involves a refashioning or reworking: today's audio storytellers are all bringing some-thing fresh to the older forms and practices, often through a hybridization of styles and/or the addition of a contemporary sensibility that reflects the present-day structure of feeling. That sensibility is often postmodernist in nature. *Love + Radio*'s inclusion of audio snippets that sound like outtakes or errors is a means of calling attention to the constructedness of radio. Many audio storytellers ironically incorporate of pop music. Autobiographical personal documentary podcasts like *StartUp* are self-reflexive and rely on metanarrative framing devices.

Another key component of Jay David Bolter and Richard Grusin's the-ory of remediation is that new media contain a twin logic of immediacy and hypermediacy.[85] On the one hand, new media strive to erase all traces of mediation (immediacy). On the other hand, new media always multiply themselves, and users are increasingly made aware of the medium itself (hypermediacy). In the case of podcasts (and internet radio, more generally), immediacy can be seen in the modern computer and smartphone interfaces making the listener's engagement with the radio device seem more transparent than an older stand-alone radio with its bulky box and dials and antenna. Interactive touchscreen mobile apps make listening more seamless, permitting listeners increasing control over what they listen to and when, where, and how they listen to it. This gives the experience a more personal and direct feel. Moreover, compared to analog technology or live broadcasting, digital editing software lets podcast editors erase recording imperfections in ways that, at least potentially, could conceal and deny the very process of production.

Despite these trends in favor of transparent immediacy, there is something of a fascination among contemporary podcast producers with hypermediacy and pointing to the signs of mediation. Numerous audio storytelling series fetishize telephone technology, podcasts like *Heavyweight* incorporating regular bits involving phone calls to staffers or friends. *Radiolab* presents its credits and sponsorship messages as telephone voicemail messages. Both *Radiolab* and *Love + Radio* audibly draw attention to conversations that are happening over phone or videochat—even seemingly enhancing the distortion in the mix, when they could reduce or remove it in postproduction. *Love + Radio* places the interviewer into nonnarrated stories even when their presence is unnecessary. Also, regardless of the industry's discursive push to separate podcasting from broadcasting, it is remarkable how many of the podcasts and podcast institutions place emphasis on "radio" in their logos and names: *Radiolab*, *Love + Radio*, *Radio Diaries*, Radiotopia, and so on. Also, for listeners, the close association of podcasting and headphone listening is a place where the mediating presence is actually enhanced rather than diminished. Thus, podcasting denies mediation in some key ways, while it multiples and stresses mediation in others.

Returning to the concept of sociability, nonnarrated podcasts like *Love + Radio* immerse its audience in the inner thoughts and minutiae of the everyday lives of individuals. Episodes are structured as deep dives into the realities of ordinary people's everyday existence. These programs record the thoughts of a diverse range of ordinary people—as so many blogs and chat rooms and social media feeds and other text-based digital media also do—while also capturing the texture and feel of that experience through the aurality of speech (tone, pitch, rhythm, volume, etc.).

This is one critical place where podcasting links up with broadcasting, and it is also where podcasting lays bare the contemporary structure of feeling: the embrace of long-form storytelling. Much has been said in the internet era about the supposed deterioration of attention spans, the tabloidization of news, the spread of a bite-size media "snack culture," and so on.[86] The basic thesis of these laments is that there has been a paradigm shift in our culture toward shorter and shorter snippets of media and information—and, especially in the realm of journalism and politics, this implies a move away from more serious, "objective" styles of social discourse to more formulaic, melodramatic narratives that are devoid of context or grounded cultural analysis. The result is an overall dumbing down of society, and more often than not, the internet and the always-on lifestyle it is said to engender are fingered as the culprit.

The popularity of audio storytelling, though, belies this notion that we are a vulgarized culture obsessed with soundbites and memes and hashtags. For the most part, this is lengthy programming that, due to its linearity, requires close attention from the listener. In print and digital journalism, beginning around 2010, there has been a similar (re)turn to extensive, deeply reported articles, in a movement alternately called *long-form journalism* or *slow journalism*. Spearheaded by websites and apps like Longreads, Narratively, and Longform, these are in-depth stories, or "deep dives," that take considerable time to research and write as well as read.[87] Contrary to the instant gratification and fast-paced "always-on lifestyle" of modern digital culture, this long-form journalism is pointing to a resurgence of attentive, "deep reading."[88]

Perhaps as a response to the speeding up of life in other areas of media and culture, more and more people are turning to podcasts that ask them to slow down and *listen*. Indeed, cultural critics have observed a growing cultural appreciation for *slowness* in the realm of serialized podcasts, especially.[89] Added to that, the mode of address in podcasting tends toward the casual and colloquial. This is content that demands close attention, and listeners are said to build up levels of dedication and trust with the podcast hosts whispering in their ears that are unrivaled in other media.[90] Joe Lambert has suggested that slowing down is a central feature of digital storytelling: it demonstrates a commitment to listening, to stopping and paying attention, to each other, and to the world.[91] Attentively listening to these podcasted voices and their intimate disclosures of everyday life—what Kate Lacey calls "listening out"—represents a significant form of social and political participation online.[92] For as much as podcasting is framed as an inward-focused, even antisocial activity—we tend to listen alone, on headphones, to only our self-selected slate of programs—there is also an aspect of audio storytelling that is encouraging a more contemplative, empathetic, highly sociable form of *listening out* to the world and its people, and to their feelings, concerns, and politics.

Conclusion

Radio: The Stealth Medium

The radio is *one-sided* when it should be two-sided. It is only a distribution apparatus, it merely dispenses. And now to say something positive, that is, to uncover the positive side of the radio with a suggestion for its re-functionalization: radio must be transformed from a distribution apparatus into a communications apparatus. The radio could be the finest possible communications apparatus in public life, a vast system of channels. That is, it could be so, if it understood how to receive as well as to transmit, how to let the listener speak as well as hear, how to bring him into a network instead of isolating him. Following this principle the radio should step out of the supply business and organize its listeners as suppliers. . . . Radio must make exchange possible.
 —Bertolt Brecht, 1932[1]

The German playwright and socialist activist Bertolt Brecht envisioned a two-way radio that would be a truly democratic, participatory communications medium. Writing in the early 1930s, Brecht was critiquing the one-to-many transmission model of broadcasting that had been established only a decade prior, and its centralized structure with a stark division between an expert-oriented production culture and a primarily domestic culture of reception. Like untold Marxist-inspired critics then and since, Brecht was particularly concerned about the social and political role of radio within the public sphere—and living in a nation where radio was controlled by the state, he was also expressly concerned about radio as a national public institution. To him, radio had yet to find its *"purpose in life"*; it was merely an "acoustic department store" that copied its form and content from existing cultural fields such as popular music, the theater, newspapers, and academic lectures.[2] Instead, Brecht conceived of a multidirectional radio that would be "a communications apparatus for the general benefit of the public," one in which "the audience is not only to *be instructed* but also must instruct."[3] He envisioned radio stations organizing discussions between businesses and their consumers, public debates about the economy, civic forums where citizens could question their local government representatives, as well as collectively created theatrical performances that could be both entertaining and educational. Brecht dreamed of what we today might call a crowdsourced radio, a medium that could promote radical social change by incorporating the audience in a meaningful way and, importantly, draw its content from across the scope of everyday life. His notion of "radio as a communications apparatus" parallels John Dewey's theory of "communication [as] participation in a common world"—the primary purpose of communication being to enable people to participate in a shared social space, in which there should be no limitations put on participation or restrictions placed on the topics of conversation.[4]

Brecht's two-way radio integrating everyone and everything was a utopic ideal. He did not specify how participation on the air should be organized for it to be truly equal or to prevent it from deteriorating into chaotic noise, nor did he establish how content could be produced without some consistent programming strategies in place and skilled cultural intermediaries employed to oversee them. (In fact, Brecht states that "radio's formal task is to . . . make the interests [of the public] interesting," suggesting that he had formal conventions and producer-facilitators in mind, not completely raw, self-produced works.)[5] Regardless, two-way radio *was* technologically possible. After all, radio in the United States had been a two-way medium prior to the institutionalization of broadcasting in the 1920s. Since that potential future for wireless radio was foreclosed, Brecht's utopian vision of a more participatory media culture has been echoed repeatedly over the decades by critics writing about different media in diverse places across the globe. Finally, in the 1990s, the internet seemed to answer the call—the World Wide Web system, in particular. Here was a communications medium that was not only structured as a truly democratic participatory public platform, but the medium itself was discursively positioned as an embodiment of democracy. The promotional rhetoric of the internet, writes Martin Spinelli, maintains that "the Internet is not just a corrective to democracy, but *is* democracy."[6]

Certainly, the internet makes a reality of "the vast system of channels" that Brecht imagined. Internet radio has very few of terrestrial radio broadcasting's technical restrictions on bandwidth, essentially no government regulation (outside music licensing), and comparatively low production and distribution costs.[7] At present, anyone in the United States with a high-speed internet connection or a smartphone has instant access to literally thousands of radio streams online, compared to at most a few dozen broadcast stations over the air. Moreover, many of the individuals and institutions profiled here in *Sound Streams* have adopted ideas similar to Brecht's, mobilizing discourses of radio as participatory and as facilitating of a more socially and politically empowered public. The Internet Multicasting Service was built around notions of open government and civic engagement. Pseudo's interactive ChatRadio concept sought to the make the online listening audience the focal point of its talk radio programming. Audioblogging and its direct descendant, podcasting, put radio directly in the hands of ordinary people, emphasizing an unscripted style of talk centered around personal experience and self-expression. Professional producers like those behind *The Brian Lehrer Show* have taken advantage of the affordances of the internet to expand audience participation in traditional talk radio formats. Even automated music radio platforms like Pandora have turned listeners into coproducers of sorts, as they shape their own listening experience according to their individual needs and desires. In nearly each and every case, internet radio innovators have strived to create at least what they personally perceived to be a more participatory form of broadcasting.

These discourses of democratization, participatory community, and the media-empowered citizen have, of course, often been false or at least compromised. Nevertheless, an overarching theme in *Sound Streams* has been to highlight the multitude of continuities that are present between existing "old" and emerging "new" media. Thus, it is worth restating that radio has gone through numerous convergences and transferences over the medium's century-long history, and variations on Brecht's utopian vision for radio occurred at each of these moments. Radio's convergence with the internet has hardly been the first time that a better possible world for radio—and society—was supposedly upon us. For instance, the expansion of AM stations and the arrival of FM in the late 1940s and early 1950s—paired with the national networks' transition to television—vastly expanded the listening options available to most Americans, as well as shifted much of the programming focus from syndicated national shows to local stations serving local needs. The further expansion of FM stations in the 1960s and 1970s was accompanied by a wave of experimental and countercultural programming, plus a boom in educational and noncommercial public and college radio, including the creation of the National Public Radio (NPR) network. Surges of optimism over more and more diverse radio programming always eventually lulled, however, as excessive advertising, consolidated ownership, rising production costs, shrinking audiences, and so forth led to bland, standardized programming and decreasing variety. Then satellite radio, or HD digital radio, or (ahem) internet radio would come along, cloaked in the promise of righting all the previous wrongs. I am not endorsing such a one-dimensional, cynical picture, merely recapitulating a widely circulated discourse. My point, in fact, is that there have historically been many different phases and *practices* of radio, each with interrelated yet different configurations of programming, different economic and industrial structures, different groups of producers and production cultures, different audiences and relationships with the listeners, and so forth. And these different phases and practices have not only overlapped, they have coexisted—even if one mode comes to dominate the popular narrative about what radio "is" at a given time and place.

It is not uncommon to read sweeping statements about the supposed "state" of radio today. Case in point: "Radio today is one of three things: talk show host and their listeners hollering at each other, preacher trying to get into your pocket, or computers playing music."[8] These may be prevailing trends at certain locations along the radio dial, but the fact of the matter is that the contemporary radio landscape in the United States is expansive. For every preprogrammed Top 40 hit music radio station or conservative talk radio station there is a community radio station channeling the eclectic spirit of 1970s freeform radio or a low-power (LPFM) station microbroadcasting to a local neighborhood about neighborhood politics. Listen widely and generously, and you will hear considerable variety on America's airwaves.

To return to Brecht's contention that radio should be transformed into a participatory "communications apparatus," it arguably *has* already been just that for a long while now. At least that is the case in a country like the United States where there is a prevailing emphasis on localism.[9] As I highlighted in the study of local talk radio in chapter 4, especially, broadcast radio has historically been more inclusive of the audience than any other mass communication medium in the United States. Surely, power and control of "the apparatus" are not as evenly distributed as progressive reformers like Brecht would want to see. However, the distinctive conversationality of radio talk and the widespread focus on quotidian activities and personal expression in radio's content are symbolic of how radio is the quintessential "people's medium."[10] Indeed, this connects to my concept of radio's sociability and how it produces a social space that structures our everyday lives.

Granted, in many locations across the country, the medley of radio stations within broadcast range is not as heterogenous as it could or should be. Rural regions are lucky to be serviced by one or two genuinely local radio stations: most of what they receive over the air consists of syndicated network programming and repeaters of stations from larger urban areas. Nevertheless, the comprehensive broadcast radio landscape in the United States is far more diverse than even most seasoned radio scholars tend to acknowledge. This is where the internet comes in, too: internet radio creates, as the radio industry market research firm Edison Research grandiosely describes it, "the infinite dial."[11] The internet expands the radio programming that the average listener receives by a thousandfold, making accessible a good many previously distant regional stations plus far-away and foreign broadcasters and scores more online-only radio stations, music streams, and podcasts.

Remediation has been a central concept threaded throughout *Sound Streams*, and accordingly internet radio is less a "revolution" than it is simply a new transmission platform for the already expansive options in the radio universe. The internet is a particularly suitable medium for alternative voices and production styles/formats that do not fit the contemporary radio industry's regulatory, economic, or aesthetic structures for one reason or another. But that does not mean these online radio productions are radically *new*. In fact, it is notable that many popular press and trade journalists analogized internet radio and podcasting to pirate radio when the practices were first emerging in the 1990s and early 2000s. The issue was not that the programming they offered was groundbreaking or unprecedented, but rather that internet technology was being used to create a space for these previously marginalized and/or illicit radio practices, both to flourish and to become legitimate. The internet helped level the playing field between radio's different practices and forms and its varied producers and industry sectors.

Internet radio entails a convergence of media technology alongside a convergence of radio industries and conventions. It erases many of the divisions (real or perceived) that have existed between commercial radio, public radio, pirate radio, amateur radio, and so on. At least to the audience, all of this programming appears more or less equally as "radio" online, especially on enormous web radio portals or podcatcher apps that inventory thousands of audio programs side by side. For example, go to the TuneIn site today and visit the "News" homepage: it presents multiple NPR, CBC, and BBC syndicated news programs and podcasts sitting alongside productions from podcast networks and digital media outlets such as Vox and Crooked Media; podcasts from legacy print media institutions including the *New York Times*, *The Economist*, the *Wall Street Journal*, the *New Yorker*, and the Associated Press; audiocasts from television news channels like CNN, MSNBC, and Fox News; a slew of local broadcast radio simulcasts; and many completely independent productions like Dan Carlin's *Common Sense* podcast.[12] There is a dizzying convergence of media platforms and logics at work here, yet they all present together in this one space as radio.

To a large extent, *Sound Streams* has been fueled by a desire to reclaim the term *radio*. I established at the outset of the book that one of the overarching goals was to construct a portrait of radio's convergence with the internet that strikes a balance between continuity and change. Rather than a manifestly altered or new medium, radio on the internet is more akin to a conglomeration—a convergence—of every type of radio that exists anywhere or that has ever existed. That is not to suggest that there is nothing at all original, innovative, or unique about internet radio and podcasting. Much of my study has emphasized the changes and new opportunities that they present—along with new challenges. Nevertheless, radio as a subject has always had to deal with a multiplicity of objects that makes constructing a coherent narrative difficult. While I am hesitant to make too many universal claims about radio's ontology or to postulate a false unity, I nevertheless believe it to be a worthy endeavor to treat radio as a coherent object in order to create a historically informed study of radio's politics and the continuously unfolding efforts to shape and control the medium.

Particularly if we look beyond the technology itself and instead focus on the culture of radio and the narrative strategies, or discourses, used by participants, it is quite clear that across the generations there is a continuity to the role that both producers and audiences expect radio to play in society. Whether it is the amateur wireless of the 1910s or Golden Age dramas or modern narrative nonfiction podcasts, radio has a special capacity for facilitating community and human connectedness and for bringing the listener a sense of participation in a community. This *sociability* transcends a particular technology or platform. What we call a media technology matters: new names tend to create insularity, limiting dialogue to a few industry insiders and "experts" who share common assumptions and specific norms of practice—commitment to a paradigm, basically. I fear that if we jettison "radio" for "podcast" or "audio" or whatever neophilic marketing term gets cooked up next, then we will lose the connection to the fundamental questions that have given the medium meaning. Renaming a medium also divides up the audience— "I am a *podcast* fan, not a *radio* fan"—which means dividing up the community the media support. This potentially weakens its potential for social change.

Michele Hilmes has called radio a "'stealth' medium," in that it is an incomparably flexible, portable, and adaptable mode of mass communication.[13] Broadcast radio signals are rather easy and inexpensive to produce as well as receive. They are also difficult to block and can transcend official boundaries.[14] This stealthiness transferred to the internet, where in the 1990s radio and radio-like audio features were in many cases the first audiovisual components added to basic text-and-graphics websites—in the process making the World Wide Web a fully multimedia platform and, in no small way, stimulating the development of the modern streaming media industry.

Hilmes also suggests that radio is a stealth medium in another sense, namely that radio as an object of academic study has historically been overshadowed in the media studies field by film, television, and visual media. Radio has been marked with an inferior cultural status. This "public 'forgetting'" and denigration can be linked to a number of factors, including the perception of radio as an older, defective technology in the wake of television's emergence; the medium's association with youth, racial and ethnic minorities, and popular culture (especially rock and roll and popular music); and the medium's "persistent ephemerality" and lack of materiality.[15] Likewise, Jonathan Sterne convincingly points out how sound has historically been placed in a subordinate position to the visual within cinema and media studies.[16] Since the late 1990s, though, sound has experienced a considerable increase in attention as an object of research across the humanities and social sciences. In media studies, sound research has extended from its origins in film sound and film music to both radio and popular music, so much so that "radio studies" and "popular music studies" are now thriving subfields.[17]

The revitalized interest in sound and sound media since the late 1990s— coalescing in the interdisciplinary field of *sound studies*—has perhaps not coincidentally corresponded with the rise of the web, cheap ubiquitous computing, and other forms of networked digital media (e.g., smartphones, satellite and HD radio, digital television). It is at this moment of digital media convergence—of essentially all old and new media circulating across multiple media platforms, especially the internet, and in the process losing their individual distinctiveness, at least at the purely technological level— that sound media like radio have shed their taken-for-granted status and begun to be the focus of serious scholarly attention again (some would say for the very first time). With this book, I have sought to merge the simultaneously emerging fields of sound studies and internet studies. I hope *Sound Streams* has amplified the heretofore understudied history of internet radio along with hidden areas in the broader history of the internet, while also drawing attention to the larger sociohistorical context of contemporary media convergence.

The Rise of Podcast Networks: PRX's Radiotopia

Sound Streams is a history, though of course our present and future are nothing if not the sum total of our past. I wish to conclude this book with some brief thoughts on the rapid professionalization of podcasting and the recent growth of podcast networks like PRX's Radiotopia, and what the emergence of these media institutions in the past few years—since around 2014—might mean for the future of radio innovation and artistry. In my analysis of *Welcome to Night Vale* (chapter 6), I discussed how podcasting is still discursively framed as an independent and even insurgent media practice, even as the traditional radio industry and other legacy media institutions have invested heavily in the podcasting field and the practice has become highly professionalized and increasingly consolidated. Indeed, *Night Vale* is something of an outlier today among popular podcasts for remaining a truly independent project (and yet, its creators have also begun to strike out into podcast network territory with their "Night Vale Presents" brand). Most of the ambitious feature-documentary programs mentioned in chapter 7 are attached to either a traditional radio station/ network or an upstart podcast network. *Radiolab* is produced and distributed by WNYC, the nation's largest public radio station—which in 2015 launched a quasi-podcast network, WNYC Studios.[18] *This American Life* and its *Serial* spinoff were produced by Chicago public broadcaster WBEZ (and *TAL* was long distributed by PRI), becoming an independent organization in 2015—albeit with distribution assistance from PRX.[19]

Following the broadcast network model, podcasters realized they could increase advertising and fund-raising, streamline distribution, cut production costs, and attract bigger audiences if they formed networks of interconnected podcasts. In addition to Radiotopia and WNYC Studios, notable podcast networks include Gimlet Media, the commercial venture of former *This American Life / Planet Money* producer Alex Blumberg and the subject of *StartUp* Season 1; Audible, Amazon's audiobook service that also produces a slate of original podcast series; Earwolf and Wolfpop (now merged together into the Earwolf brand), both owned by the podcast ad-sales company Midroll Media, which is a subsidiary of the newspaper and broadcasting conglomerate E.W. Scripps Company; the now-defunct Panoply, a subsidiary of the online magazine *Slate*, itself a subsidiary of the media conglomerate Graham Holdings; the 20th Century Fox–backed Wondery; the comedy-focused independent networks Maximum Fun and Nerdist; the tech culture network 5by5; the educational and documentary network HowStuffWorks; and Infinite Guest and SoundWorks, the podcasting arms of the public radio producer-distributors American Public Media (APM) and Public Radio International (PRI), respectively. The podcast network model received a considerable mark of validation when, in February 2019, the audio streaming platform Spotify acquired Gimlet for a reported $230 million—and then quickly followed that up with the purchase of true crime podcast network Parcast.[20]

The *Love + Radio*, *Fugitive Waves*, and *Radio Diaries* podcasts examined in chapter 7 are all members of PRX's Radiotopia podcast network. The Public Radio Exchange (PRX) is a Boston-area nonprofit public media company that was initially founded in 2002 as an internet-based digital distribution platform meant to open up the public radio marketplace to a more diverse pool of producers and content.[21] "Making public radio more public," its tagline reads. The basic premise is that independent producers can directly upload their radio pieces to PRX's online system, easing access for radio makers operating outside the major public radio production-distribution outlets (NPR, APM, PRI) and allowing them to control all the rights and licensing to their work. Meanwhile, local public radio stations can strengthen their programming by acquiring a wider range of content for their broadcasts, which especially benefits smaller stations that cannot afford to produce much original content of their own. Shortly after its inception, PRX also expanded into technology development and content production. In 2005, it created the "Pubcatcher" tool, which enabled public radio stations to manage and customize their podcast subscription feeds.[22] More recently, it built stand-alone mobile apps for multiple public radio programs, including *Radiolab* and *This American Life*, and with its "public benefit corporation" partner RadioPublic developed the podcast website platform Podsite. In terms of programming, PRX debuted *The Moth Radio Hour* in 2009, which today is syndicated for broadcast to hundreds of public radio stations and also podcasted.[23] It has subsequently produced and distributed additional podcasts, including *Snap Judgment, Reveal, Transistor, How to Be Amazing with Michael Ian Black* (a partnership with Audible), *HowSound,* and *Esquire Classic* (a partnership with *Esquire* magazine). Since 2008 it has also programmed PRX Remix, a satellite radio channel turned syndicated series turned mobile app. The "experimental radio stream" is curated by Roman Mars of the *99% Invisible* podcast and pulls together news, features, and interviews clips from various independent producers and podcasts.[24] Although it operates as an independent organization, PRX is deeply embedded in the public radio industry: it was cofounded by producer Jay Allison and the public radio trade group Station Resource Group with significant funding from the Corporation for Public Broadcasting (CPB), the National Endowment for the Arts (NEA), and other major private donors to NPR, including the MacArthur Foundation and the Ford Foundation.[25]

PRX launched the Radiotopia podcast network in February 2014 with an initial lineup of seven podcast series: *99% Invisible, Benjamen Walker's Theory of Everything, Strangers, The Truth, Fugitive Waves, Radio Diaries,* and *Love + Radio.*[26] By 2019, it had expanded its roster to some twenty podcasts, including popular hits *The Memory Palace, Ear Hustle, Song Exploder, ZigZag,* and *Criminal,* and the network claimed more than nineteen million monthly downloads of its programs.[27] Radiotopia was the brainchild of *99% Invisible* producer/host Roman Mars and PRX CEO Jake Shapiro, who together sought to develop a "collective of the best story-driven shows on the planet," emphasizing "digital-first audio programming."[28] These are programs that are either podcast-only or broadcast only in small regional markets and which have found much larger audiences online via podcast (such as Mars's own *99% Invisible,* which began in 2010 producing short segments for the local broadcasts of *Morning Edition* on San Francisco's KALW but quickly became a podcast phenomenon—its success confirmed by a 2012 Kickstarter campaign that earned $170,000, at the time the highest-earning journalism campaign in the site's history).[29] All of the programs in the Radiotopia lineup are "high-quality shows" fitting into the radio storytelling modes described in chapter 7. While they utilize "nontraditional approaches to telling stories," according to Shapiro, they are nevertheless "NPR-ready" in the sense that they adopt themes and aesthetics familiar to regular public radio listeners and they could potentially end up on broadcast radio, at least with a little editing and repackaging.[30]

Radiotopia stands apart from a number of the other start-up podcast networks, such as Gimlet, in that it uses a nonprofit model. Its podcast producers are "members," not staffers, who retain editorial control and financial ownership of their shows, as well as work out of their own facilities rather than a network studio. The PRX-backed Radiotopia staff provides these independent producers with marketing, advertising sales, and distribution support—not to mention brand recognition—in exchange for a share of any revenue their shows generate. Radiotopia's funding model closely resembles public radio's "three legged stool" approach: revenue comes from sponsorships (35 percent), donations (about 50 percent), and grants and philanthropy (about 15 percent).[31] The main difference being that sponsorship in public broadcasting consists of "underwriting," whereas podcasts do not need to adhere to the FCC rules and can run straight-up advertisements, in any quantity and in any style or placement. The podcast network's donations, too, typically come in the form of crowdfunding campaigns instead of pledge drives (Radiotopia, following Mars' 99% *Invisible* example, has experienced great success via Kickstarter)—although like public radio, Radiotopia has begun to push for sustaining memberships.[32]

Audiences would be hard pressed to find any audible differences between the supposedly noncommercial podcasts from Radiotopia (or NPR and its affiliates) and the podcasts from a for-profit network—to the point that they run ads from many of the same companies like MailChimp, Squarespace, and Audible. (That said, the FCC's broadcast regulations against advertising are "sieve-like," and underwriting on many noncommercial stations has gotten increasingly ad-like.)[33] Thus, aesthetically as well as institutionally, the lines between commercial and noncommercial as well as between radio and podcast content have gotten exceedingly blurry in recent years.

Mars, Shapiro, and their Radiotopia associates frequently compare the podcast network to an indie music record label.[34] In a promotional video for Radiotopia's highly successful 2014 Kickstarter crowdfunding campaign, for instance, Mars states, "It's kind of like an indie label for the best producers to make their best work."[35] Elsewhere, Alex Blumberg has spoken of his desire to make Gimlet the "HBO of podcasts," evoking a premium cable television model for podcasting.[36] In the Radiotopia video, Mars also uses film and television industry jargon, such as the practices of piloting and "greenlight[ing]" new shows. All of these analogies suggest that, far from disrupting the status quo, podcast networks are replicating well-established media industry institutional structures and economic models. Moreover, despite the allusions to music, film, and television, the primary functions of a podcast network are precisely those of a broadcast radio station/network: managing the business and finance aspects of radio production and distribution while reducing costs by clustering multiple similar programs together, leaving producers free to concentrate on the editorial and creative elements.

Podcasting's ongoing consolidation into formalized networks is a mixed blessing. On the one hand, it has fostered a burgeoning field of innovative audio content and original podcast-only or podcast-first programming. No longer is podcasting merely a repository for recycled over-the-air programming or second-rate extensions of existing media properties—which is primarily how the established radio industries and other professional media industries treated podcasting for the first decade of its existence. This environment is opening up opportunities for programs like *Love + Radio* that are too niche, too unconventional, and/or too explicit for broadcast radio. On the other hand, the oft-heralded amateur or DIY podcasts are being crowded out of the space. And disappearing with these independents are many of the utopian visions for podcasting as a democratizing force. There may not be any restrictions on creating a podcast and putting it online; however, it is becoming increasingly difficult to find an audience among the crowded podcast landscape. Of the hundreds of thousands of podcast titles on iTunes, 95 percent have fewer than two thousand listeners.[37] For a podcast to be "sustainable" (i.e. attract advertising that will cover production costs and potentially make the producers a profit), its producers need to release new content every two to three weeks and consistently receive around 100,000 downloads—and as podcasts get more popular, that threshold is increasing to 250,000 downloads.[38] Therefore, less than the top 1 percent of podcasters are actually currently making a living off of it.[39]

It is not my suggestion that all podcasters need or even want commercial success. After all, many amateur podcasters approach the practice as a hobby, and thus the simple pleasure of making radio and socially interacting with friends online is a satisfactory end in itself.[40] Moreover, throughout the past century, nearly all new mass media have followed a trajectory where they start open and full of revolutionary promise but then, as they gain social acceptance and economic viability, consolidate into increasingly closed systems. This is what Tim Wu terms "the Cycle."[41] The development of podcasting over the past fifteen years looks remarkably similar to the development of terrestrial radio in its first fifteen years of widespread use, during the 1910s and early 1920s: initially embraced by technologists and hobbyists who used it mainly for informal, interpersonal communication, it eventually turned into a media space for more widely consumed and entertainment-focused content controlled by professionalized (and commercialized) institutions. Such developments, it seems, are the natural outcomes of media growth, and we must not cynically treat them as inherently negative. Indeed, early adopters tend to be a small, relatively elite group. Broadening out may bring patterns of commercialization and professionalization, but it also garners larger, more heterogeneous audiences and a more diverse pool of producers—a potential boon for a more democratic mediated public sphere.

Podcast networks like Radiotopia have created previously unavailable opportunities for independent producers like *Love + Radio*'s Nick van der Kolk and his collaborators to make at least a modest living off their podcasts. (That *Love + Radio* was poached by the well-heeled upstart podcast subscription service Luminary in 2019 suggests that the producers are doing well for themselves financially.)[42] These are shows that could not find a permanent place on broadcast radio, and that would not be financially viable without the resources and support of a larger entity like Radiotopia. "The stereotype of podcasting is entirely true," says Lea Thau, producer of the former Radiotopia podcast *Strangers*, "which is that it takes up all your time and makes you no money."[43] Indeed, the *Love + Radio* story from chapter 7 is an almost archetypal example of a radio storytelling podcast during the past decade: started essentially as a creative side project for a young radio producer working a day job elsewhere in the radio industry (remember: van der Kolk worked for PRX, Chicago Public Media, and *Snap Judgment*), it received intermittent support from established organizations (alt.NPR, WBEZ) but never enough to make it a full-time project. At the time of its launch in 2005, NPR was only paying producers $100–$125 per podcast episode for alt.NPR contributions.[44] That is hardly a sustainable income, especially considering that audio storytelling pieces are among the most time- and labor-intensive types of radio to produce. "An unscripted story may sound less produced, but the opposite is often true. It takes a lot of work to make something that doesn't feel like it was produced," explains Joe Richman.[45] The average life-cycle of a *Radiolab* episode, as an example, is approximately six weeks—and it has a staff of a dozen or more.[46] Radio storytelling, as well as most radio drama, requires the kind of economic and administrative support only a more formally organized institution like a podcast network can provide, at least if a series is to survive over the long term.

Therein also lies a key problem, which is that this one corner of "creative audio" podcasting is dominated by already established public radio professionals. While networks like Radiotopia have helped producers such as van der Kolk, Thau, Jonathan Mitchell (*The Truth*), Benjamen Walker (*Theory of Everything*), Nate DiMeo (*The Memory Palace*), or Phoebe Judge and Lauren Sophrer (*Criminal*) turn their moonlighting side projects into full-time careers, these individuals were all already working within public radio. They merely made a lateral move into podcasting full time. Podcast networks, then, may be functioning as gatekeepers that are replicating existing radio industry hierarchies and keeping truly fresh faces out of the podcasting space. Since the production skills required to create the "high quality," "sound rich" storytelling programs that are increasingly in demand with audiences are rare and not widely taught, the search for new voices is unlikely to turn up people not already active in the cloistered world of public radio. Adam Ragusea has been critical of the lack of diversity in contemporary public radio, pointing out that there is an "internship barrier," in which the main way to gain employment in public radio is through unpaid internships, temporary jobs, and volunteer labor.[47] It is primarily upper-middle-class (and white) young people who are in a position to make such sacrifices. Few US universities teach the audio editing and postproduction skills needed for feature-documentary and drama soundwork, in particular; those skills mainly need to be learned on the job or be self-taught. However, few aspiring radio producers are in a position to endure the one- to four-year "wilderness period" of interning or underemployment it requires to get a foot in the proverbial door of the radio/podcast industry. Until these internship barriers and digital production gaps are more directly addressed at a structural level, the industry's inclusion problem will endure.[48]

It would seem internet radio, podcasting in particular, is approaching a crossroads: as it enters into mainstream popularity and profitability, its promise of media democratization risks being undermined. Spotify's 2019 acquisition of numerous podcast networks and technology platforms—Gimlet Media, Parcast, Anchor, and a promise to purchase more companies in the near future—confirms the streaming music giant is attempting to establish itself as the go-to source for on-demand audio of all varieties.[49] It is also bringing a vertical integration model to podcasting, in which Spotify controls the content from production through distribution and consumption. New podcast services like the subscription-based Luminary, which launched in April 2019, offer a distribution model that pulls podcasts off the open web and places them behind a paywall. Both moves signal that the platformization of podcasting is in full swing, and that the relatively open podcasting industry of today might soon be left behind for walled gardens.[50] Such developments are deeply concerning from a political economic perspective for, among other reasons, increasing the possibility of anticompetitive behaviors and decreasing inclusion and diversity through heightened barriers to entry. The question stands then: Will radio remain two-sided, or revert back to being one-sided again?

Methodological Notes on Interdisciplinarity and Developing a Convergent Methodology

Internet history and sound media research traverse multiple academic disciplines and fields of study, and this appendix elaborates on the interdisciplinary, multimethod approach utilized for *Sound Streams*. I used three main methodologies in the research and writing of this book: discourse analysis, ethnography, and textual analysis. Discourse analysis of archival and other primary sources analyzed the ideological and rhetorical frames within which technologists and media producers made sense of their actions. Participant observation of a converged newsroom and original interviews with producers illustrated the ways in which the work of making radio has changed as a result of media convergence. Textual analysis of soundworks demonstrated the ways in which the programming and form of radio shifted through its convergence with networked digital media culture.

My main methodology is *discourse analysis*, examining a wide range of archival and primary sources from the period 1993–2019. These included everything from popular press newspaper and magazine articles, trade press reporting, media interviews, and press releases to blog posts, message board discussions, SEC filings, and patent applications. Internet history, due partly to its recentness and its ephemerality, rarely benefits from the formal collections available in public libraries and archives, and the researcher must get creative with sources and techniques. I also conducted original interviews with individuals connected to a few case studies. These varied materials enabled me to evaluate the discourses of internet radio's innovators and early adopters: the computer engineers developing the technologies, the start-up entrepreneurs marketing them to the public, the established radio professionals implementing the internet into their existing routines, the media critics and audiences figuring out how the new media fit into their daily lives. I probed how these different individuals and communities envisioned what internet radio was (and is) *for*—what they considered to be internet radio's purposes and effects: how it should be used and by whom; how it fit into the contemporary media environment; what its defining characteristics were; how it was like or unlike broadcast radio and other media. Doing so allowed me to establish the rhetorical situation and context of these utterances: Whose interests did this discourse advance? What ideological positions and assumptions did it disclose? What discourses—and whose voices—were being suppressed or contained, or left completely silent?

The concept of *media convergence* that sits at the analytical center of this book also extended into my methodology. As James Hay and Nick Couldry suggest, media convergence must be understood not just at the level of the medium itself (e.g., content, technology, economy), nor through the surrounding media culture (e.g., fandom, production cultures), but also in terms of media research, theory, and methodology.[1] Convergence's intersection of media forms and industries has created an intersection of different disciplines in the arts and sciences. Media convergence thereby involves not just the merger and hybridization of the media themselves but also the intersection of perspectives and approaches to studying the media. In that spirit, I have sought to integrate approaches from multiple disciplines and fields of scholarship related to the study of media, technology, and culture, including media and cultural studies, communication studies, science and technology studies, cultural history, oral history, cultural anthropology, and the digital humanities.

Sound Streams will no doubt be most closely associated with *media studies* and the interdisciplinary field of *sound studies*. This book was written during a period of great expansion in radio studies and sound media, more broadly. Academic research on North American radio and sound recording history has swelled since the 2000s.[2] Beyond their core subject matter, these studies share a critical-cultural approach to media history that carefully considers ideological and cultural contexts. My approach in *Sound Streams* is especially indebted to the work of Michele Hilmes, who in *Radio Voices* studies the early twentieth-century emerging medium of radio as a complex "social practice grounded in culture," providing an analysis of discursive practices as they occurred across sites of production, texts, reception, and sociohistorical context, including the political landscape of the era.[3] Here and elsewhere, including in *Network Nations*, Hilmes pays particularly close attention to the discourses by which radio producers made sense of broadcasting and its place in the world, and how these discourses ideologically constructed the institutions they built and the programming they made.[4]

My research turns attention to technologists and the origins of the communication technologies underlying internet radio. Here I am influenced by the work of Jonathan Sterne, who takes a genealogical approach (in the tradition of Michel Foucault) to charting the emergence of sound media technologies and practices in disparate contexts over long periods of time. This approach underlines resemblances among otherwise diverse institutions and practices, plus the importance of analyzing how and why people actually developed and used these media technologies in the first place.[5] An overarching theme in *Sound Streams* has been to situate how the historical actors who pioneered internet radio understood what they were doing *in that moment*, especially how they conceived of their practices in relation to traditional radio broadcasting.

Particularly central to my methodology is Julie D'Acci's *interdisciplinary circuit of media study*, or "integrated approach" model.[6] This cultural studies approach to media study turns its attention broadly to the production of knowledge, meaning, and value within a society. It focuses on issues of consciousness (culture as politics and everyday life), identity and representation, difference and the cultures of neglected or underrepresented groups (culture as plural), and cultural form and texts. D'Acci's integrated approach emphasizes four sites for analysis: cultural artifact, production, reception, and sociohistorical context. For D'Acci each of these sites is "conjunctural," in the sense that they are interrelated. The economic, cultural, social, and subjective discourses can be mobilized at each moment on the circuit.[7] In other words, this approach calls for a convergence of disciplinary theories and methods.

Methods from *science and technology studies* (STS), in particular the *social shaping of technology* (SST) and *social construction of technology* (SCOT) perspectives, influence my thinking about what constitutes a "medium," plus spur my general interest in the social and cultural dimensions of technological development. In addition to Sterne, whose work bridges media studies and history of science and technology, the research of Susan Douglas and Brian Winston is especially instructive here. Both seek to challenge technologically deterministic conceptions of media "revolution," revealing instead how media technologies develop slowly and through complex social processes.[8] This approach highlights the roles of ideology and power, and how the common sense about a medium develops culturally rather than in a purely top-down fashion, which often happens when scholars focus singularly on ownership and industry economics. Winston's model of development and change in media technology, described back in chapter 3, was particularly useful in my research. A subtly recurring thread in *Sound Streams* was the numerous failures and flops: the many technologies and firms that came and went, and the role their discursive positioning played in their demise.

Social constructivist models have been influential in some of the more prominent histories of computers and the internet, such as Paul E. Ceruzzi's *A History of Modern Computing* and Janet Abbate's *Inventing the Internet*.[9] The approach's focus on how people and institutions (and neither technology nor society alone) jointly shape the development and incorporation of a technology helps trace the ways in which media become ordinary, domesticated elements of everyday life. This move helps explicate the ways symbolic and material meanings of technology develop culturally.

It is my hope that *Sound Streams* will make contributions to each of these fields and disciplines. Internet radio being a converged medium, this book is very much intended to be a work of *internet studies*, not only media studies or radio studies. I hope readers will appreciate insights provided here into the history of the early World Wide Web—to date a remarkably underexamined area of internet history. Internet radio is inherently multimedia, and, moreover, streaming media started with radio in the 1990s. Radio made the text-and-graphics-based web a fully multimedia platform, spurring its rapid growth during the dot-com era and establishing many of the institutions and conventions that streaming video (television and film) would later adopt in the 2000s. Thus, *Sound Stream's* history of internet radio reveals many heretofore unexplored aspects of the multimedia internet, especially the role of audio in the history of modern computing.

A final approach that informs my mixed-methods model is that of *new media studies*. As a field, cultural studies eschews disciplinarity. Nevertheless, one of cultural studies' major contributions to the study of *new media history*—and in particular, the history of *emergent media*—has been to place an emphasis on discourse and ideology, and the ways in which different social groups imagine and value media. As communication historians like James Carey and John Durham Peters have shown, the introduction of new media is routinely met with ambivalence, the media becoming objects onto which people project their hopes and anxieties about the changing nature of social life.[10] This means various groups struggle not just over who will control a new medium but also how it should be used (or not used), what values and meanings it should represent, and so on. It also means that new media are always accepted and understood in relation to existing media, as well as existing sociocultural dynamics. "New media" is itself a nebulous concept and there is no single "new media studies" approach; rather, new media studies is a conglomeration of many varied approaches derived from multiple disciplines.[11] Of particular interest to me are histories of emergent media and studies of the discursive construction of "newness" in new media, including work by Sterne, Lisa Gitelman, Thomas Streeter, and Jay David Bolter and Richard Grusin.[12]

A problem with more celebratory accounts of the internet and networked digital media culture is the radical break and fundamental newness that theorists claim for these "new media." This discourse often takes the form of negative comparisons made to older media; new media achieve their newness through direct comparison with, and usually the denigration of, older media. As Carolyn Marvin and other media historians have described, though, new technologies are always in tension with existing technologies. New media must be conceptually accepted and understood by a society before they can be implemented, and in practice this means new innovations are articulated through existing technologies, ideologies, and practices. "New practices do not so much flow directly from technologies that inspire them as they are improvised out of old practices that no longer work in new settings," writes Marvin.[13] Jay David Bolter and Richard Grusin challenge linear narratives of media history, arguing that new media always negotiate with other existing media in their formation, a process they call *remediation*.[14] I built on their theory of remediation throughout this book.

This perspective positions moments of media convergence as important sites of cultural struggle. The struggle occurs at the level of the technological and economic, as well as socially and culturally. Lisa Gitelman and Geoffrey Pingree argue that periods of media emergence and transition are always accompanied by uncertainty; they are key moments in which society negotiates the risk and potential of the new media. This process reveals much about the culture itself and the society's ideas about its own past, present, and future. Gitelman and Pingree highlight the concept of "framing," and how new media "frame our collective sense of time, place, and space," "our understanding of the public and the private," "our apprehension of 'the real,'" and our "relation to competing forms of representation."[15] Thus, moments of media change like radio's convergence with the internet are important sites to study, not just for understanding the status of the radio medium itself but also for understanding the role of radio within our contemporary culture and how radio reflects our view of ourselves.

Notes

Introduction

1. *Silicon Valley*, Season 2, Episode 3, "Bad Money," written and directed by Alec Berg, HBO, April 26, 2015, video.

2. My thinking on *practice* draws on the work of Pierre Bourdieu and Nick Couldry, the latter approaching media studies from a sociology of culture perspective. Studying the notion of practice in this way shifts attention from the economic, technological, or textual, instead guiding research toward discourse and the social functions of media in everyday life (i.e., what people do with or in relation to media and what people say about the media). Nick Couldry, *Listening beyond the Echoes* (Boulder, CO: Paradigm, 2006), 33–48.

3. Erving Goffman, *Forms of Talk* (Philadelphia: University of Pennsylvania Press, 1981), 171–72.

4. Zizi Papacharissi, *Affective Publics: Sentiment, Technology, and Politics* (New York: Oxford University Press, 2015).

5. Ed Madison and Ben DeJarnette, *Reimaging Journalism in a Post-truth World* (Santa Barbara, CA: Praeger, 2018), 82–83.

6. John Hartley, *Digital Futures for Cultural and Media Studies* (Malden, MA: Wiley-Blackwell, 2012), 59–93; Rosalind Coward, *Speaking Personally: The Rise of Subjective and Confessional Journalism* (Houndmills, UK: Palgrave Macmillan, 2013), 12.

7. Christian Fuchs, "Social Media and the Public Sphere," *tripleC* 12.1 (2014): 75.

8. Henry Jenkins, *Convergence Culture: Where Old and New Media Collide* (New York: NYU Press, 2006), 1–24.

9. Janet Staiger and Sabine Hake, "Preface," in *Convergence Media History*, ed. Janet Staiger and Sabine Hake (New York: Routledge, 2009), ix–xi.

10. Todd Gitlin, *Inside Prime Time*, paperback ed. (New York: Pantheon, 1985), 63–85.

11. James Hay and Nick Couldry, "Rethinking Convergence/Culture: An Introduction," *Cultural Studies* 25.4–5 (2011): 473.

12. Henry Jenkins, "Convergence? I Diverge," *MIT Technology Review*, June 2001, 93.

13. The theory of disruptive innovation has become received wisdom within the high-tech industry since its coinage in the mid-1990s, mistakenly deployed whenever technologies or markets change. Clayton M. Christensen, Michael E. Raynor, and Rory McDonald, "What Is Disruptive Innovation?," *Harvard Business Review*, December 2015.

14. Jay David Bolter and Richard Grusin, *Remediation: Understanding New Media* (Cambridge, MA: MIT Press, 1998), 15.

15. Ibid.

16. Ibid., 224.

17. Web 2.0 does not represent a specific technology or set of technologies; rather it describes changes in web software design and business models facilitating enhanced creativity, information sharing, and collaboration among users. Practices like "user-generated content," "peer production," and "collective intelligence" are fundamental aspects of Web 2.0. Tim O'Reilly, "What Is Web 2.0?: Design Patterns and Business Models for the Next Generation of Software," in *The Social Media Reader*, ed. Michael Mandiberg (New York: NYU Press, 2012), 32–52.

18. Mark Deuze, "Media Industries, Work and Life," *European Journal of Communication* 24.4 (2009): 467–80.

19. Susan J. Douglas, *Inventing American Broadcasting, 1899–1922* (Baltimore: Johns Hopkins University Press, 1987).

20. Michele Hilmes, *Radio Voices: American Broadcasting, 1922–1952* (Minneapolis: University of Minnesota Press, 1997), 34–74.

21. Kristen Haring, *Ham Radio's Technical Culture* (Cambridge, MA: MIT Press, 2006).

22. Tom McCourt and Eric W. Rothenbuhler, "Burnishing the Brand: Todd Storz and the Total Station Sound," *Radio Journal* 2.1 (2004): 3–14.

23. Michael B. Schiffer, *The Portable Radio in American Life* (Tucson: University of Arizona Press, 1991).

24. Noah Arceneaux, "CB Radio: Mobile Social Networking in the 1970s," in *The Mobile Media Reader*, ed. Noah Arceneaux and Anandam Kavoori (New York: Peter Lang, 2012), 55–68.

25. Christopher H. Sterling and Michael C. Keith, *Sounds of Change: A History of FM Broadcasting in America* (Chapel Hill: University of North Carolina Press, 2008).

26. Ibid., 136–39.

27. I am indebted here to Jonathan Sterne et al.'s notion of podcasting as a *practice* rather than a new *medium* or *format*: "The Politics of Podcasting," *Fibreculture Journal* 13 (2008), http://thirteen.fibreculturejournal.org/fcj-087-the-politics-of-podcasting/

28. Kate Lacey, "Ten Years of Radio Studies: The Very Idea," *Radio Journal* 6.1 (2008): 24.

29. Richard Berry, "Will the iPod Kill the Radio Star? Profiling Podcasting as Radio," *Convergence* 12.2 (2006): 144.

30. Andrew Crisell, *Understanding Radio* (New York: Methuen, 1986); Martin

Shingler and Cindy Wieringa, *On Air: Methods and Meanings of Radio* (New York: Oxford University Press, 1998).

31. Paddy Scannell, "What Is Radio For?," *Radio Journal* 7.1 (2009): 89–95; Chris Priestman, "Narrowcasting and the Dream of Radio's Great Global Conversation," *Radio Journal* 2.2 (2004): 77–88.

32. On the difference between *object* and *thing* in philosophy, see Ian Bogost, *Alien Phenomenology, or What It's Like to Be a Thing* (Minneapolis: University of Minnesota Press, 2012), 24.

33. Scannell, "What Is Radio For?," 89.

34. The texts on media phenomenology that I primarily drawing upon: Roger Silverstone, *Television and Everyday Life* (New York: Routledge, 1994); Paddy Scannell, *Radio, Television, and Modern Life: A Phenomenological Approach* (Cambridge, MA: Blackwell, 1996); Paddy Scannell, *Television and the Meaning of Live* (Malden, MA: Polity, 2014).

35. Jonathan Sterne, *The Audible Past: Cultural Origins of Sound Reproduction* (Durham, NC: Duke University Press, 2003), 223.

36. Ibid., 213.

37. Brian Winston, *Media Technology and Society: A History from the Telegraph to the Internet* (New York: Routledge, 1998).

38. Casey Man Kong Lum, "Notes toward an Intellectual History of Media Ecology," in *Perspectives on Culture, Technology, and Communication: The Media Ecology Tradition*, ed. Casey Man Kong Lum (New York: Hampton, 2006).

39. Alternately, this network of technologies, institutions, and practices could be described as the conjuncture of economic, cultural, social, and subjective discourses reflected in Julie D'Acci's "circuit of media study," or integrated approach, model of media studies. Juile D'Acci, "Cultural Studies, Television Studies, and the Crisis in the Humanities," in *Television after TV: Essays on a Medium in Transition*, ed. Lynn Spigel and Jan Olsson (Durham, NC: Duke University Press, 2004), 418–46.

40. Michele Hilmes, "The New Materiality of Radio: Sound on Screens," in *Radio's New Wave*, ed. Jason Loviglio and Michele Hilmes (New York: Routledge, 2013), 60.

41. Crisell, *Understanding Radio*, 56–57.

42. Echoing Hilmes, I am excluding music from this categorization, unless it is music composed explicitly for radio use (e.g., jingles, theme songs, original interstitial music) or music that is otherwise recontextualized for radio consumption.

43. In other words, it is not simply the audio of an event intended for another medium, such as a play performed for the theater. A stage play is designed to be heard *and* seen. A radio play, even of the same production, would differ considerably, since any visual content/context would need to be conveyed audibly.

44. *Sociability* is not my term, nor am I the first to identify it as the primary characteristic of radio. I am building here on a body of radio studies scholarship, including works by Paddy Scannell, cited throughout this introduction, and David Hendy: *Radio in the Global Age* (Malden, MA: Polity, 2000), 184.

45. Crisell, *Understanding Radio*, 11. The blindness-of-radio concept originates from a 1936 essay by Rudolf Arnheim: "In Praise of Blindness; Emancipation from

the Body," in *Radio*, trans. Margaret Ludwig and Herbert Read (London: Faber & Faber, 1936), 133–203.

46. I am discussing radio and television here together under the rubric of broadcasting. Though I am referring to radio, a majority of the scholarly literature on liveness refers to television. Nevertheless, television is an offspring of radio, and the live transmission model is shared by both media. Thus, I mostly use the theory on television liveness interchangeably for radio.

47. Paddy Scannell, "Editorial," *Media, Culture & Society* 23.6 (2001): 701.

48. Mary Ann Doane, "Information, Crisis, Catastrophe," in *Logics of Television: Essays in Cultural Criticism*, ed. Patricia Mellencamp (Bloomington: Indiana University Press, 1990), 238.

49. John Durham Peters, "Witnessing," *Media, Culture & Society* 23.6 (2001): 719.

50. Daniel Dayan and Elihu Katz, *Media Events: The Live Broadcasting of History* (Cambridge, MA: Harvard University Press, 1992), viii.

51. Scannell, *Radio*, 86.

52. John Ellis, *Visible Fictions: Cinema, Television, Video* (New York: Routledge, 1992), 132.

53. Asa Kroon Lundell, "The Design and Scripting of 'Unscripted' Talk: Liveness versus Control in a TV Broadcast Interview," *Media, Culture & Society* 31.2 (2009): 271.

54. Jane Feuer, "The Concept of Live Television: Ontology as Ideology," in *Regarding Television: Critical Approaches—an Anthology*, ed. E. Ann Kaplan (Frederick, MD: University Publications of America, 1983), 12–22.

55. Ibid., 14.

56. Ellis, *Visible Fictions*, 135.

57. Raymond Williams, *Television: Technology and Cultural Form*, Routledge Classics ed. (New York: Routledge, 2003), 90.

58. Kate Lacey, *Listening Publics: The Politics and Experience of Listening in the Media Age* (Malden, MA: Polity, 2013), 101.

59. Circa April 2019, there were an estimated seven hundred thousand podcast series and twenty-nine million episodes. Ross Winn, "2019 Podcast Stats & Facts (New Research from Apr 2019)," *Podcast Insights*, May 7, 2019, https://www.podcastinsights.com/podcast-statistics/

60. Samuel Greengard, *The Internet of Things* (Cambridge, MA: MIT Press, 2015), 89.

61. Norie Neumark, "Different Spaces, Different Times: Exploring Possibilities for Cross-Platform 'Radio,'" *Convergence* 12.2 (2006): 214–16.

62. Derek Kompare, "Flow," in *Keywords for Media Studies*, ed. Laurie Ouellette and Jonathan Gray (New York: NYU Press, 2017), 72–74.

63. Williams, *Television*, 95.

64. Scannell, *Radio*, 152.

65. Ibid., 172.

66. Paddy Scannell, "Review Essay: The Liveness of Broadcast Talk," *Journal of Communication* 59.4 (2009): E3.

67. Crisell, *Understanding Radio*, 3–18, 45–66.

68. Walter Ong calls this "secondary orality," since it is more deliberate and self-conscious than "primary orality": *Orality and Literacy: The Technologizing of the Word* (London: Methuen, 1982), 136.

69. Crisell, *Understanding Radio*, 56–67.

70. Nick Couldry, "Liveness, 'Reality,' and the Mediated Habitus from Television to the Mobile Phone," *Communication Review* 7.4 (2004): 360.

71. Karin van Es, *The Future of Live* (Malden, MA: Polity, 2017).

72. Lance Sieveking, *The Stuff of Radio* (London: Cassell, 1934).

73. Priestman, "Narrowcasting," 83.

74. Graeme Turner, "'Liveness' and 'Sharedness' Outside the Box," *FlowTV* 13.11 (2011): par. 6, http://flowtv.org/2011/04/liveness-and-sharedness-outside-the-box/

75. Scannell, *Radio*, 153.

76. James Fallows, "How to Save the News," *The Atlantic*, June 2010, http://www.theatlantic.com/magazine/archive/2010/06/how-to-save-the-news/308095/

Chapter 1

1. *Geek of the Week*, "Dr. Marshall T. Rose," hosted by Carl Malamud, Internet Multitasking Service, March 31, 1993, audio, http://town.hall.org/radio/Geek/033193_geek_ITR.html

2. Tim Berners-Lee, "Information Management: A Proposal," CERN, March 1989, http://www.w3.org/History/1989/proposal.html

3. John Schwartz, "With the Frontier Growing Up, Pioneer Malamud Closes His 'Net Shop," *Washington Post*, May 13, 1996, F19.

4. *Computer Chronicles*, "The Internet," Stewart Cheifet Productions, 1993, video, https://archive.org/details/episode_1134

5. Union Metrics, "On the Future of the Internet and Everything," accessed May 26, 2019, http://unionmetrics.tumblr.com/post/73313589203/on-the-future-of-the-internet-and-everything

6. CERN, "The Birth of the Web," accessed May 26, 2019, http://home.web.cern.ch/topics/birth-web

7. Steven Baker, "Multicasting for Sound and Video," *Unix Review* 12.2 (1994): 23.

8. Carl Malamud, interview by author, July 27, 2015.

9. John Markoff, "Turning the Desktop PC into a Talk Radio Medium," *New York Times*, March 4, 1993, D18.

10. IMS's radio program files were made available through a website hosted by the National Center for Supercomputing Applications (NCSA) at University of Illinois Urbana-Champaign, but this would not have been the primary way most users accessed the files in 1993. Kevin M. Savetz, "Plug In, Log On, Tune In: Internet Talk Radio," *Microtimes*, May 31, 1993, 181.

11. Michael Scott, "One File Reading: The Sound of the Internet," *Vancouver Sun*, October 16, 1993, D8.

12. Baker, "Multicasting," 23.

13. Stephen Casner and Stephen Deering, "First IETF Internet Audiocast," *ACM SIGCOMM Computer Communications Review* 22.3 (1992): 1.

14. Malamud, interview by author.

15. Elliott Karpilovsky, Lee Breslau, Alexandre Gerber, and Subhabrata Sen, "Multicast Redux: A First Look at Enterprise Multicast Traffic," *WREN'09: Proceedings of the ACM SIGCOMM*, August 21, 2009, 1, http://www.research.att.com/export/sites/att_labs/techdocs/TD_7QANV3.pdf

16. Malamud, interview by author.

17. David Bank, "Agent of Change," *San Jose Mercury News*, January 8, 1995, 4E.

18. Carl Malamud, "You Can Join in Public Radio's Reinvention, On-line," *Current*, October 18, 1993, 13.

19. Carl Malamud, "Internet Talk Radio: Flame of the Internet," slideshow presentation at IETF Meeting, March 1993, https://public.resource.org/archive/1993_03_01_itr.pdf

20. *Computer Chronicles*, video.

21. Malamud, "You," 13.

22. Ibid.

23. Markoff, "Turning," A1.

24. Ralph Engelman, *Public Radio and Television in American: A Political History* (Thousand Oaks, CA: Sage, 1996), 130–32.

25. Markoff, "Turning," A1.

26. Steve Behrens, "Flatow's Talk Show Marks a Digital First," *Current*, May 31, 1993, 11.

27. Eric S. Raymond, "The Cathedral and the Bazaar," *First Monday* 3.3 (1998), http://firstmonday.org/article/view/578/499

28. Malamud, IETF slideshow.

29. Carl Malamud, "Internet Town Hall," mission statement, February 1, 1993, https://public.resource.org/archive/1993_02_01_itr.pdf

30. Josh Hyatt, "Radio Waves to the Future," *Boston Globe*, January 23, 1994, 75.

31. IMS's early programming (1993–94) archive is located online: http://museum.media.org/radio/

32. "HarperAudio Joins Internet 'Radio Show,'" *ABA Newswire*, February 21, 1994, 6.

33. Internet Multicasting Service, "'Your Place in Cyberspace,'" press kit, December 2, 1994, 2, https://public.resource.org/archive/1994_12_02_itr.pdf

34. Carl Malamud, letter to Clayton Boyce, February 4, 1993, https://public.resource.org/archive/1993_02_04_npc.pdf

35. IMS press kit, 2.

36. Ibid., 3; Carl Malamud, *A World's Fair for the Global Village* (Cambridge, MA: MIT Press, 1997), 145, 190.

37. National Science Foundation Office of Legislative and Public Affairs, "The National Science Foundation's Global Schoolhouse Project," press release, April 26, 1993, https://public.resource.org/archive/1993_04_26_gsh.pdf; UN50 Committee, "UN Closing Ceremony Looks to Future," press release, June 25, 1995, https://public.resource.org/archive/1995_06_25_un.pdf

38. Peter H. Lewis, "Internet Radio Station Plans to Broadcast around the Clock," *New York Times*, September 19, 1994, A15; Wendy Grossman, "Broadcasting: Music in the Wires," *The Guardian* (UK), March 10, 1995, 5.

39. Malamud, interview by author.

40. Grossman, "Broadcasting," 5.

41. Behrens, "Flatow's Talk Show," 1.

42. Malamud, interview by author.

43. Bank, "Agent of Change," 4E.

44. IMS press kit, 3.

45. Bank, "Agent of Change," 1E.

46. Ibid., 4E.

47. Jared Sandberg, "Live from Las Vegas . . . It's Talk Radio, TV, Rock on the Internet," *Wall Street Journal*, May 4, 1994, B5.

48. Internet Multicasting Service, "Internet Multicasting to Provide Three Full Days of Cyberspace Broadcasts at Networld+Interop Event in Las Vegas," press release, May 3, 1994, https://public.resource.org/archive/1994_05_03_cyber.pdf

49. Lewis, "Internet Radio Station," 5.

50. Deb Roy and Carl Malamud, "Integration of a Large Text and Audio Corpus Using Speaker Identification," Association for the Advancement of Artificial Intelligence (AAAI), technical report SS-97-03 (1997): 1, http://www.aaai.org/Papers/Symposia/Spring/1997/SS-97-03/SS97-03-002.pdf

51. Malamud, interview by author.

52. Thomas A. Stewart and Joyce E. Davis, "The Netplex: It's a New Silicon Valley," *Fortune*, March 7, 1994, 98–102.

53. Janet Abbate, *Inventing the Internet* (Cambridge, MA: MIT Press, 1999), 113–25.

54. Stewart and Davis, "The Netplex," 98.

55. John Markoff, "Building the Electronic Superhighway," *New York Times*, January 24, 1993, A6.

56. Stewart and Davis, "The Netplex," 102.

57. Carl Malamud, "Internet Multicasting Service," slideshow presentation for Sun Microsystems, June 18, 1993, https://public.resource.org/archive/1993_06_18_ims.pdf

58. Thomas A. Stewart and Patty de Llosa, "Boom Time on the New Frontier," *Fortune*, September 27, 1993, 153–61.

59. Malamud, *A World's Fair*, 145–46.

60. Bank, "Agent of Change," 4E.

61. Andrew Ross, "Earth to Gore, Earth to Gore," in *Technoscience and Cyberculture*, ed. Stanley Aronowitz, Barbara Martinsons, and Michael Menser (New York: Routledge, 1996), 118–19.

62. Tim Jordan, *Hacking: Digital Media and Technological Determinism* (Malden, MA: Polity, 2008), 12.

63. Scott, "One File Reading," D8.

64. IMS press kit; Ann H. Greenberg, "Internet Access to Large Government Data Archives: The Direct Edgar Access System," proposal to National Science Foundation, July 7, 1993, 23, https://public.resource.org/archive/1993_07_07_edgar.pdf

65. Carl Malamud and Marshall T. Rose, "An Experiment in Remote Printing," *Interoperability Report* 7.9 (1993): 27–28; Carl Malamud and Marshall T. Rose,

"Principles of Operation for the TPC.INT Subdomain: General Principles and Policy," IETF Network Working Group RFC 1530, October 1993, https://tools.ietf.org/html/rfc1530

66. Peter H. Lewis, "Internet Users Get Access to S.E.C. Filings Fee-Free," *New York Times*, January 17, 1994, D2.

67. Mark H. Anderson and Jared Sandberg, "SEC Will Take Over a Project Offering Free Internet Access to Corporate Filings," *Wall Street Journal*, August 28, 1995, C15.

68. Malamud, interview by author.

69. Patrick L. Doddy, "Teen-ager at Campaign Helm: Trunkey's Top Man Only 17-Years Old," *Wheaton Daily Journal*, October 3, 1976, 1, https://public.resource.org/archive/1976_10_03.pdf

70. Carl Malamud (@carlmalamud), "my feeling on working with our government as a free-lance civil servant," Twitter, June 8, 2016, https://twitter.com/carlmalamud/status/740717816878407680

71. Indiana University, academic record of Carl Malamud, January 22, 1998, https://public.resource.org/archive/1983_12_31_iu.pdf; Nancy Scola, "Washington's I.T. Guy," *American Prospect*, July–August 2010, 23.

72. Carl Malamud, "The MBA Computer Literacy Project 1983: Understanding the Computer Potential at Indiana University," report for Indiana University Master of Business Administration, September 1, 1983, https://public.resource.org/archive/1983_09_01_literacy.pdf

73. Greenberg, "Internet Access," 23.

74. Prentice Hall, "What *Is* an Internet Explorer?," brochure, October 17, 1991, https://public.resource.org/archive/1991_10_17_new.pdf; Carl Malamud, "FYI, ANSI Petition Drive," memo, October 9, 1992, https://public.resource.org/archive/1992_10_09_ansi.pdf

75. Marshall McLuhan and Quentin Fiore, *The Medium Is the Massage: An Inventory of Effects* (New York: Bantam, 1967), 63.

76. Carl Malamud, "Lifting Every Voice," *St. Petersburg Times*, March 7, 1993, 1D.

77. IMS press kit, 3.

78. Malamud, ITH mission, 4.

79. Malamud, "Lifting," 6D.

80. Malamud, ITH mission, 4.

81. Malamud, interview by author.

82. "Science Friday, 1993: The Future of the Internet," *Science Friday*, November 26, 2010, audio, http://www.sciencefriday.com/segment/11/26/2010/science-friday-1993-the-future-of-the-internet.html

83. Hyatt, "Radio Waves," 75.

84. Malamud, "You," 14.

85. Ibid.

86. Malamud, ITH mission, 4.

87. IMS press kit, 6; Dave Briscoe, "Clinton Could Go Global on Cyberstation," *Toronto Star*, January 9, 1994, E9.

88. IMS press kit, 6.

89. Malamud, "Lifting," 6D.

90. Behrens, "Flatow's Talk Show," 13.

91. John Schwartz, "Superhighway Routed through Capitol Hill: Network Plans to Deliver Sound Bites as Bytes," *Washington Post*, September 19, 1994, A3.

92. IMS press kit, 7

93. Markoff, "Turning," D18.

94. *TechNation*, "Internet Radio and 'Wired' Magazine," hosted by Moira Gunn, March 31, 1993, audio, http://town.hall.org/radio/TechNation/033193_tech_ITR.html

95. Tony Waltham, "Weekly Radio Show to Be New Service on Internet," *Post Database*, January 6, 1993, 3.

96. Markoff, "Turning," D18.

97. Schwartz, "With," F19.

98. Malamud, interview by author.

99. Jared Sandberg, "On-Line Internet Expo Will Promote Cyber Space to the Whole Wired World," *Wall Street Journal*, March 14, 1995, B6; John Schwartz, "A World's Fair for Modem Times," *Washington Post*, March 15, 1995, B1; John Markoff, "Coming Soon to Computer's Everywhere, a World's Fair," *New York Times*, December 25, 1995, A56.

100. David Morris, "Robin Hood of Web 'Steals' from Data-Rich, Gives to Public," *Saint Paul Pioneer Press*, June 30, 1998, 7A.

101. Julie K. L. Dam, "Welcome, Cybernauts!," *Time*, March 11, 1996, 53.

102. Sandberg, "On-Line," B6.

103. Carl Malamud, "Internet 1996 World Exposition," speech, Singapore, May 14, 1995, https://public.resource.org/archive/1995_05_14_singapore.pdf

104. Malamud, *A World's Fair*, 177–78.

105. Jared Sandberg, "Wild Horses Couldn't Drag Mick's On-Line Fans from This Concert," *Wall Street Journal*, November 18, 1994, B1.

106. Eric Boehlert, "Radio Biz Finds New Way to Network," *Billboard*, March 26, 1994, 119.

107. Mark Hudis, "Radio: Waves of the Future," *Mediaweek*, January 27, 1997, 42.

108. Nicholas Garnham, *Emancipation, the Media, and Modernity: Arguments about the Media and Social Theory* (New York: Oxford University Press, 2000), 75–78.

109. "WXYC'S Simulcast," WXYC, November 12, 2004, http://www.ibiblio.org/wxyc/about/first/index.shtml

110. Michael Dekker, "KJHK Broadcasts Go Live on the Internet," *Lawrence Journal-World*, December 13, 1994, 1.

111. Jack Schofield, "Netwatch," *The Guardian* (UK), December 15, 1994, 5; David Beran and Jennie Ruggles, "This Is Radio Web," *Gavin*, September 13, 1996; Donna Petrozzello, "Online Services: Radio on the Internet," *Broadcasting & Cable*, January 23, 1995, 158–59.

112. Tim Ross, "WWW.WXYC.ORG," *In/Audible*, Fall 2004, 20–23, 34.

113. Anthony Nguyen, email message to author, July 6, 2015.

114. Gary Hawke, email message to author, July 25, 2015.

115. Paul Jones, interview by author, August 11, 2015.

116. Robert K. Merton, *The Sociology of Science: Theoretical and Empirical Investigations* (Chicago: University of Chicago Press, 1973), 371.

117. SunSITE's online archive: http://www.ibiblio.org/sunsite/sunsiteworld.html

118. Michael Shoffner, interview by author, August 9, 2015.

119. Ross, "WWW.WXYC.ORG," 21.

120. Ibid.

121. John Selbie, interview by author, July 25, 2015.

122. Ibid.

123. "KJHK on the Internet Signals Radio's Future," *Kansas Alumni Magazine*, February–March 1995, 12.

124. Robert Burcham, email message to author, July 5, 2018.

125. Jordan, *Hacking*, 6–7.

126. Jones, interview by author.

127. Selbie, interview by author.

128. Ibid.

129. Tim Dorcey, "CU-SeeMe Desktop Video Conferencing Software," *Connexions* 9.3 (1995), http://ftp.icm.edu.pl/packages/cu-seeme/html/DorceyConnexions.html

130. Mark Hodges, "Videoconferencing for the Rest of Us," *Technology Review*, February–March 1996, 17–19.

131. Mark Gibbs, "Foraging in the Team Toolbox," *Network World*, January 15, 1996, SS7; Selbie, interview by author.

132. Ross, "WWW.WXYC.ORG," 21.

133. *The State of Things*, "WXYC/Foreign Exchange," hosted by Frank Stasio, WUNC, November 24, 2004, audio, http://www.ibiblio.org/wunc_archives/sot/?p=108

134. Ross, "WWW.WXYC.ORG," 22.

135. Michael Shoffner, "Golden Days," WXYC, November 2, 1994, http://www.ibiblio.org/wxyc/about/first/golden.html

136. Jones, interview by author.

137. Steven McClung, "College Radio Station Web Sites: Perceptions of Value and Use," *Journalism & Mass Communication Educator*, Spring 2001, 63.

138. Petrozzello, "Online Services," 159.

139. Shoffner, interview by author.

140. Selbie, interview by author.

141. Shoffner, interview by author.

142. While the SunSITE-WXYC team had a particular outcome in mind, the technological development was still driven more by curiosity than any entrepreneurial desire to make a profit or build a marketable product. David Bell, *Science, Technology and Culture* (New York: Open University Press, 2006), 33.

143. Jones, interview by author.

144. *The State of Things*, "WXYC/Foreign Exchange."

145. Selbie, interview by author.

146. Ross, "WWW.WXYC.ORG," 23; Jones, interview by author.

147. Selbie, interview by author.

148. "KJHK on the Internet," 13.

149. Selbie, interview by author.

150. Ross, "WWW.WXYC.ORG," 23.

151. "KJHK On the Internet," 12.

152. John Burgess, "Internet Creates a Computer Culture of Remote Intimacy," *Washington Post*, June 28, 1993, A1.

153. Stewart and de Llosa, "Boom Time," 153–61.

154. Brad Templeton, "Re: Press Release: WXYC Broadcasting on Internet (fwd)," email message, December 10, 1994, http://www.ibiblio.org/wxyc/about/first/templeton.html

155. Chris Campbell, "10 Years of Streaming," email message, November 7, 2004.

156. Robert Burcham, "Protocols for Internet Radio," *Dr. Dobb's: The World of Software Development*, July 1, 1995, par. 4, http://www.drdobbs.com/protocols-for-internet-radio/184409771

157. Shoffner, interview by author.

158. Jordan, *Hacking*, 5.

159. Burcham, "Protocols."

160. John Selbie, "Cyber Radio 1—Internet Radio Broadcast Software," Usenet Non-Commercial Radio newsgroup message, October 13, 1995, https://groups.google.com/forum/#!search/wrek$20selbie/rec.radio.noncomm/SwIQ0BvPC_0/Q2E6ykK11awJ

161. "We Got Here First," WREK, October 12, 2004, https://web.archive.org/web/20041012191117/http://www.wrek.org/wreknet-first.html; Selbie, interview by author.

162. Scott Donaton, "TV Will Be 'Electronic Highway' for Time," *Advertising Age*, February 1, 1993, 1.

163. Peter W. Mitchell, "Digital on the Air," *Stereo Review*, November 1993, 103.

164. Times Mirror Center for The People & The Press, "Technology in the American Household: Americans Going Online . . . Explosive Growth, Uncertain Destinations," October 16, 1995, 25, http://www.people-press.org/1995/10/16/americans-going-online-explosive-growth-uncertain-destinations/

165. Susannah Fox and Lee Rainie, "The Web at 25 in the U.S.: Part 1," Pew Research Center, February 27, 2014, http://www.pewinternet.org/2014/02/27/part-1-how-the-internet-has-woven-itself-into-american-life/

166. Laura Santhanam, Amy Mitchell, and Kenny Olmstead, "The State of the News Media 2013: Audio by the Numbers," Pew Research Center, March 19, 2013.

167. Hudis, "Radio," 42.

168. Andy Marx, "Measure Streaming? Keep On Dreaming," *Internet World*, November 1, 1999, 43.

169. David Bank, "Cyberstation Goes on the Air in Show-Biz Fashion," *San Jose Mercury News*, May 4, 1994, 12D.

170. Ibid.

171. Bank, "Agent," 1E.

172. Fritz Nelson, "That Was Now, This Is Then," *Network Computing*, December 16, 2003, 20.

Chapter 2

1. Geoff Orens, "Cuban, Mark," in *2001 Current Biography Yearbook*, ed. Clifford Thompson (New York: H.W. Wilson, 2001), 99.

2. Kara Swisher and Evan Ramstad, "Yahoo! Plans to Buy Internet Broadcaster," *Wall Street Journal*, April 1, 1999, A3; Saul Hansell, "Obsessively Independent, Yahoo Is the Web's Switzerland," *New York Times*, August 23, 1999, C1.

3. Associated Press, "Perusing Yahoo's Most Expensive Acquisitions," *AP English Worldstream*, May 20, 2013.

4. Rebecca Buckman, "Broadcast.com Soars 3-½ Times from IPO Price," *Wall Street Journal*, July 20, 1998, B6.

5. David Barboza, "Broadcast.com Soars in Opening Day Frenzy," *New York Times*, July 1998, 18, D1.

6. Broadcast.com, "Form 424B4," SEC file no. 333-5287, filed July 17, 1998.

7. Yahoo! Inc., "Form 8-K," SEC file no. 000-28018, filed July 20, 1999.

8. Broadcast.com, *1998 Annual Report*, SEC file no. 000-24591, filed April 27, 1999.

9. Richard Tedesco, "Video Streaming: The Not Ready for Prime Time Medium," *Broadcasting & Cable*, May 25, 1998, 22.

10. Broadcast.com, "Form 424B4."

11. Clea Simon, "The Web Catches and Reshapes Radio," *New York Times*, January 16, 2000, AR15.

12. Katy Bachman, "Cozying Up to the Web," *Mediaweek*, February 21, 2000, 5.

13. RealNetworks Inc., *2000 Annual Report*, SEC file no. 000-23137, filed April 2, 2001.

14. Marcus Wohlsen, "1995, When CD-ROMs and Microsoft Ruled," *Wired*, October 25, 2012.

15. *Internet History Podcast*, "Episode 30: Chapter 5, Supplemental 4—Real Networks CEO Rob Glaser," interview by Brian McCullough, September 8, 2014, audio, http://www.internethistorypodcast.com/2014/09/real-networks-ceo-rob-glaser/

16. "About the Trustee: Overview," Glaser Progress Foundation, accessed May 26, 2019, http://www.glaserprogress.org/overview/about_trustee.asp

17. Robert H. Reid, *Architects of the Web: 1,000 Days That Built the Future of Business* (New York: Wiley, 1999), 74.

18. Ibid., 77.

19. Ibid.

20. *Internet History Podcast*, "Episode 30."

21. Rob Walker, "Between Rock and a Hard Drive," *New York Times Magazine*, April 23, 2000, 76.

22. Reid, *Architects of the Web*, 77; Walker, "Between Rock," 76.

23. Jeff Leeds, "Air America, Home of Liberal Talk, Files for Bankruptcy Protection," *New York Times*, October 14, 2006, B7.

24. Progressive Networks, "RealAudio Press Conference," NAB Show, Las Vegas, NV, April 10, 1995, audio, https://web.archive.org/web/19980705015657/http://www6.real.com/corporate/pressroom/pr/launchconf/index.html

25. Reid, *Architects of the Web*, 79–80.

26. Progressive Networks, "RealAudio Product Announcement," press release, April 10, 1995, http://web.archive.org/web/19980215091103/http://www8.real.com/corporate/pressroom/pr/prodannounce.html; Michael Krantz, "Groovin' to the Internet," *Mediaweek*, April 10, 1995, 8; Evelyn Tan Powers, "Radio Days Now on the Internet," *USA Today*, April 10, 1995, 1D.

27. Progressive Networks, "Progressive Networks Introduces Version 2.0 of the RealAudio System," press release, October 30, 1995, http://web.archive.org/web/19980215090941/http://www8.real.com/corporate/pressroom/pr/2.0release.html

28. Jan Ozer, "Sound Blasts the Web," *PC Magazine*, March 26, 1996, 103–4.

29. Progressive Networks, "RealAudio Product Announcement"; Progressive Networks, "Progressive Networks Ships RealAudio System," press release, July 25, 1995.

30. Jan Ozer, "Get Real about Internet Sound with RealAudio Version 2.0," *PC Magazine*, August 1996, 51.

31. Lee Gomes, "The Internet Gets a Voice," *San Jose Mercury News*, September 25, 1995, 1E.

32. Thomas A. Powell, "A Medley of Products Sound Out the 'Net," *Communications Week*, August 19, 1996, IA2.

33. David Pogue, "Taking Their Lumps of Coal," *New York Times*, December 25, 2003, G1.

34. Progressive Networks, "Progressive Networks Launches Timecast," press release, April 29, 1996, http://web.archive.org/web/19980215090749/http://www8.real.com/corporate/pressroom/pr/timecast.html; Progressive Networks, "MCI and Progressive Networks Launch First Internet Broadcast Network Designed to Reach Large-Scale Audiences," press release, August 5, 1997, http://web.archive.org/web/19980215090411/http://www8.real.com/corporate/pressroom/pr/rn/index.html; RealNetworks, Inc., *1999 Annual Report*, SEC file no. 000-23137, filed March 30, 2000.

35. Ozer, "Sound," 113, 116.

36. Azeem Azhar, "Battlelines Drawn in Internet Radio Ratings War," *The Guardian* (UK), October 26, 1995, 4; Jan Ozer, "Streaming Media: Getting Back in the Game," *PC Magazine*, September 22, 1998, 50.

37. Joshua Quittner, "Radio Free Cyberspace," *Time*, May 1, 1995, 91; Azhar, "Battlelines Drawn," 4.

38. Michael Cunningham, "Don't Touch That Dial!," *Irish Times*, May 22, 1995, 8.

39. M. Sharon Baker, "Progressive Networks Plugs Sound Software into Internet," *Puget Sound Business Journal*, May 12, 1995, 6; Brett Atwood, "Higher-Quality RealAudio 3.0 Debuts," *Billboard*, September 28, 1996, 84; Richard Tedesco, "Online Services: Progressive Debuts RealAudio 3.0," *Broadcasting & Cable*, September 23, 1996, 52–54.

40. Progressive Networks, "Version 2.0."

41. Ozer, "Get Real," 51.

42. Progressive Networks, "Progressive Networks Announces 'Live Rea-

lAudio' System," press release, August 30, 1995, http://web.archive.org/web/19980215091014/http://www8.real.com/corporate/pressroom/pr/livemariners.html

43. Donna Petrozzello, "Radio on the Internet: ABC Radio Enters WWW," *Broadcasting & Cable*, August 21, 1995, 38; Progressive Networks, "ABC RadioNet First to Fully Integrate Live RealAudio," press release, September 7, 1995, http://web.archive.org/web/19980215091007/http://www8.real.com/corporate/pressroom/pr/radionet.html

44. Harry A. Jessell and Paige Albiniak, "Progressive Networks' Glaser: First Audio, Now Video," *Broadcasting & Cable*, February 10, 1997, 26–28; Joshua Quittner, "They've Gotta Have It," *Time*, February 24, 1997, 60.

45. Progressive Networks, "MCI."

46. Variations of this basic Broadcast.com origin story were repeated in numerous profiles of Cuban and Wagner. See Richard Murphy, "Making a Killing on the Internet," *Success: The Leading Magazine for Entrepreneurs*, May 1999, 54–59; Orens, "Cuban, Mark," 97–100; Elizabeth A. Rathbun, "Yahoo! Broadcast's Mark Cuban: Crusader for Convergence," *Broadcasting & Cable*, October 18, 1999, 57; Devin Leonard, "Mark Cuban May Be a Billionaire, but What He Really Needs Is Respect," *Fortune*, October 15, 2007, 172–82.

47. T. C. Doyle, "The Cuban Story: From Neighborhood VAR to Internet Czar," *VARbusiness*, October 25, 1999, 54.

48. Jaeb's involvement in AudioNet is largely undocumented, though the acquisition of Cameron Audio Networks (aka Cameron Broadcasting Systems) is outlined in Broadcast.com's 1998 SEC prospectus ("Form 424B4"). Jaeb's version of the story is told in a 2012 podcast interview: *Mixergy*, "The Unknown Founder Who Got 10% of Broadcast.com—with Chris Jaeb," interview by Andrew Warner, September 28, 2012, audio, http://mixergy.com/interviews/chris-jaeb-broadcast-interview/

49. Progressive Networks, "Progressive Networks Ships."

50. Evan Ramstad, "AudioNet Scores by Tapping a Mundane Medium," *Wall Street Journal*, February 26, 1998, B8.

51. AudioNet's press release incorrectly places the event on Friday, September 1, 1995; the game actually took place on Saturday, September 2, 1995. AudioNet, "First FullTime SimulNetCast Radio Station on the Net," press release, September 1, 1995, http://blogmaverick.com/2015/01/16/the-worlds-first-streaming-radio-station-and-first-live-sporting-event-on-the-net/

52. Murphy, "Making a Killing," 56.

53. Jaeb quoted in *Mixergy*, "Unknown Founder."

54. AudioNet, "First"; Brett Atwood, "KPIG First 24-Hour Online Radio," *Billboard*, September 2, 1995, 101.

55. Chuck Taylor, "Radio Biz Poised to Gain from the Online Explosion," *Billboard*, May 8, 1999, 85; Yahoo! Inc., "Form 8-K."

56. David Raths, "Portland Businesses, Radio Stations Join Move to Web Audio," *Portland Business Journal*, June 7, 1996, 9.

57. Mark Cuban, "Roadshow," slideshow presentation, uploaded September 8, 2009, https://www.slideshare.net/mcuban/roadshow-1968443

58. Ramstad, "AudioNet Scores," B8.

59. Michael Wilke, "Online Radio Startups Music to Net Users' Ears," *Advertising Age*, March 25, 1996, 40; Andrea Adelson, "Remaking Radio as Internet Voice," *New York Times*, September 23, 1996, D7; Mark Hyman, "Do You Love the Orioles and Live in L.A.? Play-by-Play Broadcasts of Faraway Games Are All Over the Web," *Business Week*, May 12, 1997, 108.

60. Broadcast.com, "Form 424B4"; Broadcast.com, *1998 Annual Report*.

61. Richard Tedesco, "Internet Ratings: Sony Soars, Heaven's Gate Flashes in Web Ratings," *Broadcasting & Cable*, June 9, 1997, 54.

62. Dan Routman, "Super Bowl XXXI to Feature Live Webcasts," *PR Newswire*, press release, January 20, 1997; Richard Tedesco, "Super Bowl Website Preps for Game Day," *Broadcasting & Cable*, January 20, 1997, 55.

63. Alan Goldstein, "CEO of Dallas-Based Broadcast.com Reflects on Future after IPO," *Dallas Morning News*, August 31, 1998.

64. Jeff Bounds, "Fiddling for Wall Street's Signal: Web-Radio Pioneer Audio-Net Hears IPO in the Distance," *Dallas Business Journal*, February 23, 1997, 1.

65. Murphy, "Making a Killing," 56.

66. Sandy Chen, Arial Friedman, Darren Landy, Mark Stencik, and Joey Shammah, "Case 1: Broadcast.com," in Allan Afuah and Christopher L. Tucci, *Internet Business Models and Strategies: Text and Cases* (New York: McGraw Hill Irwin, 2001), 187.

67. Nancy Lawson, "Finding a New Web Frequency," *Baltimore Business Journal*, October 7, 1996.

68. Mark Cuban, "Is This a Proposal from 1995 or 2015?," *Blog Maverick*, January 16, 2015, http://blogmaverick.com/2015/01/16/is-this-a-proposal-from-1995-or-2015/

69. Richard Tedesco, "Web Will Touch All Bases: Major League Baseball in Power Plan Online," *Broadcasting & Cable*, May 11, 1998, 72–73.

70. R. Lee Sullivan, "Radio Free Internet," *Forbes*, April 22, 1996, 44–45.

71. Josh Daniel, "Now Hear This," *Texas Monthly*, May 1996, 30.

72. Steven Vonder Haar and Kimberly Weisul, "Yahoo! Goes Multimedia with Broadcast.com Deal," *Interactive Week*, April 5, 1999.

73. Alan Goldstein, "Tough Surfing: Yahoo Still Trying to Ride Broadcast.com's Wave to the Future," *Dallas Morning News*, July 25, 2001; Brian Bergstein, "Yahoo! to End Broadcasting Services," *Associated Press Online*, June 26, 2002.

74. "A Talk with the King of Streaming: Mark Cuban," *Internet World*, October 1, 1999, 34, 40.

75. Robert D. Hof and Elizabeth Lesly, "Don't Surf to Us, We'll Surf to You," *Business Week*, September 9, 1996, 109.

76. Andrea Petersen, "What Is a Portal—and Why Are There So Many of Them? Once Gateways to the Web, They Keep Expanding," *Wall Street Journal*, December 10, 1998, B8.

77. George Anders, "The Race for 'Sticky' Web Sites—behind the Deal Frenzy, a Quest to Hang on to Restless Clickers," *Wall Street Journal*, February 11, 1999, B1.

78. Henry Jenkins, Sam Ford, and Joshua Green, *Spreadable Media: Creating Value and Meaning in a Networked Culture* (New York: NYU Press, 2013), 4.

79. Hudis, 42; Tom Girard, "'Net Radio Mixes Ad Models," *Daily Variety*, June 27, 2000, A8; Broadcast.com, "Form 424B4."

80. Chuck Taylor, "Real-Time Audio Livens Radio Station Web Sites," *Billboard*, June 8, 1996, 6.

81. Taylor, "Radio Biz," 85.

82. Rebecca Ann Lind and Norman J. Medoff, "Radio Stations and the World Wide Web," *Journal of Radio Studies* 6.2 (1999): 203–21; Mary Jackson-Pitts and Ross Harms, "Radio Websites as a Promotional Tool," *Journal of Radio Studies* 10.2 (2003): 270–82; Scott R. Hamula and Wenmouth Williams Jr., "The Internet as a Small-Market Radio Station Promotional Tool," *Journal of Radio Studies* 10.2 (2003): 262–69.

83. Wilke, "Online Radio Startups," 40.

84. Megan Sapnar Ankerson, *Dot-Com Design: The Rise of a Usable, Social, Commercial Web* (New York: NYU Press, 2018), 90.

85. Taylor, "Real-Time," 106.

86. Michele Hilmes, *Only Connect: A Cultural History of Broadcasting in the United States*, 4th ed. (Boston: Wadsworth, 2014), 51–52.

87. Taylor, "Real-Time Audio," 106; Orens, "Cuban, Mark," 99.

88. Murphy, "Making a Killing," 56.

89. Goldstein, "CEO."

90. Susan Pulliam and Jared Sandberg, "Internet Stocks' Lightning Surge Is a Result of Surprise Profits," *Wall Street Journal*, November 3, 1995, C1.

91. David Wessel, "Fed Chairman Pops the Big Question: Is Market Too High?," *Wall Street Journal*, December 6, 1996, A3.

92. Toni Sant, *Franklin Furnace and the Spirit of the Avant-Garde: A History of the Future* (Chicago: Intellect, 2011), 118; Casey Kait and Stephen Weiss, *Digital Hustlers: Living Large and Falling Hard in Silicon Alley* (New York: Regan, 2001), 239.

93. Andy Cohen, "The Jupiter Mission," *Sales & Marketing Management*, April 2000, 56.

94. Paul M. Eng and Mark Lewyn, "On-Ramps to the Info Superhighway," *Business Week*, February 7, 1994, 108.

95. Claire Tristan, "King of the Gods," *MC: Technology Marketing Intelligence*, August 1998, 26.

96. Daniel Roth, "My, What Big Internet Numbers You Have!," *Fortune*, March 15, 1999, 114–18.

97. Judith Messina, "Web-Gazing Pays for Research Firm," *Crain's New York Business*, August 11, 1997, 19.

98. Eric W. Pfeiffer, "Start Up: The Story of a Prodigy—What Ever Happened to America's First Cutting-Edge Online Service?," *Fortune*, October 5, 1998, 19.

99. Howard Rheingold, *The Virtual Community: Homesteading on the Electronic Frontier*, rev. ed. (Cambridge, MA: MIT Press, 2000), 295–97.

100. Kait and Weiss, *Digital Hustlers*, 244–45.

101. "The New York Cyber Sixty," *New York*, November 13, 1995, 49.

102. Kait and Weiss, *Digital Hustlers*, 242. Another Pseudo executive, Dennis Adamo, states they were getting a 30 percent royalty on billings of $750,000/month, which would place Pseudo's monthly earnings closer to $225,000/month. *Internet*

History Podcast, "Episode 135: The Pseudo.com Story with Dennis Adamo," interview with Brian McCullough, March 19, 2017, audio, http://www.internethistory-podcast.com/2017/03/the-pseudo-com-story-with-dennis-adamo/

103. David Kirkpatrick, "Suddenly Pseudo," *New York*, December, 20, 1999; Charles Platt, "Steaming Video," *Wired*, November 2000.

104. Kirkpatrick, "Suddenly Pseudo."

105. Rick Ayre and Thomas Mace, "Online Services: Adapting to the Web," *PC Magazine*, June 11, 1996, 136.

106. Eric Gwinn, "Prodigy Likes Its Chat Rooms a Little Noisy," *Chicago Tribune*, January 20, 1995.

107. Kirkpatrick, "Suddenly Pseudo."

108. Sant, *Franklin Furnace*, 118. The original show was live broadcast on Thursdays 11:00 p.m. EST.

109. Leslie Goff, "In Depth: Silicon Alley," *Computerworld*, April 22, 1996, 81.

110. Kevin Kelly, *New Rules for the New Economy* (New York: Viking Penguin, 1998).

111. Frank Owen, "Let Them Eat Software," *Village Voice*, February 6, 1996, 28; Richard Florida, *The Rise of the Creative Class* (New York: Basic, 2002).

112. Andrew Ross, *No-Collar: The Humane Workplace and Its Hidden Costs* (Philadelphia: Temple University Press, 2004), 21–24.

113. Rich Tedesco, "Pseudo Radio Gaining Momentum," *Broadcasting & Cable*, July 22, 1996, 46.

114. Program schedule and show descriptions available through Internet Archive Wayback Machine capture of Pseudo website April 13, 1997. All programming information referenced in this chapter pulled from these descriptions: https://web.archive.org/web/19970413215115/http://www.pseudo.com/netcast/themepage/index.html

115. Wilke, "Online Radio Startups," 40.

116. Pseudo Programs Inc., "Internet Radio Shows," advertisement, *Back Stage*, April 19, 1996, 7.

117. Jesse Kornbluth, "Surfing the Net for God: 'Reverend Billy' Offers Guidance to 'Webchildren,'" *SF Gate*, August 6, 1996.

118. Trip Gabriel, "Where Silicon Alley Artists Go to Download," *New York Times*, October 8, 1995, A49.

119. Kirkpatrick, "Suddenly Pseudo"; Platt, "Steaming Video."

120. Kirkpatrick, "Suddenly Pseudo."

121. Gabriel, "Silicon Alley Artists," A49.

122. Ira Breskin, "Pseudo Programs' Joshua Harris: Observer of the On-Line Game Now Weighs in as Player," *Investor's Business Daily*, May 8, 1996, 2.

123. Marshal Rosenthal, "Radio's New Frontier," *Daily Variety*, August 30, 1999, A14.

124. "Tune In to Internet Radio," *MediaWeek*, April 19, 1999, 94; Elizabeth A. Rathbun, "Broadcasters Told to Make Internet Gains," *Broadcasting & Cable*, April 26, 1999, 44.

125. David Kushner, "Listen Up, Talk Radio, This is the Internet Speaking," *New York Times*, April 13, 1997, 39.

126. Ibid.

127. Breskin, "Pseudo Programs," 2.

128. Tedesco, "Pseudo," 46; Kushner, "Listen Up, Talk Radio," 40.

129. Kushner, "Listen Up, Talk Radio," 39–40.

130. Austin Bunn, "Machine Age: Psmoke and Fire," *Village Voice*, July 7, 1998, 35.

131. Ethan Brown, "The Sound of the City: Totally Wired," *Village Voice*, September 15, 1998, 132.

132. Mike Rinzel, interview by author, January 6, 2015.

133. Austin Bunn, "Machine Age: Plugging NetTV," *Village Voice*, January 20, 1998, 31.

134. Rinzel, interview by author.

135. Andrew Smith, *Totally Wired: The Wild Rise and Crazy Fall of the First Dotcom Dream* (London: Simon & Schuster UK, 2012), 41–42. The *Launder My Head* video is accessible online: https://vimeo.com/22906987

136. Rinzel, interview by author.

137. Bruce Haring, "'Streaming Media' Ready to Roar," *USA Today*, August 5, 1998, 4D.

138. Jennifer L. Rewick, "Pseudo Programs' Rapid Expansion Produces Unexpected Growing Pains," *Wall Street Journal*, December 13, 1999.

139. Rinzel, interview by author.

140. Pseudo Programs Inc., "Beep! Beep! Pseudo Picks Up Speed with Road Runner," *Business Wire*, press release, March 3, 1999; Anya Sacharow, "Sprite Sponsors Pseudo, for Real," *Brandweek*, July 6, 1998, 36; Steven Vonder Haar, "Cast the Net," *Brandweek*, March 15, 1999, 20.

141. Pseudo Programs Inc., "Pseudo Launches Internet Television on Excite," *PR Newswire*, press release, October 6, 1998; Pseudo Programs Inc., "EchoStar and Pseudo Programs Announce Interactive Television Broadcast Agreement," *Business Wire*, press release, January 8, 1999.

142. Pseudo Programs Inc., "Coming Soon: The QB Club Channel on Pseudo.com," *New York Times*, advertisement, October 18, 1999, C19; Pseudo Programs Inc., "Pseudo Programs, Inc. and NFL Quarterback Club Huddle to Launch Premier Internet TV Football Player Lifestyle Channel," *PR Newswire*, press release, October 12, 1999.

143. Jeanne Cummings and Phil Kuntz, "Republican Convention 2000: Despite Little News or Interest, Conventions Become More Lavish," *Wall Street Journal*, July 31, 2000, A24; Leslie Wayne, "Rather Be in Philadelphia? G.O.P. Convention's Online," *New York Times*, August 3, 2000, G7.

144. Richard Tedesco, "Fifth Estater: Breaking New Ground, Again," *Broadcasting & Cable*, February 7, 2000, 60.

145. Dennis Stillwell, "1999: The Year 'E' Took US By Storm," *Journal of Commerce*, December 29, 1999, 10; Catherine Tymkiw, "Bleak Friday on Wall Street," *CNNMoney*, April 14, 2000, http://money.cnn.com/2000/04/14/markets/markets_newyork/

146. Ken Kerschbaumer, "Interactive Media: The Online Blues," *Broadcasting & Cable*, September 25, 2000, 52–53; Justin Oppelaar, "Pseudo Shutters; 'Net Set Shudders," *Variety*, September 25, 2000, 35–36.

147. *60 Minutes II*, "The Dot-Com Kids," interview with Bob Simon, CBS, February 15, 2000, transcript, http://www.cbsnews.com/news/the-dot-com-kids/

148. Vonder Haar, "Cast the Net," 17.

149. Amanda Griscom, "Pseudo's Surveillance Palace: Happy New Year, Pod People!," *Village Voice*, December 21, 1999, 41; David Usborne, "Josh Harris Was the King of New York's Silicon Alley," *The Independent* (UK), March 16, 2001, 1, 7; Jim Hanas, "Second Acts: The Bubble Boy," *Radar*, February 2008, 40–43.

150. *We Live in Public*, directed by Ondi Timoner, IndiePix Films, 2010.

151. Aymar Jean Christian, *Open TV: Innovation beyond Hollywood and the Rise of Web Television* (New York: NYU Press, 2018), 47.

152. Patricia G. Lange, "Videos of Affinity on YouTube," in *The YouTube Reader*, ed. Pelle Snickars and Patrick Vonderau (Stockholm: National Library of Sweden, 2009), 71.

153. Arbitron/Edison Media Research, "Internet Study VI: Streaming at a Crossroads," February 2001, 1.

154. Chuck Taylor, "Clear Channel/InXsys Launch Net Radio Site," *Billboard*, July 24, 1999, 3.

155. Thomas Streeter, *The Net Effect: Romanticism, Capitalism, and the Internet* (New York: NYU Press, 2011), 69–92.

156. Michael C. Keith and Christopher H. Sterling, "Disc Jockeys (DJs or Deejays)," in *Encyclopedia of Radio*, vol. 1, ed. Christopher H. Sterling (New York: Routledge, 2004), 760.

157. Michele Hilmes, *Radio Voices: American Broadcasting, 1922–1952* (Minneapolis: University of Minnesota Press, 1997), 39.

158. John Markoff, "Turning the Desktop PC into a Talk Radio Medium," *New York Times*, March 4, 1993, D18.

159. Quittner, "Radio," 91.

160. Stephen Duncombe, *Notes from Underground: Zines and the Politics of Alternative Culture* (New York: Verso, 1997), 11–12; Bob Ostertag, *People's Movement, People's Press: The Journalism of Social Justice Movements* (Boston: Beacon, 2006), 9.

161. Mark Cuban, "Nabkeynote," slideshow presentation for NAB Show 1999, Las Vegas, NV, published September 8, 2009, https://www.slideshare.net/mcuban/nabkeynote

162. Stewart and de Llosa, "Boom Time," 153–61.

163. David Bank, "Agent of Change," *San Jose Mercury News*, January 8, 1995, 4E.

164. R. Lee Sullivan, "Internet is a Pirate of the Airwaves," *BRW*, April 29, 1996, 48.

165. Hilmes, *Radio Voices*, 34–74.

166. John Burgess, "Internet Creates a Computer Culture of Remote Intimacy," *Washington Post*, June 28, 1993, A1.

167. Leonard Wiener, "Tinkering with Radio on the Web," *U.S. News & World Report*, April 1, 1996, 72.

168. Mark Cuban, "Another Interview about Streaming Media from 1999," *Blog Maverick*, August 24, 2014, http://blogmaverick.com/2014/08/24/another-interview-about-streaming-media-from-1999/

169. Adelson, "Remaking Radio," D7.

170. Hilmes, *Radio Voices*, 42–43.

Chapter 3

1. Greg Milner, *Perfecting Sound Forever: An Aural History of Recorded Music* (New York: Faber and Faber, 2009), ix.

2. John Cheever, "The Enormous Radio," *New Yorker*, May 17, 1947, 28–33.

3. Douglas Kahn, "Introduction: Histories of Sound Once Removed," in *Wireless Imagination: Sound, Radio, and the Avant-Garde*, ed. Douglas Kahn and Gregory Whitehead (Cambridge, MA: MIT Press, 1992), 1–29; Jonathan Crary, *Suspensions of Perception: Attention, Spectacle, and Modern Culture* (Cambridge, MA: MIT Press, 1999), 290.

4. Jeffrey Sconce, *Haunted Media: Electronic Presence from Telegraphy to Television* (Durham, NC: Duke University Press, 2000), 93.

5. Erving Goffman, *Forms of Talk* (Philadelphia: University of Pennsylvania Press, 1981), 197–327.

6. Tim O'Reilly, "What Is Web 2.0? Design Patterns and Business Models for the Next Generation of Software," in *The Social Media Reader*, ed. Michael Mandiberg (New York: NYU Press, 2012), 32–52.

7. Dave Winer, "The Unedited Voice of a Person," *Scripting News*, January 1, 2007, http://scripting.com/2007/01/01.html#theUneditedVoiceOfAPerson

8. Michele Hilmes, *Radio Voices: American Broadcasting, 1922–1952* (Minneapolis: University of Minnesota Press, 1997), 34–46.

9. Jonathan Sterne et al., "The Politics of Podcasting," *Fibreculture Journal* 13 (2008), http://thirteen.fibreculturejournal.org/fcj-087-the-politics-of-podcasting/

10. John B. Horrigan and Maeve Duggan, "Home Broadband 2015," Pew Research Center, December 21, 2015, http://www.pewinternet.org/2015/12/21/home-broadband-2015/

11. "Federal Communications Commission Releases Data on High-Speed Services for Internet Access," Federal Communications Commission, August 9, 2001, https://transition.fcc.gov/Bureaus/Common_Carrier/News_Releases/2001/nrcc0133.html

12. Stefan Fatsis, "Getting in the Game: Sports in Cyberspace is Seen as a Sure Thing, and Nobody Wants to Be Left Behind," *Wall Street Journal*, March 28, 1996, R6.

13. Mark Berniker, "ESPN Enters the Internet Zone," *Broadcasting & Cable*, April 10, 1995, 60.

14. Richard Tedesco, "SportsZone Nets More NBA Games," *Broadcasting & Cable*, December 9, 1996, 136.

15. Richard Tedesco, "CBS Sports Comes to the 'Net," *Broadcasting & Cable*, March 10, 1997, 72–73.

16. Donna Petrozzello, "ABC Radio Enters WWW," *Broadcasting & Cable*, August 21, 1995, 38.

17. Scott Kirsner, "Profits in Site?: When Will Online Journalism Ventures Begin to Make Money?," *American Journalism Review*, December 1997, 40–44.

18. Jared Sandberg, "It Isn't Entertainment That Makes the Web Shine; It's Dull Data," *Wall Street Journal*, July 20, 1998, A1, A6.

19. John Durham Peters, *The Marvelous Clouds: Toward a Philosophy of Elemental Media* (Chicago: University of Chicago Press, 2015), 7.

20. Richard Linnett, "6 Shops in Talk Soup," *Adweek*, August 23, 1999, 6.

21. Jane L. Levere, "On-Line Radio Tries the All-Talk Format," *New York Times*, September 13, 1999, C5; Mark Walsh, "Everything Old Is New Again," *Crain's New York Business*, July 3, 2000, 3.

22. "Drought in Advertising Prompts eYada.com to Shutter Operations," *Wall Street Journal*, July 10, 2001, B7; Jose Alvear, "eYada No More," *Streaming Media*, July 10, 2001.

23. Meyrowitz was forthcoming about his use of established broadcast practices: Jose Alvear, "Q&A with Robert Meyrowitz, President and CEO of eYada.com," *Streaming Media*, April 25, 2000.

24. Program schedule and show descriptions available through Internet Archive Wayback Machine capture of eYada website from June 7, 2001: https://web.archive.org/web/20010607022743/http://www.eyada.com/channel_main.cfm?channelid=10

25. Henry Jenkins, Sam Ford, and Joshua Green, *Spreadable Media: Creating Value and Meaning in a Networked Culture* (New York: NYU Press, 2013), 49.

26. TalkNetRadio.com, "TalkNetRadio.com Launches Registration Site for Do-It-Yourself Internet Radio; Internet Users Worldwide Invited to Register to Create Their Own Internet Talk Programs," *BusinessWire*, press release, January 5, 1999.

27. Ibid., par. 2.

28. Ibid., par. 3.

29. Jose Alvear, "GiveMeTalk.com Lets Users Create Internet Talk Shows," *Streaming Media*, February 23, 2000.

30. Thomas Claburn, "Shouting Match," *Ziff Davis Smart Business for the New Economy*, September 2000, 66.

31. GiveMeTalk!, "William Gross Named CEO of GiveMeTalk!, Do-It-Yourself Streaming Media," *PRNewswire*, press release, April 12, 2000, par. 2.

32. Matt Richtel, "Want Your Own Talk Show? You Can Get One, on the Net," *New York Times*, January 14, 1999; "Make Your Own Internet Talk Radio Show on Your Mac," *Mac Observer*, February 25, 2000.

33. Ben Hammersley, "Audible Revolution," *The Guardian* (UK), February 11, 2004; Andy Goldberg, "The People's Radio," *The Independent* (UK), December 8, 2004, 11.

34. Doc Searls, "DIY Radio with PODcasting," *Doc Searls' IT Garage*, September 28, 2004, https://web.archive.org/web/20051219003155/http://www.itgarage.com/node/462; Russell Kay, "Podcasting," *Computerworld*, October 3, 2005, 34.

35. Andy Bowers, "The Year of the Podcast: A Status Report on a Much-Hyped New Medium," *Slate*, December 30, 2005, http://www.slate.com/articles/podcasts/2005/12/the_year_of_the_podcast.html

36. "'Blog' Picked as Word of the Year," *BBC News*, December 1, 2004, http://news.bbc.co.uk/2/hi/technology/4059291.stm; Leslie Walker, "Bloggers Gain Attention in 2004 Election," *Washington Post*, November 4, 2004, E1.

37. Wallys W. Conhaim, "iPod Sprouts Successful Changes in Media Landscape," *Information Today*, March 2005, 25–26.

38. Apple Inc., "Apple Takes Podcasting Mainstream: Discover, Subscribe, Manage & Listen to Podcasts Rights in iTunes 4.9," press release, June 28, 2005, https://www.apple.com/pr/library/2005/06/28Apple-Takes-Podcasting-Mainstream.html

39. I would suggest that we are on the cusp of a *fourth wave* in 2020, as the platformization of podcasting is notably reshaping the industry, the style and form of the content, and also the listening experience through increased professionalization and economic consolidation. For more on industry shifts see John L. Sullivan, "The Platforms of Podcasting: Past and Present," *Social Media + Society* (2019), DOI:10.33767/osf.io/4fcgu.

40. Tiziano Bonini, "The 'Second Age' of Podcasting: Reframing Podcasting as a New Digital Mass Medium," *Quaderns del CAC* 41 (2015): 21–30.

41. Dave Winer, "Payloads for RSS," *DaveNet*, January 11, 2001, http://scripting.com/davenet/2001/01/11/payloadsForRss.html

42. Hammersley, "Audible," par. 2.

43. This basic story was circulated through dozens of reports about podcasting in 2004–5 and beyond. Some examples: Byron Acohido, "Radio to the MP3 Degree: Podcasting," *USA Today*, February 9, 2005, 1B; Celeste Biever, "And Now for the Podcast . . . ," *New Scientist*, February 12, 2005, 24; Brian X. Chen, "Aug. 13, 2004: 'Podfather' Adam Curry Launches *Daily Source Code*," *Wired*, August 13, 2009.

44. Daniel Terdiman, "Adam Curry Gets Podbusted," *CNET*, December 2, 2005, http://www.cnet.com/news/adam-curry-gets-podbusted/

45. Sinclair Target, "The Rise and Demise of RSS," *Two-Bit History*, December 18, 2018, https://twobithistory.org/2018/12/18/rss.html. Dave Winer still insists that RSS is an essential component of podcasting: "A podcast is a series of digital media files made available over the open web through an RSS feed with enclosures." From "What Is a Podcast?," *Scripting News*, April 9, 2019, http://scripting.com/2019/04/09/135711.html

46. Bradley Johnson, "Device Costs $299, but Content is Free," *Advertising Age*, October 7, 1996, 48; Brett Atwood, "Net Audio Freed from the Computer," *Billboard*, February 15, 1997, 62; Audio Highway, "Audio Highway and National Public Radio Ink Pact," *PRNewswire*, press release, August 11, 1997; Audiohighway.com, "Form 424A," SEC file no. 333-59823, filed August 14, 1998.

47. Trudi M. Rosenblum, "Downloading Audio from the Internet: The Future of Audio?," *Publishers Weekly*, March 2, 1998, 27–28; Lisa Hamm-Greenawalt, "Listen to the Sound of Content Being Sold," *Internet World*, October 15, 1999, 52–54.

48. Eliot Van Buskirk, "Bragging Rights to the World's First MP3 Player," *CNET*, January 25, 2005, http://www.cnet.com/news/bragging-rights-to-the-worlds-first-mp3-player/

49. Sterne et al., "The Politics of Podcasting," pars. 33–34.

50. Brian Winston, *Media Technology and Society, a History: From the Telegraph to the Internet* (New York: Routledge, 1998), 1–15.

51. He also introduces the third transformation of "the 'law' of suppression of radical potential." That last theory is not directly relevant to my discussion here.

52. Hamm-Greenawalt, "Listen to the Sound," 53.

53. Jacob Aron, "Right Track, Wrong Tunes," *New Scientist*, February 18, 2012, 42–43; Andrew Cunningham, "A Brief History of USB, What It Replaced, and What Has Failed to Replace It," *Ars Technica*, August 17, 2014, http://arstechnica.

com/gadgets/2014/08/a-brief-history-of-usb-what-it-replaced-and-what-has-failed-to-replace-it/

54. Aaron Barlow, *The Rise of the Blogosphere* (Westport, CT: Praeger, 2007), 152.

55. danah boyd, "A Blogger's Blog: Exploring the Definition of a Medium," *Reconstruction* 6.4 (2006), http://reconstruction.eserver.org/Issues/064/boyd.shtml

56. Chris Taylor, "Pssst. Wanna See My Blog?," *Time*, February 11, 2002, 68.

57. Jill Walker Rettberg, *Blogging*, 2nd ed. (Malden, MA: Polity, 2014), 6–8; Scott Rosenberg, *Say Everything: How Blogging Began, What It's Becoming, and Why It Matters* (New York: Crown, 2009), 17–45. *Justin's Links* archives located online: http://www.links.net/

58. Walker Rettberg, *Blogging*, 9.

59. Rosenberg, *Say Everything*, 46–73.

60. Ibid., 48.

61. DaveNet archive located online: http://scripting.com/davenet/about.html

62. Rosenberg, *Say Everything*, 58–64; Walker Rettberg, *Blogging*, 8–12.

63. *Scripting News* archive located online: http://scripting.com/archive.html

64. Walker Rettberg, *Blogging*, 8.

65. History of RSS, including Dave Winer's role, outlined in Ben Hammersley, *Developing Feeds with RSS and Atom* (Sebastopol, CA: O'Reilly Media, 2005), 1–12.

66. Winer, "Payloads"; Chris Pfaff, "Podcasting: The Next Syndication Play," *Produced By*, Winter 2005–6, 48–52.

67. Pfaff, "Podcasting," 49.

68. Winer, "Payloads."

69. Tristan Louis suggested on a software developer online discussion board in October 2000 that RSS be revised to carry sound files: "For example, it could serve up a radio feed related to this story." From "Some Suggestions for RSS .92," *Discussion of XML News / Announcement / Syndication / Resource Discovery Formats* (Yahoo! Groups), October 13, 2000, message, https://groups.yahoo.com/neo/groups/syndication/conversations/topics/698

70. Dave Winer, "History of the Radio UserLand News Aggregator," *UserLand Software*, June 11, 2002, http://radio.userland.com/userGuide/reference/aggregator/newsAggregator

71. Laura Gordon-Murnane, "Saying 'I Do' to Podcasting," *Searcher: The Magazine for Database Professionals*, June 2005, 44–51.

72. Winer, "Payloads."

73. Stephen Humphries, "'Podcast' Your World: Digital Technology for iPod Does for Radio What Blogs Did for the Internet," *Christian Science Monitor*, December 10, 2004, 12. Early podcaster Dave Slusher of *Evil Genius Chronicles* said, "As a concept, I see it as most closely equivalent to doing for internet audio content what TiVo does for TV." In Daniel Terdiman, "Podcasts: New Twist on Net Audio," *Wired*, October 8, 2004.

74. Winer, "Payloads."

75. Andrew Sullivan, "The Blogging Revolution: Weblogs Are to Words What Napster Was to Music," *Wired*, May 1, 2002.

76. boyd, "A Blogger's Blog," pars. 40–46.

77. The "podfather" title got attached to Curry during the first wave of mainstream media attention heaped upon podcasting in the spring 2005—rumors being that Curry himself devised the title and aggressively promoted it. For an early reference to Curry as "the podfather": Ken Belson, "An MTV Host Moves to Radio, Giving Voice to Audible Blogs," *New York Times*, May 2, 2005, C3.

78. Winer, "Payloads."

79. Dawn C. Chmielewski, "Podcasting Power: New Technology Delivers Personalized Broadcasts to Your MP3 Player," *San Jose Mercury News*, January 31, 2005, 1E.

80. Colleen Walsh, "The Podcast Revolution: Two Berkman Fellows Helped to Make It Happen," *Harvard Gazette*, October 27, 2011, http://news.harvard.edu/gazette/story/2011/10/the-podcast-revolution/

81. Jeffrey A. Dvorkin, "Chris Lydon's Gamble: Rolling the Dice with Public Radio," NPR Ombudsman, accessed March 20, 2016, http://www.npr.org/yourturn/ombudsman/2001/010319.html. Lydon hosted *The Connection* starting in 1994. The widely syndicated series continued with other hosts until 2005, at which time it was canceled and replaced with *On Point*, still broadcasting today. Lydon returned to WBUR in 2014 with the weekly program *Open Source*. For Lydon's history with WBUR see Ed Siegel, "Christopher Lydon Returns to WBUR with 'Open Source,'" *The ARTery*, November 20, 2013, https://www.wbur.org/artery/2013/11/20/lydon-open-source

82. Internet Archive Wayback Machine capture of christopherlydon.org shortly after March 2001 launch: https://web.archive.org/web/20010401114958/http://www.christopherlydon.org/

83. Christopher Lydon, "A God for Bloggers," *Christopher Lydon Interviews* . . . , June 21, 2003, http://blogs.harvard.edu/lydondev/2003/06/21/a-god-for-bloggers/

84. Dave Winer, "Chris Lydon Speaks of Ralph Waldo Emerson and Weblogs," *DaveNet*, July 1, 2003, http://scripting.com/davenet/2003/07/01/chrisLydon-SpeaksOfRalphWal.html

85. Bob Doyle, "The First Podcast," *EContent*, September 2005, 33; C. Walsh, "The Podcast Revolution," par. 22.

86. Dave Winer, "Chris Lydon's Weblog for the Ears," *DaveNet*, July 31, 2003, http://scripting.com/davenet/2003/07/31/chrisLydonsWeblogForTheEar.html. Lydon's audioblog series never had an official name; it was simply referred to as *Christopher Lydon Interviews*. . . . An archive of the interview segments is online: http://blogs.harvard.edu/lydondev/all-the-lydon-interviews-in-one-download/

87. Michael Geoghegan and Dan Klass, "Podcasting 101," *Entertainment and Sports Lawyer* 24.4 (2007): 20–23; Kay, "Podcasting," 34.

88. Audio of Dave Winer's *Morning Coffee Notes* and other miscellaneous podcasts were archived at http://morningcoffeenotes.com/ but the site is no longer online.

89. *Trade Secrets* online archive: http://secrets.scripting.com/

90. Dave Winer, "How Podcasting Got Its Name," *Scripting News*, April 7, 2013, http://threads2.scripting.com/2013/april/howPodcastingGotItsName

91. *Daily Source Code*, Episode 1, hosted by Adam Curry, August 13, 2004, audio.

92. Chmielewski, 1E; Dave Slusher, "Harold Gilchrist on the Birth of Pod-

casting," *Evil Genius Chronicles* February 18, 2005, http://evilgeniuschronicles.org/2005/02/18/harold-gilchrist-on-the-birth-of-podcasting/

93. Dave Winer, "A Podcast about Podcasting," *Scripting News*, September 30, 2015, audio, http://scripting.com/2015/09/30/aPodcastAboutPodcasting.html

94. Jodi Dean, *Blog Theory: Feedback and Capture in the Circuits of Drive* (Malden, MA: Polity, 2010), 43.

95. Walker Rettberg, *Blogging*, 33–34.

96. Ibid., 30.

97. Dean, *Blog Theory*, 74.

98. "Broadcast Your Own Show," *Personal Computer World*, October 2005; Dan Tynan, "30 Things You Didn't Know You Could Do on the Internet," *PC World*, July 2005, 80.

99. Garth Kidd, "Audioblogging," *Deadly Bloody Serious*, November 19, 2002, https://web.archive.org/web/20021223122619/http://www.deadlybloodyserious.com/outlines/AudioBlogging.html; Harold Gilchrist, "My Contribution to Audioblogging/Podcasting," *Audioblogging 2.0*, December 4, 2005, http://radioweblogs.com/0100368/stories/2005/02/19/myContributionToAudiobloggingpodcasting.html

100. Geoghegan and Klass, par. 9; Doug Kaye, "Podcasting History," *Blogarithms: Doug Kaye's Blog*, December 2, 2005, https://blogarithms.com/2005/12/02/podcasting-history/

101. Johnny Dee, "The Cult www.chuckpalahniuk.net," *The Guardian* (UK), September 13, 2003, 28. Many of Palahniuk's 2003 audioblog recordings, along with transcripts, are accessible via an Internet Archive Wayback Machine capture of *The Cult's* "Chuck's AudioBlogs" web page from November 19, 2003: https://web.archive.org/web/20031119035904/http://www.chuckpalahniuk.net/blog.php

102. Chuck Palahniuk, "Sunday, September 14th, 2003," *The Cult: www.chuckpalahniuk.net*, captured October 2, 2003, transcript, https://web.archive.org/web/20031002005204/http://www.chuckpalahniuk.net/author/audioblog_transcripts.htm

103. Zach Dundas, "A Hazardous Outing: For Years, Fight Club Author Chuck Palahniuk Managed to Keep His Personal Life under Wraps. Whoops," *Williamette Week*, October 1, 2003, 9; Erik Hedegaard, "A Heartbreaking Life of Staggering Weirdness," *Rolling Stone*, June 30, 2005, 124, 128, 130.

104. Audblog, "audblog Is the First Audio Blogging Service Created Directly for the Blogging Community," press release, February 15, 2003. The very first Audblog posts, created by the company's founder, Noah Glass, were published on a blog called *wackydelicious*; they are available via an Internet Archive Wayback Machine capture from July 21, 2003: https://web.archive.org/web/20030721231628/http://wackywacky.blogspot.com/

105. Evan Williams, "How Odeo Happened," *Evhead*, February 25, 2005, http://www.evhead.com/2005/02/how-odeo-happened.asp. The exact date of audioBLOGGER's release is unclear, though it was active as early as April 2003, only two months after Audblog's initial launch. The earliest Internet Archive Wayback Machine capture of the site is from April 1, 2003: https://web.archive.org/web/20030401215932/http://audioblogger.com/

106. "audioBLOGGER FAQ," audioBLOGGER.com, captured August 2, 2003, https://web.archive.org/web/20030802083756/http://www.audioblogger.com/faq.html

107. Arik Hesseldahl, "Google Goes Blog-Crazy," *Forbes*, February 18, 2003, http://www.forbes.com/2003/02/18/cx_ah_0218google.html

108. Wheaton was a prolific blogger with ties to the California internet technology community. He was friends with Audblog's developers and audioblogged regularly between June 2003 and April 2005, before he transitioned to the podcast *Radio Free Burrito*. Wheaton's audioblog archive is online: http://wwdnaudblog.blogspot.com/

109. Peter J. Howe, "Computer, Microphone, iPod Make Broadcasting Personal," *Knight Ridder Tribune Business News*, December 20, 2004. By the June 2005 launch of iTunes 4.9 with its dedicated podcast features, the number of regular podcast series was up to three thousand, and by July 2005 it more than doubled to an estimated six to seven thousand.

110. Williams, "How Odeo Happened," par. 23.

111. Miguel Helft, "Yearning for Freedom . . . from Venture Capital Overlords," *New York Times*, November 24, 2006, C5.

112. John Markoff, "For a Start-Up, Visions of Profit in Podcasting," *New York Times*, February 25, 2005, C1, C4; Dom Sagolla, *140 Characters: A Style Guide for the Short Form* (Hoboken, NJ: Wiley, 2009), xvii.

113. Kevin J. Delaney, "When a Tech Start-Up's Dreams Turn Prosaic," *Wall Street Journal*, April 3, 2007, B1; Nicholas Carlson, "The Real History of Twitter," *Business Insider*, April 13, 2011, http://www.businessinsider.com/how-twitter-was-founded-2011-4

114. Carlson, "Real History"; Nick Bilton, *Hatching Twitter: A True Story of Money, Power, Friendship, and Betrayal* (New York: Penguin, 2013); Dhiraj Murthy, *Twitter: Social Communication in the Twitter Age* (Malden, MA: Polity, 2013).

115. Sagolla, *140 Characters*, xvii–xviii.

116. Audblog's Noah Glass was a big supporter of Dorsey's text status idea, as well as its initial project manager. It is actually Glass who came up with the name twttr (soon changed to Twitter). Williams abruptly fired Glass in the fall of 2006, and his role in the creation of Twitter has largely been concealed. Carlson, "Real History"; John C. Abell, "March 21, 2006: Twitter Takes Flight," *Wired*, March 21, 2011.

117. Alice E. Marwick, *Status Update: Celebrity, Publicity, and Branding in the Social Media Age* (New Haven, CT: Yale University Press, 2013), 209.

118. Ibid.; Dean, *Blog Theory*, 61–90.

119. Manuel Castells, *Networks of Outrage and Hope: Social Movements in the Internet Age* (Malden, MA: Polity, 2012), 2.

120. Zizi Papacharissi, *Affective Publics: Sentiment, Technology, and Politics* (New York: Oxford University Press, 2015), 68.

121. Admittedly, it is impossible to say whether Twitter would have been anywhere near as popular if it was a sound-based platform.

122. Martin Shingler and Cindy Wieringa, *On Air: Methods and Meanings of Radio* (New York: Oxford University Press, 1998), 34–41.

123. Professional radio/podcast producers often use "amateurish" as a euphemism for unlistenable. That cultural divide between the "podcasters" (the scrappy amateurs) and the "pro-casters" (the professional broadcasters) is discussed in Jason Loviglio, "What I Learned at Podcast Movement 2015," *Antenna: Responses to Media & Culture*, September 1, 2015, http://blog.commarts.wisc.edu/2015/09/01/what-i-learned-at-podcast-movement-2015/

124. Goldberg, "The People's Radio," 11.

Chapter 4

1. Jay David Bolter and Richard Grusin, *Remediation: Understanding New Media* (Cambridge, MA: MIT Press, 1998), 5.

2. Ibid., 28.

3. Jeffrey S. Wilkinson, August E. Grant, and Douglas J. Fisher, *Principles of Convergent Journalism*, 2nd ed. (New York: Oxford University Press, 2012).

4. Ian Hutchby, *Media Talk: Conversation Analysis and the Study of Broadcasting* (New York: Open University Press, 2006), 28–31.

5. The term *talkback radio* is most widely used in Australia. John Tebbutt, "Imaginative Demographics: The Emergence of a Radio Talkback Audience in Australia," *Media, Culture & Society* 28.6 (2006): 857–82.

6. Jason Loviglio, *Radio's Intimate Public: Network Broadcasting and Mass-Mediated Democracy* (Minneapolis: University of Minnesota Press, 2005), 38–69.

7. Susan J. Douglas, "Letting the Boys Be Boys: Talk Radio, Male Hysteria, and Political Discourse in the 1980s," in *Radio Reader: Essays in the Cultural History of Radio*, ed. Michele Hilmes and Jason Loviglio (New York: Routledge, 2002), 485–504.

8. James F. Hamilton, "Historical Forms of User Production," *Media Culture & Society* 36.4 (2014): 491–507.

9. Andrew Crisell, *Understanding Radio* (New York: Methuen, 1986), 65–66.

10. WNYC is a subsidiary of the nonprofit corporation New York Public Radio (NYPR), a system of eight non-commercial radio stations that includes the classical music radio station WQXR (105.9 FM) and the four stations in the NPR member network New Jersey Public Radio. New York Public Radio, "About Us," accessed May 26, 2019, http://www.nypublicradio.org/about/

11. Steven Lee Myers, "New York, Signing Off, to Sell Its Radio and TV Stations," *New York Times*, March 22, 1995, http://www.nytimes.com/1995/03/22/nyregion/new-york-signing-off-to-sell-its-radio-and-tv-stations.html

12. New York Public Radio, *Media Kit: Q3 2017*, accessed May 26, 2019, https://static1.squarespace.com/static/53ff2c53e4b0e1f6ca3c017d/t/5a340369e4966b79a0770f2c/1513358196242/Media+Kit+Q3+2017.pdf; David Hinckley, "Public Station WNYC Is Dominating the Ratings for Talk Radio," *New York Daily News*, July 11, 2014.

13. Ben Sisario, "WNYC to Open New Division to Feed a Market for Podcasts," *New York Times*, October 13, 2015, B1.

14. NPR has 999 member stations in 2019. However, for the main period discussed in this chapter, during the 1990s and 2000s, the number of stations was

closer to 700. For an overview of NPR stations and the network's structure: "About NPR," NPR, accessed May 26, 2019, https://www.npr.org/overview

15. Richard Dean, "Reporting Online: 1995: Beginning of NPR.org," in *This Is NPR: The First Forty Years* (San Francisco: Chronicle, 2010), 182–83.

16. Glen Sansone, "NPR, MN Public Radio Build Network, Anger Stations," *CMJ New Music Report*, April 26, 1999, 4.

17. Dean, "Reporting Online," 182.

18. WNYC cited the prohibitive cost of RealAudio servers and "restrictions in broadcasting rights for network programming—like NPR, PRI, and BBC broadcasts" for its initial delay in offering audio streams/downloads online: "Talking Back to WNYC.org," *WNYC Program Guide*, March–April 1998, 2.

19. "Onward and Upward," *WNYC Program Guide*, May–June 1999, 5; "From New York to the World," *WNYC Program Guide*, July–August 1999, 2.

20. Laura Walker, "New Year, New News," Memo to WNYC Radio Board of Trustees, January 12, 2000, New York Public Radio Archives; "WNYC.org Offers Audio Streaming," *WNYC Program Guide*, Spring 2000, 2.

21. "Radio New Universe," *WNYC Program Guide*, January–February 1998, 2.

22. Dean, "Reporting Online," 182.

23. Laura Walker, interview by author, January 13, 2015.

24. Laura Walker, "Message from the President," *WNYC Program Guide*, Summer 2000, 2.

25. "Talking Back," 2.

26. Ibid.

27. *On the Line*'s The Soapbox message board available through an Internet Archive Wayback Machine capture from December 7, 1998: https://web.archive.org/web/19981207033528/http://www.wnyc.org/talk/ontheline/wwwboard.shtml

28. Dean, "Reporting Online," 182.

29. Ben Max, "Brian Lehrer, Interviewee: Iconic Radio Host Discusses His Career as 25th WNYC Anniversary Approaches," *Gotham Gazette*, July 1, 2014, video, http://www.gothamgazette.com/index.php/government/5129-brian-lehrer-wnyc-interview-25-year-anniversary

30. "Talk Program Descriptions," WNYC, captured December 3, 1998, https://web.archive.org/web/19981201071959/http://www.wnyc.org/program/index.html

31. *On the Media*, "Everyone's a Critic," hosted by Brooke Gladstone, WNYC, January 6, 2006, audio, http://www.wnyc.org/story/128890-everyones-a-critic/

32. Michael Kane, "Brian Lehrer: Radio Host Guides the Talk of the Town," *New York Post*, June 29, 2009, 34; Batya Ungar-Sargon, "Hello, Brian Lehrer? This Is All of New York On the Line," *Tablet*, July 5, 2013, https://www.tablet-mag.com/jewish-arts-and-culture/books/136873/hello-brian-lehrer; David Wallis, "WNYC's Brian Lehrer on the Art of Silencing an Agitated Caller," *Observer* (New York), February 5, 2016, http://observer.com/2016/02/radios-brian-lehrer-on-the-art-of-silencing-a-crazed-caller/; Clyde Haberman, "Asking the Tough Questions, Often While Unseen, for Almost a Quarter-Century," *New York Times*, December 30, 2013, A14.

33. WNYC, "WNYC Radio's The Brian Lehrer Show Wins Peabody Award," press release, April 2, 2008, http://www.wnyc.org/press/wnyc-radios-the-brian-

lehrer-show-wins-peabody-award/; "The Brian Lehrer Show (WNYC Radio)," Peabody Awards, accessed May 26, 2019, http://www.peabodyawards.com/award-profile/the-brian-lehrer-show

34. Brian Lehrer, "The Brian Lehrer Show—2007 Peabody Award Acceptance Speech," filmed at 66th Annual Peabody Awards, New York, published September 21, 2015, video, https://www.youtube.com/watch?v=WUkYTD_ED94

35. *The Brian Lehrer Show*, "Our One," featuring Brian Lehrer and Nuala Mc-Govern, WNYC, February 19, 2009, audio, http://www.wnyc.org/story/29800-our-one/

36. Scott Chappell, "Brian Lehrer: Journalist," *Stated Magazine*, 2014, http://www.statedmag.com/articles/interview-brian-lehrer-of-wnyc-public-radio.html

37. Michael Purdy, "What Is Listening?," in *Listening in Everyday Life: A Personal and Professional Approach*, ed. Michael Purdy and Deborah Borisoff (Lanham, MD: University Press of American, 1991), 3.

38. Ungar-Sargon, "Hello, Brian Lehrer?," par. 10.

39. Wallis, "WNYC's Brian Lehrer," par. 2.

40. Andrew Belonsky, "WNYC Host Brian Lehrer on Wikileaks, Loughner, and the Need for Nation-Building," *Death and Taxes*, January 26, 2011, par. 37.

41. Jo Piazza, "Brian Lehrer Marks 25 Years of Having a Show on Public Radio's WNYC," *Wall Street Journal*, October 20, 2014.

42. Kate Lacey, *Listening Publics: The Politics and Experience of Listening in the Media Age* (Malden, MA: Polity, 2013), 7–8.

43. Regarding how *Lehrer*'s producers' attention to the live broadcast trumps any concerns for time-shifted listening or social media content, Lehrer says: "The bottom line, though, is that we are a live show in the middle of business hours, and we never sacrifice live, real-time programming." In "Answers from Brian Lehrer of WNYC," *New York Times*, December 3, 2008, par. 8, https://cityroom.blogs.ny-times.com/2008/12/03/answers-about-new-york-public-radio-part-1/

44. "The 60-Second Interview: Brian Lehrer, Host of the Brian Lehrer Show on WNYC," *Politico*, March 4, 2014, http://www.politico.com/media/story/2014/03/the-60-second-interview-brian-lehrer-host-of-the-brian-lehrer-show-on-wnyc-002684#ixzz4CJeI6FA0

45. Belonsky, "WNYC Host Brian Lehrer," par. 15.

46. Megan Ryan, interview by author, January 9, 2015.

47. "Answers from Brian Lehrer," par. 9.

48. The following observations are all derived from my fieldwork, unless otherwise noted. The broadcast I observed was on Friday, January 9, 2015. Audio of the complete episode is online: http://www.wnyc.org/story/the-brian-lehrer-show-2015-01-09. I also observed the WNYC newsroom, including *The Brian Lehrer Show* production team, at various times during the week of January 5–9, 2015.

49. Nick Couldry, "Liveness, 'Reality,' and the Mediated Habitus from Television to the Mobile Phone," *Communication Review* 7.4 (2004): 360.

50. Dan Gillmor, *We the Media: Grassroots Journalism by the People, for the People*, paperback ed. (North Sebastopol, CA: O'Reilly Media, 2006), xix–xxix. Lehrer praises citizen journalism in Carlson, "Real History," par. 12, and Belonsky, "WNYC Host Brian Lehrer," par. 6.

51. Jose van Dijck and Thomas Poell, "Understanding Social Media," *Media and Communication* 1.1 (2013): 2.

52. *The Brian Lehrer Show*, "How Many SUVs Are on Your Block?," featuring Brian Lehrer and Jeff Howe, WNYC, July 26, 2007, audio, http://www.wnyc.org/story/25493-how-many-suvs-are-on-your-block/

53. *The Brian Lehrer Show*, "SUV Counting: The Results," featuring Brian Lehrer, Jim Colgan, and Jamie Kitman, WNYC, August 2, 2007, audio, http://www.wnyc.org/story/25545-suv-counting-the-results/

54. *The Brian Lehrer Show*, "Are You Being Gouged?," featuring Brian Lehrer and Jim Colgan, WNYC, September 24, 2007, audio, http://www.wnyc.org/story/25927-are-you-being-gouged/

55. *The Brian Lehrer Show*, "Are You Being Gouged?: The Results," featuring Brian Lehrer, Jim Colgan, David Leonhardt, and Jay Rosen, WNYC, October 11, 2007, audio, http://www.wnyc.org/story/26067-screening-the-war/

56. Jeremy Egner, "Crowdsourcing: Enlisted Legmen, Formerly Known as the Audience," *Current*, July 28, 2008, http://current.org/files/archive-site/news/news-0813crowdsourcing.shtml

57. Penny O'Donnell, "Journalism, Change, and Listening Practices," *Continuum* 23.4 (2009): 503–17; Matt Sienkiewicz, "Start Making Sense: A Three-Tier Approach to Citizen Journalism," *Media Culture & Society* 36.5 (2014): 691–701.

58. *The Brian Lehrer Show*, "Crowdsourcing: Hillary Clinton's Schedule," featuring Brian Lehrer and Andrea Bernstein, WNYC, March 19, 2008, audio, http://www.wnyc.org/story/27207-crowdsourcing-hillary-clintons-schedule/

59. Brian Lehrer, "And, a Reminder from Brian Lehrer . . . ," New York Development Wiki, captured October 4, 2009, https://web.archive.org/web/20091004030511/http://issues.wnyc.org/wiki/index.php/New_York_Development_Wiki

60. Brian Lehrer, "The Free Country Manifesto," WNYC, September 7, 2010, http://www.wnyc.org/story/93347-its-free-country-so-what/

61. Carlson, "Real History," par. 8.

62. Piazza, "Brian Lehrer Marks," par. 15.

63. Jody Avirgan, "How the Brian Lehrer Show Makes It Work Online the Other 22 Hours of the Day," NPR Training, published March 5, 2014, video, https://vimeo.com/88288433

64. Emily Alfin Johnson, "How WNYC's The Brian Lehrer Show Efficiently Serves Their Online Audience," NPR Digital Services, March 3, 2014, http://digitalservices.npr.org/post/how-wnycs-brian-lehrer-show-efficiently-serves-their-online-audience

65. Ryan, interview by author.

66. Avirgan, video.

67. Ibid.

68. This is true at least for professional radio stations and shows, by which I mean commercial broadcasters and major public service broadcasters like NPR member stations. Many smaller noncommercial stations, such as college stations and LPFM broadcasters, still lack a substantial web presence.

69. Rebecca Ann Lind and Norma J. Medoff, "Radio Stations and the World

Wide Web," *Journal of Radio Studies* 6.2 (1999): 203–21; Bradley Carl Freeman, Julia Klapczynski, and Elliott Wood, "Radio and Facebook: The Relationship Between Broadcast and Social Media Software in the U.S., Germany, and Singapore," *First Monday* 17.4 (2012), http://firstmonday.org/ojs/index.php/fm/article/view/3768

70. NPR, "NPR Digital Services Steps Up Efforts to Measure and Monetize Station Streams," press release, March 27, 2012.

71. Theories of audience exploitation on the internet are plentiful, especially among critical Marxist media scholars. This position is perhaps best represented in the work of Mark Andrejevic, who argues that audience interactivity creates value for mass media producers, most of all in the form of data that can be commodified by marketers: Mark Andrejevic, "Exploiting YouTube: Contradictions of User-Generated Labour," in *The YouTube Reader*, ed. Pelle Snickers and Patrick Vonderau (Stockholm: National Library of Sweden, 2009), 406–23; Mark Andrejevic, "Watching Television without Pity: The Productivity of Online Fans," *Television & New Media* 9.1 (2008): 24–46.

72. Jason Loviglio, "Public Radio in Crisis," in *Radio's New Wave: Global Sound in the Digital Era*, ed. Jason Loviglio and Michele Hilmes (New York: Routledge, 2013), 28.

73. New York Public Radio, "Local Content & Services Report—CPB Station Action Survey for 2015," March 11, 2016, 1, https://media.wnyc.org/media/resources/2016/Mar/11/CPB_2015_-_Sec_6_Local_Content_Services_Report_FINAL_VERSION_ixHpbfN.pdf

74. *Remix*, "WNYC and the Content Revolution," panel discussion with Laura Walker, WNYC, September 17, 2014, audio, http://www.wnyc.org/story/wnyc-and-content-revolution-remix/. I heard this digital media business language repeatedly during my interviews with executive staff at New York Public Radio. The industry press on public radio podcasting is especially rife with discourses of "monetization." Another ubiquitous term is "audio content." Journalism critics have rightly argued that "content" is "a vague, cynical word" that reduces reporting to a commodity: Jon Christian, "Dear Journalists: For the Love of God, Please Stop Calling Your Writing Content," *Slate*, May 25, 2016, http://www.slate.com/blogs/lexicon_valley/2016/05/25/content_and_its_discontents_it_s_massively_depressing_when_journalists_call.html

75. Clickbait is the practice of posting curiosity-raising headlines, quotes, images, GIFs, etc., in order to encourage people to click to see more, primarily for the purpose of generating web traffic that can then be used to raise online advertising revenue. Geert Lovink, *Social Media Abyss: Critical Internet Cultures and the Force of Negation* (Malden, MA: Polity, 2016), 6.

76. WNYC has been financially successful over the past decade, with a steadily expanding listenership (much of it online and via podcast) and a growing staff (much of the new staff being placed in digital positions or with WNYC Studios' expanding portfolio of podcasts). Moreover, *Lehrer* is one of the network's most listened-to programs and a fundraising powerhouse, which permits the show's producers considerable autonomy. Smaller stations with shrinking budgets are not likely to enjoy such flexibility, and their local talk-news programs—if they can even

afford to produce any—often do not have the resources to utilize the internet as proactively as *Lehrer* does.

77. Kane, "Brian Lehrer," par. 11.

Chapter 5

1. "Apple Worldwide Developer Conference 2015," filmed June 8, 2015, video, 98:50, http://www.apple.com/live/2015-june-event/

2. The significance of these events to Apple's corporate identity and their place in popular culture can be seen in *Steve Jobs* (2015, director Danny Boyle): the biopic film is structured entirely around incidents taking place immediately before three of these product launches.

3. "Apple WWDC 2015."

4. Steven Rosenbaum, "The Curation Explosion, and Why Humans Still Trump Tech," *Forbes*, July 26, 2015, http://www.forbes.com/sites/stevenrosenbaum/2015/07/26/the-curation-explosion/#66a9bbb6231c

5. Apple Inc., "Introducing Apple Music—All the Ways You Love Music. All In One Place," press release, June 8, 2015, https://www.apple.com/pr/library/2015/06/08Introducing-Apple-Music-All-The-Ways-You-Love-Music-All-in-One-Place-.html

6. For 2015, streaming made up 34.3 percent of revenues, digital downloads 34.0 percent, physical media sales 28.8 percent, synch licenses 2.9 percent: Joshua P. Friedlander, "News and Notes on 2015 RIAA Shipment and Revenue Statistics," RIAA, March 22, 2016, http://www.riaa.com/wp-content/uploads/2016/03/RIAA-2015-Year-End-shipments-memo.pdf

7. Jeremy Wade Morris, *Selling Digital Music, Formatting Culture* (Berkeley: University of California Press, 2015), 152.

8. Hugh McIntyre, "Streaming Is the Future of Music, but It's Shaky," *Forbes*, March 16, 2016, http://www.forbes.com/sites/hughmcintyre/2016/03/16/streaming-is-the-future-of-music-but-its-shaky/#1fdc2e273464. By 2019, streaming accounted for 75 percent of U.S. recorded music revenue: Patricia Hernandez, "Streaming Now Accounts for 75 percent of Music Industry Revenue," *The Verge*, September 20, 2018, https://www.theverge.com/2018/9/20/17883584/streaming-record-sales-music-industry-revenue

9. The industry distinction between "streaming radio" and "music streaming" services usually comes down to level of user control. Radio services are "noninteractive" and marked by livestreaming; in the case of Pandora, user control is limited to selecting a "seed" artist/song and limited song skips. Music services are on-demand and fully customizable by the user.

10. For licensing reasons, Pandora only operates in the United States. Across all radio and music-streaming services, Pandora is used by 30 percent of Americans—making it the biggest online "audio brand," above Spotify, iHeartRadio, Apple Music, etc.: Edison Research, *The Infinite Dial 2019*, March 6, 2019, https://www.edisonresearch.com/wp-content/uploads/2019/03/Infinite-Dial-2019-PDF-1.pdf; Janko Roettgers, "Pandora Stock Ticks Up on Growth of Paying Subscriptions,

Revenue in Q1," *Variety*, May 3, 2018, https://variety.com/2018/digital/news/pandora-q1-2018-earnings-1202797588/

11. Ted Striphas, "Algorithmic Culture," *European Journal of Cultural Studies* 18.4–5 (2015): 395–412.

12. Julio Ojeda-Zapata, "New Radio Service in Town: Turn On Your Computers; Twin Cities Company Launches Net.Radio," *Saint Paul Pioneer Press*, November 4, 1995, 1A.

13. NetRadio Corporation, *2000 Annual Report*, SEC file no. 000-27575, filed April 2, 2001; NetRadio Corporation, "Form 10-Q," SEC file no. 000-27575, filed November 14, 2001.

14. Lisa Napoli, "Tuning Into the Music of Developing Nations," *New York Times*, March 16, 1998, D8.

15. Evan George, "Little Radio's Big Problem," April 23, 2007, *Los Angeles Downtown News*, http://www.ladowntownnews.com/news/little-radio-s-big-problem/article_9837dc03-a219-5685-bad1-089aa1d112c6.html; John Ortved, "The Death of East Village Radio," *New Yorker*, June 12, 2014, http://www.newyorker.com/culture/culture-desk/the-death-of-east-village-radio. Little Radio and East Village Radio started out as pirate radio stations, but following run-ins with the FCC made operations legit by transitioning online. Little Radio existed from 2004 until 2011–12. East Village Radio operated from 2003 to 2014, then resurrected in 2015 as a channel on start-up Dash Radio.

16. Ernest Jasmin, "Logged On, Tuned In: Regional Online Radio Stations Reach Worldwide," *News Tribune*, August 3, 2008; Vauhini Vara, "Can an Ohio Radio Station Reinvent Itself Yet Again?," *Wall Street Journal*, March 14, 2007, B1. KEXP was forced online-only in 2006 but returned to terrestrial broadcasting. WOXY transitioned from its FM signal in Oxford, Ohio, to a web-only home at WOXY.com in 2004; it shut down in 2010.

17. Kurt Hanson, "'Bloody Sunday' Decimates Internet Radio," *RAIN News*, February 1, 2016, http://rainnews.com/kurt-hanson-bloody-sunday-decimates-internet-radio/; Paul Riismandel, "Live365 to Broadcasters: We're Shutting Down Jan. 31," *Radio Survivor*, January 20, 2016, http://www.radiosurvivor.com/2016/01/20/live365-broadcasters-shutting-jan-31/; March Schneider, "Pioneering Web Radio Broadcaster Live365 Returns," *Billboard*, January 4, 2017, https://www.billboard.com/articles/business/7646952/pioneering-web-radio-broadcaster-live365-returns

18. Rob Sheffield, "The Golden Age of Wired-ness," *Rolling Stone*, June 8, 2000, S25–S26.

19. US Congress, Digital Millennium Copyright Act (Public Law 105-304), passed October 28, 1998, https://www.congress.gov/105/plaws/publ304/PLAW-105publ304.pdf

20. Ted Kalo, "SiriusXM's Grandfathered Rates for Music Must End," *Billboard*, May 8, 2015, http://www.billboard.com/articles/business/6554072/ted-kalo-congress-royalty-board-fair-play-pay-act-sirius-xm. The rate increased from 9 percent to 15.5 percent of gross revenue in 2018: Anne Steele, "Sirius XM Satellite Radio to Pay Higher Royalty Rate Starting in 2018," *Wall Street Journal*, December 15, 2017.

21. Jordan Heller, "A Startup Indie Radio Station Gains a Toehold in an Unfriendly Universe," *Christian Science Monitor*, September 21, 2010, par. 16.

22. Tim Anderson provides an overview of the legal and economic constraints placed on webcasters in the "new music industry": "Opening Pandora's Box: The Problematic Promise of Radio on the Internet," in *Popular Music in a Digital Music Economy: Problems and Practices for an Emerging Service Industry* (New York: Routledge, 2013), 92–117. For a legal history of internet radio royalties: Kellen Myers, "The RIAA, the DMCA, and the Forgotten Few Webcasters: A Call for Change in Digital Copyright Royalties," *Federal Communications Law Journal* 61.2 (2009), 431–56.

23. Susan Visakowitz, "Radio Daze," *Billboard*, February 9, 2008, 43–45.

24. Although not the focus of this chapter, I do not want to overlook the many radio makers today producing unique music radio programs for their own websites or through streaming services like Mixcloud and Soundcloud. Many of these producers are nonprofessionals, specializing in obscure music artists/genres and simply making radio for fun or as a way to be involved in a community—the internet era equivalent of college or community radio.

25. Carl Chery, "Meet the Beats Music Editorial Team," Tumblr post, January 21, 2014, http://carlchery.tumblr.com/post/74085763576/meet-the-beats-music-editorial-team

26. Elephant Ike, "Tim Westergren Interview: Interview with the Music Genome Project Founder," *Tiny Mix Tapes*, captured March 18, 2006, https://web.archive.Org/web/20060318150700/http://www.tinymixtapes.com/interviews/tim_westergren.htm

27. Striphas, "Algorithmic Culture," 396; Robyn Caplan and danah boyd, *Who Controls the Public Sphere in an Era of Algorithms?: Mediation, Automation, Power*, Data & Society Research Institute, May 13, 2016, http://www.datasociety.net/pubs/ap/MediationAutomationPower_2016.pdf

28. David Lazer, "The Rise of the Social Algorithm," *Science*, June 5, 2015, 1090–91.

29. Lev Manovich, "The Algorithms of Our Lives," *Chronicle Review*, December 16, 2013, par. 5, http://chronicle.com/article/the-algorithms-of-our-lives-/143557

30. Christian Sandvig, "Seeing the Sort: The Aesthetic and Industrial Defense of 'The Algorithm,'" *NMC Media-N* 10.3 (2014): par. 3, http://median.newmediacaucus.org/art-infrastructures-information/seeing-the-sort-the-aesthetic-and-industrial-defense-of-the-algorithm/

31. Caplan and boyd, *Who Controls*, 4.

32. Caplan and boyd, *Who Controls*, provide an overview of this literature on the power dynamics of algorithms.

33. Tarleton Gillespie, "The Relevance of Algorithms," in *Media Technologies: Essays on Communication, Materiality, and Society*, ed. Tarleton Gillespie, Pablo J. Boczkowski, and Kirsten A. Foot (Cambridge, MA: MIT Press, 2014), 169.

34. I say "most of" because there are still licensing issues, particularly with music radio, that can block some internet radio signals from being streamed in certain locales, not to mention digital divides and other issues of unequal network access.

35. Jeremy Wade Morris and Devon Powers, "Control, Curation and Musical

Experience in Streaming Music Services," *Creative Industries Journal* 8.2 (2015): 106.

36. Patrick Burkart and Tom McCourt, "Infrastructure for the Celestial Jukebox," *Popular Music* 23.3 (2004): 349.

37. Sandvig, "Seeing the Sort," pars. 4–6.

38. Richard Tedesco, "The DJ.com Becomes Spinner.com," *Broadcasting & Cable*, July 27, 1998, 46; "On Spinner, Wide Range of Choices," *New York Times*, November 5, 1998, G10.

39. Stephanie Clifford, "Pandora's Long Strange Trip," *Inc.*, October 1, 2007, http://www.inc.com/magazine/20071001/pandoras-long-strange-trip.html; J. D. Harrison, "When We Were Small: Pandora," *Washington Post*, February 6, 2015, G01.

40. Claire Cain Miller, "How Pandora Slipped Past the Junkyard," *New York Times*, March 8, 2010, B1.

41. Savage Beast Technologies, Inc., "Savage Beast Technologies Unleashes New Music Search and Recommendation Technology That Maps Genetic Makeup of Music to an Individual's Taste," *PR Newswire*, press release, November 13, 2000, par. 1.

42. John Koetsier, "From Savage Beast to Sweet Music: How Pandora Became the Internet's Radio Station," *VentureBeat*, December 21, 2012, par. 3, http://venturebeat.com/2012/12/21/from-savage-beast-to-sweet-music-how-pandora-became-the-internets-radio-station/

43. Jennie Starr, "Pandora: Finding New Music When You Have No Time to Hit the Clubs," *Searcher*, June 2006, 59.

44. Manuel Roig-Franzia, "Pandora's Keeper," *Washington Post*, April 4, 2013, C1.

45. Ryan Stabile, "Interview with Pandora Radio CEO, Joe Kennedy," *AXS*, December 1, 2014, par. 3, http://www.axs.com/interview-with-pandora-radio-ceo-joe-kennedy-30936

46. Roig-Franzia, "Pandora's Keeper," C1. Westergren credits Kennedy here with the internet radio idea, and Kennedy confirms it elsewhere. Other sources suggest that it was an investor, Larry Marcus of Walden Venture Capital, who pushed Savage Beast toward internet radio. In either case, it was not the company's original founders who steered them toward this strategy. Simone Baribeau, "How Pandora Soothed the Savage Beast," *Fast Company*, September 10, 2012, par. 3, http://www.fastcompany.com/3001052/how-pandora-soothed-savage-beast

47. Pandora Media, "Powered by the Music Genome Project, New Pandora Service Makes It Dramatically Easier to Find and Enjoy New Music," *BusinessWire*, press release, August 29, 2005, pars. 1–2.

48. These changes in the branding are based on an analysis of multiple Pandora.com homepage crawls captured by Internet Archive Wayback Machine: https://web.archive.org/web/*/pandora.com

49. Linda Rosencrance, "Pandora.com Sings with OpenLaszlo," *Computerworld*, February 9, 2006, 32.

50. Jay David Bolter and Richard Grusin, *Remediation: Understanding New Media* (Cambridge, MA: MIT Press, 1998), 15.

51. Elephant Ike, "Tim Westergren Interview," par. 14.

52. Ariana Moscote Freire, "Remediating Radio: Audio Streaming, Music Recommendation and the Discourse of Radioness," *Radio Journal* 5.2–3 (2007), 105–8.

53. Tim Westergren, "Written Direct Testimony," delivered before US Copyright Royalty Judges, October 6, 2014, 12, https://www.crb.gov/rate/14-CRB-0001-WR/statements/Pandora/7_Written_Direct_Testimony_of_Timothy_Westergren_PUBLIC_pdf.pdf; Janko Roettgers, "Rdio Launches Personalized Radio Stations to Steal Listeners Away from Pandora," *Gigaom*, August 8, 2013, par. 7, https://gigaom.com/2013/08/08/rdio-launches-personalized-radio-stations-to-steal-listeners-away-from-pandora/

54. The legal definitions for interactive and noninteractive services were set in the DMCA (Sec. 405). The noninteractive designation was upheld a decade later in *Arista Records v. Launch Media*. LAUNCHcast, started in 2001, operated similarly to Pandora by allowing listeners to create personal radio stations based on their taste choices. A group of record labels sued, arguing that this service was "interactive" and thus not covered by a statutory royalty—meaning that LAUNCHcast would need to negotiate with each sound recording copyright holder for every song it used. The court concluded that since listeners could not control or predict what particular songs would be played, webcasters like LAUNCHcast (and Pandora) were not a substitute for purchasing a song. David Oxenford, "Court of Appeals Determines That Launchcast Is Not an Interactive Service—Thus Not Needing Direct Licenses from the Record Labels," *Broadcast Law Blog*, August 22, 2009, http://www.broadcastlawblog.com/2009/08/articles/court-of-appeals-determines-that-launchcast-is-not-an-interactive-service-thus-not-needing-direct-licenses-from-the-record-labels/

55. Jessica Grose, "Opening Pandora.com's Box," *Spin*, December 7, 2005, par. 8, https://web.archive.org/web/20060413090935/http://www.spinmag.com/features/exclusives/2005/12/051207_pandora/

56. Pandora Media, Inc., *United States Patent No. US 7,003,515 B1: Consumer Item Matching Method and System*, published February 21, 2006, http://www.google.com/patents/US7003515

57. As the MGP is a proprietary invention, the full list of "genes" and "chromosomes" has never been made public. However, Pandora representatives have revealed various aspects of the MGP in interviews and promotional materials. These genes for vocal delivery are described in Eric K. Arnold, "Oakulture: The Coup's Whiteboard Show at Pandora," *Oakland Local*, March 4, 2014, par. 6, http://oaklandlocal.com/2014/03/oakulture-the-coups-whiteboard-show-at-pandora-two-new-co-working-spaces-emphasize-the-intersection-of-tech-and-culture/

58. Rob Walker, "The Song Decoders," *New York Times Magazine*, October 18, 2009, 48.

59. Vignesh Ramachandran, "Pandora Staff—Not Computers—Analyze 10,000 Songs Each Month," *Peninsula Press*, March 27, 2012, http://archive.peninsulapress.com/2012/03/27/pandora-staff-not-computers-analyze-10000-songs-each-month/

60. Elephant Ike, "Tim Westergren Interview," par. 19.

61. Danny Bradbury, "Mapping the Music Genome: A New Service Can Match

Your Tastes with Tracks You Don't Even Know," *The Independent* (UK), September 14, 2005, 47. An item-to-item recommender system that bases suggestions on previous consumption and matches them with the behavior of similar users is known as *collaborative filtering*; it is widely utilized by music-streaming services as well as e-commerce sites like Amazon.

62. Dev Patnaik, "Exclusive Q&A with Pandora Radio Founder Tim Westergren," *Wired to Care*, March 12, 2009, http://www.wiredtocare.com/?p=1203

63. Matthew Lasar, "Digging into Pandora's Music Genome with Musicologist Nolan Gasser," *Ars Technica*, January 12, 2011, par. 11, http://arstechnica.com/tech-policy/2011/01/digging-into-pandoras-music-genome-with-musicologist-nolan-gasser/

64. Walker, "The Song Decoders," 48.

65. Jonathan Segel, "Pandora, Now a Year Down the Road from Me," *Jonathan Segel*, March 21, 2013, par. 3, https://jsegel.wordpress.com/2013/03/21/pandora-now-a-year-down-the-road-from-me/

66. Journalists and Pandora representatives cite this 10,000 songs per month number repeatedly, though the math is fuzzy. For instance, Ramachandran writes that Pandora had twenty-six analysts and analysis took ten to thirty minutes per song. If analysts were working twenty hours per week (most were said to be part time), even at the minimum of ten minutes per song they could only log 2,600 songs in a month. Even if they worked forty hours a week, they would still barely clear 5,000 songs a month.

67. Pandora Media, Inc., "Form 424B4," SEC file no. 333-172215, filed June 15, 2011, 83. According to Gasser, music analysts have "advanced music degrees," specifically backgrounds in theory and history: Lasar, "Digging," par. 42. Elsewhere, Westergren has said music analysts are "trained musicians, typically with at least a four-year degree in music theory": Tim Westergren, "The Music Genome Project," *AlwaysOn: The Insiders Network*, captured March 24, 2006, par. 7, https://web.archive.org/web/20060324042152/http://www.alwayson-network.com/printpage.php?id=P10557_0_4_0_C

68. Stabile, "Interview with Kennedy," par. 8.

69. Kinley Levack, "Tracking Down Tracks: Search Engines Evolve," *EContent*, April 2004, 8; Lasar, "Digging," par. 45. Music analysts do not cover music in every genome; they only work on those genres they have expertise in. Most analysts work across three to four genomes at most.

70. Jason Kirby, "More on Pandora: Genres, Genomes, and Musical Taste," *Scholars' Lab*, February 15, 2010, par. 17, http://scholarslab.org/digital-humanities/more-on-pandora-genres/

71. Dietmar Jannach et al., *Recommender Systems: An Introduction* (New York: Cambridge University Press, 2011), 51–52.

72. Pandora's playlist algorithms take into account information like the time of day and where listeners are geographically located, not only data about personal taste and preference: John Paul Titlow, "At Pandora, Every Listener Is a Test Subject," *Fast Company*, August 14, 2013, par. 4, http://www.fastcompany.com/3015729/in-pandoras-big-data-experiments-youre-just-another-lab-rat

73. Jennifer Smith Maguire and Julian Matthews, "Are We All Cultural Inter-

mediaries Now?: An Introduction to Cultural Intermediaries in Context," *European Journal of Cultural Studies* 15.5 (2012): 551–62.

74. Pierre Bourdieu, *Distinction: A Social Critique of the Judgement of Taste*, trans. Richard Nice (Cambridge, MA: Harvard University Press, 1984), 359.

75. Jeremy Wade Morris, "Curation by Code: Infomediaries and the Data Mining of Taste," *European Journal of Cultural Studies* 18.4–5 (2015): 459.

76. Ibid.

77. Bill Brewster and Frank Broughton, *Last Night a DJ Saved My Life: The History of the Disc Jockey* (New York: Grove, 2000), 21.

78. Steve Taylor, "'I Am What I Play': The Radio DJ as Cultural Arbiter and Negotiator," in *Cultural Work: Understanding the Cultural Industries*, ed. Andrew Beck (New York: Routledge, 2003), 90.

79. Elena Razlogova, "The Past and Future of Music Listening: Between Freeform DJs and Recommendation Algorithms," in *Radio's New Wave: Global Sound in the Digital Era* (New York: Routledge, 2013), 62.

80. Eric Rothenbuhler and Tom McCourt, "Radio Redefines Itself, 1947–1962," in *Radio Reader: Essays in the Cultural History of Radio*, ed. Michele Hilmes and Jason Loviglio (New York: Routledge, 2002), 380–84.

81. Christopher H. Sterling and Michael C. Keith, *Sounds of Change: A History of FM Broadcasting in America* (Chapel Hill: University of North Carolina Press, 2008), 127–54.

82. Tyler Gray, "Pandora Pulls Back the Curtain on Its Magic Music Machine," *Fast Company*, January 21, 2011, par. 9, http://www.fastcompany.com/1718527/pandora-pulls-back-curtain-its-magic-music-machine

83. Bill Machrone, "Out of Pandora's Box," *PC Magazine*, May 23, 2006, 60.

84. Tim Westergren, "Does the Music Genome Project Include Information on Whether Songs Are Well-Known?," *Quora*, January 27, 2011, https://www.quora.com/Does-the-Music-Genome-Project-include-information-on-whether-songs-are-well-known/answer/Tim-Westergren

85. Elephant Ike, "Tim Westergren Interview," par. 7.

86. Levack, "Tracking Down Tracks," 8.

87. Patnaik, "Exclusive Q&A," par. 21.

88. Westergren, "Written," 3.

89. Richard Middleton, *Studying Popular Music* (Bristol, PA: Open University Press, 1990), 106–7.

90. Judy Lochhead, "Introduction," in *Postmodern Music/Postmodern Thought*, ed. Judy Lochhead and Joseph Auner (New York: Routledge, 2002), 2.

91. Bourdieu, *Distinction*, 342.

92. Evan Serpick, "Digital Custom Radio: How a Project to Catalog All the World's Music Became a Web Site That Can Predict What You Like," *Rolling Stone*, February 9, 2006, 20.

93. Walker, "The Song Decoders," 48.

94. Clifford, "Pandora's Long Strange Trip," par. 13.

95. Machrone, "Out of Pandora's Box," 60.

96. Laura Woods, "Analyst Spotlight: Meet Michelle Alexander, Pandora's

Eclectic Musical Mind," *Pandora Blog*, September 27, 2012, par. 1, http://blog.pandora.com/pandora-innovators/meet-michelle-alexander/

97. Steve Hardy, "Creative Generalist Q&A: Tim Westergren," *Creative Generalist*, October 31, 2006, par. 12, http://creativegeneralist.com/2006/10/creative-generalist-qa-tim-westergren/

98. Westergren, "Written," 10.

99. Striphas, "Algorithmic Culture," 407.

100. Linda Tischler, "Algorhythm & Blues: How Pandora's Matching Service Cuts the Chaos of Digital Music," *Fast Company*, December 2005, 89.

101. Walker, "The Song Decoders," 48.

102. Gretchen Lee, "Alumni Feature: Calibrating His Life to Music," *Washington Magazine*, April 2012, par. 11, http://magazine.wustl.edu/2012/April/Pages/Steve-HoganCalibratingHisLifetoMusic.aspx

103. Starr, "Pandora," 57.

104. Tischler, "Algorhythm & Blues," 89.

105. Starr, "Pandora," 58.

106. Walker, "The Song Decoders," 48.

107. Tischler, "Algorhythm & Blues," 89.

108. Jim Harrington, "2010: Thinking Inside the Box: Tim Westergren and Pandora Reinvent Internet Radio," *San Jose Mercury News*, February 4, 2010, par. 28, http://www.mercurynews.com/business/ci_14318446

109. Elephant Ike, "Tim Westergren Interview," par. 16.

110. Erin Griffith, "Silicon Valley Might Never Solve the Web's Impossible Problem: Taste," *Pando*, October 30, 2013, par. 10, https://pando.com/2013/10/30/silicon-valley-might-never-solve-the-webs-impossible-problem-taste/

111. Titlow, "At Pandora," par. 6.

112. Reyhan Harmanci, "Tech Confessional: How Pandora's IPO Changed Everything," *BuzzFeed*, August 8, 2012, par. 3, https://www.buzzfeed.com/reyhan/tech-confessional-how-pandoras-ipo-changed-every?utm_term=.dv63gyRVx#.xaP528PgM

113. The majority of Pandora's audience uses its free service tier, which operates on the same ad-supported business model as commercial broadcast radio. Since ad rates depend on attracting large numbers of listeners and keeping them "engaged" for lengthy periods of time, there is arguably a strong incentive for Pandora to limit unfamiliar new music in its listeners' streams, instead providing them with a steady flow of familiar music that will keep them from logging off.

114. Jeff Leeds, "Radio's New Strategy: Play a Hit, Again and Again," *New York Times*, December 1, 2007, B9.

115. Titlow, "At Pandora," par. 12.

116. "Surprisal" is Eric Bischke's term: Rob Pegoraro, "Pandora's 'Music Genome Project' Explores the Cold Hard Facts of How We Interact with Music," *Boing Boing*, May 24, 2014, par. 17, http://boingboing.net/2014/05/24/pandoras-music-genome-proj.html

117. Titlow, "At Pandora," pars. 10–11.

118. Martin Edlund, "The Madonna Code: Searching for the Perfect Music Rec-

ommendation System," *Slate*, July 5, 2005, par. 8, http://www.slate.com/articles/arts/music_box/2005/07/the_madonna_code.html

119. Walker, "The Song Decoders," 48.

120. Titlow, "At Pandora," par. 23; Sarah Perez, "Pandora Takes on Spotify with Dozens of Personalized Playlists Built Using Its Music Genome," *TechCrunch*, March 28, 2018, https://techcrunch.com/2018/03/28/pandora-takes-on-spotify-with-dozens-of-personalized-playlists-built-using-its-music-genome/

121. Natasha Singer, "Listen to Pandora, and It Listens Back," *New York Times*, January 5, 2014, BU3.

122. Titlow, "At Pandora," par. 4.

123. Pegoraro, "Pandora's Music Genome Project," pars. 14–15.

124. The data analysts and engineers also serve ad sales staff by performing "behavioral targeting," customizing ads by identifying correlations between users' demographics, listening habits, and ad tolerance: Singer, "Listen to Pandora," BU3.

125. Titlow, "At Pandora," par. 28; Perez, "Pandora Takes on Spotify," par. 4.

126. Titlow, "At Pandora," par. 26.

127. Chris Priestman, "Narrowcasting and the Dream of Radio's Great Global Conversation," *Radio Journal* 2.2 (2004): 85–86; Alan Beck, *The Death of Radio? An Essay in Radio Philosophy for the Digital Age* (Kent, UK: Sound Journal, 2002), Section 7, http://www.savoyhill.co.uk/deathofradio/

128. Priestman, "Narrowcasting," 85.

129. Sterling and Keith, *Sounds of Change*, 133.

130. Andrew Barker, "B'cast Radio Faces the Music; AM/FM Feeling the Squeeze as Internet Jukeboxes Branch Out," *Variety*, November 29, 2010, 15.

131. "Apple WWDC 2015."

132. Sofie Lindblom, "What Made Discover Weekly One of Our Most Successful Feature Launches to Date?," *Spotify Labs*, November 18, 2015, par. 1, https://labs.spotify.com/2015/11/18/what-made-discover-weekly-one-of-our-most-successful-feature-launches-to-date/

133. James Geddes, "Spotify Adds New Curated 'In Residence' Radio Shows to Compete with Apple Music's Beats 1," *Tech Times*, October 28, 2015, http://www.techtimes.com/articles/100111/20151028/spotify-adds-new-curated-in-residence-radio-shows-to-compete-with-apple-music-beats-one-radio.htm

134. Rdio CEO Anthony Bay quoted in Steven Rosenbaum, "The Curation Explosion, and Why Humans Still Trump Tech," *Forbes*, par. 12, http://www.forbes.com/sites/stevenrosenbaum/2015/07/26/the-curation-explosion/#14e18c1a231c; Janko Roettgers, "Rdio Launches Personalized Radio Stations to Steal Listeners Away from Pandora," *Gigaom*, August 8, 2013, https://gigaom.com/2013/08/08/rdio-launches-personalized-radio-stations-to-steal-listeners-away-from-pandora/

135. Tim Moynihan, "Songza Is Dead, but It Lives On within Google Play Music," *Wired*, December 2, 2015, http://www.wired.com/2015/12/songza-is-dead-but-it-lives-on-within-google-play-music/

136. Pandora Media, Inc., "Pandora Delivers Next Level Discovery with Handpicked New Music Stations," press release, June 15, 2016, http://press.pandora.com/phoenix.zhtml?c=251764&p=irol-newsArticle&ID=2177702

137. Nathan Ingraham, "Pandora Adds Curated Playlists to Its On-Demand Mu-

sic Service," *Engadget*, October 30, 2017, https://www.engadget.com/2017/10/30/pandora-curated-playlists-premium-service/; Billy Steele, "Pandora Goes Full Spotify with Personalized Playlists," *Engadget*, March 28, 2018, https://www.engadget.com/2018/03/28/pandora-personalized-soundtracks/

138. John Paul Titlow, "Inside Pandora's Plan to Reinvent Itself—and Beat Back Apple and Spotify," *Fast Company*, April 26, 2016, http://www.fastcompany.com/3058719/most-innovative-companies/inside-pandoras-plan-to-reinvent-itself-and-beat-back-apple-and-sp

139. Mona Lalwani, "Ebro Darden: The DJ Who Curates the Sound of New York on Beats 1," *Engadget*, October 9, 2015, https://www.engadget.com/2015/10/09/ebro-darden-curates-the-sound-of-new-york-on-beats-1/

140. Victor Luckerson, "Radio Fans Will Love Rdio's Newest Feature," *Time*, August 12, 2015, http://time.com/3993374/rdio-live-radio/

Chapter 6

1. *Radiolab*, "Musical Language," produced by Jad Abumrad and Ellen Horne, WNYC, April 21, 2006, audio, http://www.radiolab.org/story/91512-musical-language/

2. The talk was given at the SoHo Apple Store in New York City in spring 2006 and released as a podcast extra in November 2007: *Radiolab*, "Making Radiolab," produced by Jad Abumrad and Ellen Horne, WNYC, November 6, 2007, http://www.radiolab.org/story/91746-making-radio-lab/

3. The phrase "audio storytelling" is applied so liberally in industry and popular discourse that it appears as a stand-in for podcasting writ large. See, for instance, Sarah Larson, "'Serial,' Podcasts, and Humanizing the News," *New Yorker*, February 20, 2015, http://www.newyorker.com/culture/sarah-larson/serial-podcasts-humanizing-news

4. The Creative Audio Unit is the name of the Australian Broadcasting Corporation Radio National division for "new and emerging innovations in production and storytelling." Many radio producers use the term "creative audio" to describe their work. "Sound-rich audio stories" is the term the Third Coast International Audio Festival uses to describe the documentary radio programming the organization curates. The term "sound-rich" is a widely used euphemism in radio and podcast production cultures for highly produced texts.

5. Siobhan McHugh, "Editorial: Podcasting as the New Space for Crafted Audio," *RadioDoc Review* 3.1 (2017): 1.

6. Siobhan McHugh, "How Podcasting Is Changing the Audio Storytelling Genre," *Radio Journal* 14.1 (2016): 76.

7. Jonathan Kern, *Sound Reporting: The NPR Guide to Audio Journalism and Production* (Chicago: University of Chicago Press, 2008), 341.

8. Jack Hart, *Storycraft: The Complete Guide to Writing Narrative Nonfiction* (Chicago: University of Chicago, 2011), 10.

9. Neil Verma, "The Arts of Amnesia: The Case for Audio Drama, Part One," *RadioDoc Review* 3.1 (2017): 5.

10. John Hartley and Kelly McWilliam, "Computational Power Meets Human

Contact," in *Story Circle: Digital Storytelling Around the World*, ed. John Hartley and Kelly McWilliam (Malden, MA: Wiley-Blackwell, 2009), 3.

11. Ibid., 3.

12. Michael Wilson, *Storytelling and Theatre: Contemporary Storytellers and Their Art* (New York: Palgrave Macmillan, 2006), 59.

13. "What is The Moth?," The Moth, accessed March 14, 2016, http://themoth.org/about

14. "StartUp: About," Gimlet Media, accessed March 14, 2016, https://gimletmedia.com/show/startup/about/. *StartUp* returned to Gimlet in 2015–16 for a "mini-season."

15. iTunes/Apple Podcasts is far from the only podcast distribution platform available; however, it accounts for approximately 70 percent of all podcast downloads internationally. As there currently exists no industry standard audience measurement system for podcasts, the iTunes chart rankings are widely accepted as the closest thing to a publicly available "ratings" system. Eric Blattberg, "The Measurement Challenge Facing Podcast Advertising," *Digiday*, February 5, 2015, http://digiday.com/platforms/measurement-challenge-podcast-advertising/

16. Adam Carlson, "America's Most Popular Podcast: What the Internet Did to 'Welcome to Night Vale,'" *The Awl*, July 24, 2013, http://www.theawl.com/2013/07/americas-most-popular-podcast-what-the-internet-did-to-welcome-to-night-vale; "The Making of a Podcast Phenomenon: *Welcome to Night Vale* by the Numbers," *Details*, November 6, 2013, http://www.details.com/blogs/daily-details/2013/11/the-making-of-a-phenomenon-night-vale.html

17. iTunes/Apple Podcasts charts are archived at http://www.itunescharts.net

18. Greg Evans, "FX to Develop 'Welcome to Night Vale' Podcast for TV," *Deadline*, December 8, 2017, https://deadline.com/2017/12/fx-welcome-to-night-vale-production-deal-podcast-1202222829/

19. I define *radio drama* (or *radio play*) as a fictionalized, purely acoustic performance; it is a dramatized text (including adaptations of stage plays, novels, short stories, etc.) that is crafted to be heard, its essence being sound (voice, music, noise, and silence). In this chapter, I default to the recently popularized term "audio drama," although as Neil Verma has observed, there is little discernible difference between radio drama and audio drama; they are not separate or rival practices, even if some distinctions are beginning to emerge at a textual level: Verma, "Arts," 6.

20. Rebecca Greenfield, "The Future of Media is Podcasting," *Fast Company*, February 9, 2015, http://www.fastcompany.com/3041522/pod-power

21. Jenna Scherer, "A Primer on New York Podcast Welcome to Night Vale," *Time Out New York*, October 8, 2013, http://www.timeout.com/newyork/things-to-do/a-primer-on-new-york-podcast-welcome-to-night-vale

22. Sean Kelley, "Welcome to Night Vale: 5 Marketing Lessons from America's Most Popular Podcast," *Lonelybrand*, December 17, 2013, http://lonelybrand.com/blog/welcome-to-night-vale-five-marketing-lessons-from-americas-most-popular-podcast/; Mandy Kilinskis, "Accidental Viral Marketing Lessons from 'Welcome to Night Vale,'" *Quality Logo Products: Branding Beat*, updated March 18, 2019, http://www.qualitylogoproducts.com/blog/accidental-viral-marketing-lessons-from-welcome-to-night-vale/

23. David Carr, "Big Media Wants a Piece of Your Pod," *New York Times*, July 4, 2005, C1.

24. Max Willens, "Americans Listen to 21 million Hours of Podcasts Every Day, Cutting into Radio, Research Finds," *International Business Times*, January 23, 2015, http://www.ibtimes.com/americans-listen-21-million-hours-podcasts-every-day-cutting-radio-research-finds-1792814

25. Sheri Crofts et al., "Podcasting: A New Technology in Search of Viable Business Models," *First Monday* 10.9 (2005), http://firstmonday.org/article/view/1273/1193; Richard Berry, "Will the iPod Kill the Radio Star?: Profiling Podcasting as Radio," *Convergence* 12.2 (2006): 143–62; Kris M. Markman, "Doing Radio, Making Friends, and Having Fun: Exploring the Motivations of Independent Audio Podcasters," *New Media & Society* 14.4 (2011): 547–65; Kris M. Markman and Caroline E. Sawyer, "Why Pod?: Further Explorations of the Motivations for Independent Podcasting," *Journal of Radio & Audio Media* 21.1 (2014): 20–35; Nele Heise, "On the Shoulders of Giants?: How Audio Podcasters Adopt, Transform and Re-invent Radio Storytelling," *Transnational Radio Stories* (2014), https://hamburgergarnele.files.wordpress.com/2014/09/podcasts_heise_public.pdf

26. Marika Lüders, "Conceptualizing Personal Media," *New Media & Society* 10.5 (2008): 683–702.

27. Axel Bruns, "Towards Produsage: Futures for User-led Content Production," *Proceedings: Cultural Attitudes towards Communication and Technology* (2006): 1–10; Henry Jenkins, Sam Ford, and Joshua Green, *Spreadable Media: Creating Value and Meaning in a Networked Culture* (New York: NYU Press, 2013); Yochai Benkler, *The Wealth of Networks: How Social Production Transforms Markets and Freedom* (New Haven, CT: Yale University Press, 2006); Manuel Castells, *Networks of Outrage and Hope: Social Movements in the Internet Age* (Malden, MA: Polity, 2012).

28. Markman, "Doing Radio"; Markman and Sawyer, "Why Pod?"

29. Rebecca Greenfield, "The (Surprisingly) Profitable Rise of Podcast Networks," *Fast Company*, September 26, 2014, http://www.fastcompany.com/3035954/most-creative-people/the-surprisingly-profitable-rise-of-podcast-networks

30. Neil Verma, *Theater of the Mind: Imagination, Aesthetics, and American Radio Drama* (Chicago: University of Chicago Press, 2012).

31. Howard Fink, "The Sponsor's v. the Nation's Choice: North American Radio Drama," in *Radio Drama*, ed. Peter Lewis (London: Longman, 1981), 185–243.

32. Richard J. Hand and Mary Traynor, *The Radio Drama Handbook: Audio Drama in Practice and Context* (New York: Continuum, 2011). The BBC also has a digital-only channel, BBC Radio 4 Extra (launched in 2002 as BBC Radio 7), specializing in drama and comedy.

33. The Internet Archive's "Old Time Radio" collection: https://archive.org/details/oldtimeradio

34. "Old Time Radio Live," Wisconsin Public Radio, accessed March 25, 2016, http://www.wpr.org/series/old-time-radio-live

35. Gavia Baker-Whitelaw, "'Welcome to Night Vale,' Where David Lynch Meets 'The Twilight Zone,'" *Daily Dot*, June 10, 2013, http://www.dailydot.com/entertainment/welcome-to-night-vale-podcast/; Cory Doctorow, "Welcome to Night Vale: An Appreciation of the Spookiest, Funniest Podcast," *Boing Boing*,

December 12, 2013, http://boingboing.net/2013/12/12/welcome-to-night-vale-an-appr.html; Alex Biese, "Cecil Baldwin & Joseph Fink on 'Welcome to Night Vale,'" *Asbury Park Press*, June 30, 2014, http://www.app.com/story/entertainment/arts/2014/06/27/cecil-baldwin-joseph-fink-welcome-night-vale/11547587/

36. Jay David Bolter and Richard Grusin, *Remediation: Understanding New Media* (Cambridge, MA: MIT Press, 1998), 15.

37. Radio dramas do not need to be serialized; some are single plays, including adaptations of stage plays. However, the serial, along with the series (self-contained episodes involving the same group of characters or actors), proved to be the preferred forms for fictional programming during the Golden Age.

38. Danny Gallagher, "*Welcome to Night Vale* Creators Discuss Inspiration, Podcasts, Texas Roots," *Dallas Observer Blogs*, March 12, 2014, par. 19, http://blogs.dallasobserver.com/mixmaster/2014/03/a_qa_with_welcome_to_night_val.php

39. Michele Hilmes, *Radio Voices: American Broadcasting, 1922–1952* (Minneapolis: University of Minnesota Press, 1997).

40. Like *Welcome to Night Vale*, a majority of the audio drama podcasts that have emerged in the 2010s adopt the mystery, science fiction, horror, and fantasy genres.

41. Shawn VanCour, "From Mercury to Mars: A Hard Act to Follow: *War of the Worlds* and the Challenges of Literary Adaptation," *Antenna: Responses to Media & Culture*, October 14, 2013, http://blog.commarts.wisc.edu/2013/10/14/from-mercury-to-mars-a-hard-act-to-follow-war-of-the-worlds-and-the-challenges-of-literary-adaptation-2/

42. Elissa S. Guralnick, "Radio Drama: The Stage of the Mind," *Virginia Quarterly Review* 61.1 (1985): 71.

43. *Welcome to Night Vale*, "25—One Year Later," written by Joseph Fink and Jeffrey Cranor, Night Vale Presents, June 15, 2013, audio, https://beta.prx.org/stories/229728

44. Martin Shingler and Cindy Wieringa, *On Air: Methods and Meanings of Radio* (New York: Oxford University Press, 1998), 51.

45. Andrew Crisell, *Understanding Radio* (New York: Methuen, 1986), 58–59.

46. Paddy Scannell, *Broadcast Talk* (Newbury Park, CA: Sage, 1991), 2.

47. *Welcome to Night Vale*, "13—a Story about You," written by Joseph Fink and Jeffrey Cranor, Night Vale Presents, December 15, 2012, audio, https://beta.prx.org/stories/229705; *Welcome to Night Vale*, "19A—The Sandstorm," written by Joseph Fink and Jeffrey Cranor, Night Vale Presents, March 15, 2013, audio, https://beta.prx.org/stories/229730

48. Jason Loviglio, "Sound Effects: Gender, Voice, and the Cultural Work of NPR," *Radio Journal* 5.2–3 (2008), 78; Brian Montopoli, "All Things Considerate: How NPR Makes Tavis Smiley Sound Like Linda Wertheimer," *Washington Monthly*, January–February 2003, par. 1, http://www.washingtonmonthly.com/features/2001/0301.montopoli.html

49. Noah Arceneaux, "CB Radio: Mobile Social Networking in the 1970s," in *The Mobile Media Reader*, ed. Noah Arceneaux and Anandam Kavoori (New York: Peter Lang, 2012), 55–67.

50. *RiYL*, "Episode 095: Jeffrey Cranor (of *Welcome to Night Vale*)," interview

by Brian Heater, Boing Boing, March 4, 2015, audio, http://riylcast.tumblr.com/post/112688741205/episode-095-jeffrey-cranor-of-welcome-to-night

51. Alexandra Alter, "The New Explosion in Audio Books," *Wall Street Journal*, August 1, 2013; James Atlas, "Hearing Is Believing," *New York Times*, January 11, 2015, SR8.

52. "Meet the 2011 MacArthur Fellows," MacArthur Foundation, accessed March 28, 2016, https://www.macfound.org/fellows/class/2011/

53. "MacArthur Fellows Frequently Asked Questions," MacArthur Foundation, accessed March 28, 2016, https://www.macfound.org/fellows-faq/

54. "All Fellows," MacArthur Foundation, accessed May 26, 2019, https://www.macfound.org/fellows/search/all

55. "MacArthur Fellows Program: Jad Abumrad," MacArthur Foundation, September 20, 2011, https://www.macfound.org/fellows/1/

56. Brian Howe, "Pain in the Gut," *Indy Week*, October 29, 2014, 26–27.

57. Sarah Larson, "'Invisibilia' and the Evolving Art of Radio," *New Yorker*, January 21, 2015, http://www.newyorker.com/culture/sarah-larson/invisibilia-evolving-art-radio

58. Ira Glass, "Radiolab: An Appreciation," *Transom.org*, November 8, 2011, http://transom.org/2011/ira-glass-radiolab-appreciation/

59. Around *Radiolab*'s breakout success in 2010–11, before it was as widely syndicated on broadcast radio, its podcast downloads outpaced its average terrestrial radio audience by a ratio of two to one. Rebecca Nicholson, "Radiolab: The Podcast That Makes Science Fun and Geeking Out Socially Acceptable," *The Guardian* (UK), April 22, 2011, http://www.theguardian.com/tv-and-radio/2011/apr/23/radiolab-podcast-abumrad-krulwich

60. *Radiolab*, "The Radio Lab," produced by Jad Abumrad, WNYC Studios, May 25, 2017, audio, https://www.wnycstudios.org/story/15-years/

61. *Ask Me Another*, "Jad Abumrad: Accidental Scientist," interview by Ophira Eisenberg, NPR, March 15, 2013, audio, https://www.npr.org/2013/06/14/174123830/jad-abumrad-accidental-scientist

62. Rob Walker, "The Sweet Sound of Science," *New York Times Magazine*, April 10, 2011, 42–45; "People: Jad Abumrad," Radiolab, accessed March 28, 2016, http://www.radiolab.org/people/jad-abumrad/

63. Tony Perez, "The Science of Story-Telling: A Conversation with Robert Krulwich and Jad Abumrad of Radiolab," *Tin House* 13.3 (2012): 142; Jad Abumrad, "Manifesto: Jad Abumrad," *Transom.org*, July 26, 2012, http://transom.org/2012/jad-abumrad-terrors-and-virtues/

64. Abumrad, "Manifesto"; Ellen Horne, interview by author, January 9, 2015.

65. *Radiolab*, "The Radio Lab."

66. Maddie Oatman, "Radiolab's Jad Abumrad Tries Something Completely Nuts," *Mother Jones*, July–August, 2012, http://www.motherjones.com/media/2012/05/jad-abumrad-radiolab-robert-krulwich-in-the-dark

67. Howe, "Pain in the Gut," 27.

68. Ibid.

69. Perez, "Science of Story-Telling," 142.

294 · *Notes to Pages 192–95*

70. Howe, "Pain in the Gut," 27.

71. Horne, interview by author. Horne departed *Radiolab* in September 2015, after twelve years, for an executive producer position at Audible.

72. Abumrad, "Manifesto."

73. Horne, interview by author.

74. Ben Crandell, "'Radiolab' Hosts Talk about Their Experimental Show," *McClatchy-Tribune Business News*, January 26, 2012.

75. Mike Janssen, "*Radio Lab*: Where Big Ideas Become Audio Art," *Current*, October 23, 2006, http://current.org/files/archive-site/radio/radio0619radiolab.shtml. The episode, "Look Out . . . Martians!," was first aired on October 31, 2003, and later was reworked into the 2008 "War of the Worlds" episode (Season 4, Episode 3).

76. Crandell, "Radiolab Hosts Talk"; Janssen, "*Radio Lab*."

77. Marah Eakin, "Radiolab's Jad Abumrad and Robert Krulwich on Their Favorite Radiolab Episodes," *A.V. Club*, September 26, 2013, http://www.avclub.com/article/radiolabs-jad-abumrad-and-robert-krulwich-on-their-103356. The "Robert and Jad romance story" is also recounted by Abumrad and Krulwich in a *Radiolab* podcast "extra" from 2008. In it, they state that this very first segment they ever produced together was sent to Ira Glass for a Flag Day episode of *This American Life*. This does not quite match up with the piece Abumrad describes in the *A.V. Club* interview (there is no banter, and no "strange skit about rabbits"). Nevertheless, "It was horrible. It was really horrible," Glass remarked, adding, "It's just amazing that you were able to put together such a wonderful program after that." The podcast also includes audio of that rejected Abumrad-Krulwich *TAL* submission: *Radiolab*, "Jad and Robert: The Early Years," WNYC, May 6, 2008, audio, http://www.radiolab.org/story/91820-jad-and-robert-the-early-years/

78. Perez, "Science of Story-Telling," 142–43.

79. Since Season 13 (2014–15), *Radiolab* has increasingly pivoted away from science topics to broader studies of culture and human nature.

80. Marc Fisher, "With 'Radio Lab,' Krulwich and Co. Will Stretch the Shape— and Sound—of Reporting," *Washington Post*, October 1, 2006, http://www.washingtonpost.com/wp-dyn/content/article/2006/09/29/AR2006092900236.html

81. Patricia Burstein, "Laughing All the Way to the Bank," *New York*, February 15, 1988, 24.

82. Amy Farber, "Historical Echoes: Not-So-Classical Opera Explains Interest Rates," *Liberty Street Economics*, August 24, 2012, http://libertystreeteconomics.newyorkfed.org/2012/08/historical-echoes-not-so-classical-opera-explains-interest-rates.html#.Vvq9QhIrJPM

83. Burstein, "Laughing All the Way," 24.

84. Eric Benson, "228 Minutes with Jad Abumrad and Robert Krulwich," *New York*, July 30, 2012, http://nymag.com/news/intelligencer/encounter/jad-abumrad-robert-krulwich-2012-7/

85. John Biewen, "Introduction," in *Reality Radio: Telling True Stories in Sound*, ed. John Biewen and Alexa Dilworth (Chapel Hill: University of North Carolina Press, 2010), 5.

86. Ibid., 5.

87. Jad Abumrad, "No Holes Were Drilled in the Heads of Animals in the Making of This Radio Show," in *Reality Radio: Telling True Stories in Sound*, ed. John Biewen and Alexa Dilworth (Chapel Hill: University of North Carolina Press, 2010), 45.

88. Ibid., 44.

89. *Radiolab*, "Patient Zero," produced by Jad Abumrad, WNYC, November 14, 2011, audio, http://www.radiolab.org/story/169879-patient-zero/. Episode first aired as Season 10, Episode 4; updated and rebroadcast in 2014 as Season 13, Episode 3.

90. Josh Richmond, "Noisecasting: The Search for Podcasting's Bleeding Edge," *The Timbre*, February 10, 2015, par. 1, http://thetimbre.com/noisecasting-search-podcastings-bleeding-edge/

91. The Old Time Radio Researcher's Group home page: https://archive.org/details/OTRR_Home_Page

Chapter 7

1. "Behind the Scenes with Nick van der Kolk and Laura Kwerel," Third Coast International Audio Festival, accessed May 26, 2019, https://www.thirdcoastfestival.org/article/nick-van-der-kolk-and-laura-kwerel-bts

2. These are some of the story themes listed on the website for StoryCenter (formerly the Center for Digital Storytelling), the nonprofit organization operated by Joe Lambert, one of the cofounders of the digital storytelling movement: http://www.storycenter.org/stories/

3. Sarah Kozloff, *Invisible Storytellers: Voice-Over Narration in American Fiction Film* (Berkeley: University of California Press, 1989), 33.

4. Ira Glass, "Foreword," in Jessica Abel, *Out on the Wire: The Storytelling Secrets of the New Masters of Radio* (New York: Broadway, 2015), x.

5. I have written about these wartime radio documentary programs in Andrew J. Bottomley, "The Ballad of Alan and Auntie Beeb: Alan Lomax's Radio Programmes for the BBC, 1943–1960," *Historical Journal of Film, Radio, and Television* 36.4 (2015), 8–12. See also Michele Hilmes, *Network Nations: A Transnational History of British and American Broadcasting* (New York: Routledge, 2012), 120–32.

6. The Transom website—"a showcase and workshop for new public radio"—regularly features nonnarrated radio pieces and production techniques: http://transom.org/tag/non-narrated/

7. "About Radio Diaries," Radio Diaries, accessed March 15, 2016, http://www.radiodiaries.org/about/; "About Us," Kitchen Sisters, accessed March 15, 2016, http://www.kitchensisters.org/about.htm

8. "Company Profile," Sound Portraits, accessed March 15, 2016, http://www.soundportraits.org/about/

9. David Isay, "A Sound Portraits Manifesto," *Transom.org*, July 1, 2003, http://transom.org/2003/david-isay-sound-portraits/

10. Jessica Abel, *Out On the Wire: The Storytelling Secrets of the New Masters of Radio* (New York: Broadway, 2015), 41.

11. "About Radio Diaries."

12. Joe Richman, "Manifesto," *Transom.org* (2014, April 15), http://transom.org/2014/joe-richman/

13. Radio Diaries, "About Radio Diaries."

14. The Kitchen Sisters, "Manifesto," *Transom.org*, September 18, 2012, http://transom.org/2012/kitchen-sisters-kitchen-vision/

15. *Love + Radio* deviates occasionally from the nonnarrated mode, with select episodes featuring participation from the show's producers and, occasionally, even exposition.

16. "WXBC Bard Radio," *The Bard Free Press*, February 21, 2005, 9.

17. Some of the material originally produced for WXBC's *This One Time* was repurposed under the *Love + Radio* banner.

18. "Love + Radio," Need Supply Co., February 14, 2016, http://blog.needsupply.com/2016/02/14/love-radio/; Adrianne Mathiowetz, "Where I Am Now," PRX, July 10, 2014, https://blog.prx.org/2014/07/adrianne-mathiowetz-now/

19. Tim Feran, "NPR Increases Programming Available Online," *Knight Ridder Tribune Business News*, May 25, 2006; NPR, "Mobilcast Brings NPR Podcasts to Mobile Phones," *BusinessWire*, press release, March 27, 2006.

20. Mathiowetz, "Where I Am Now."

21. Baker departed *Love + Radio* in 2018 to start his own "audio production house," Phenomephon, as well as direct Marvel's audio drama series *Wolverine: The Long Night*, a Stitcher Original.

22. "Episodes," Love + Radio, accessed May 22, 2019, http://loveandradio.org/category/episodes/

23. "Love + Radio," NPR, accessed March 16, 2016, http://www.npr.org/rss/podcast.php?id=510064

24. "WBEZ-FM to Bring Back 'Love + Radio' Podcasts," *Chicagoland Radio & Media*, May 21, 2012, http://chicagoradioandmedia.com/news/2502-wbez-fm-to-bring-back-love-radio-podcasts

25. PRX, "PRX Launches Radiotopia," press release, February 4, 2014, http://media.prx.org/PRXRadiotopia_release_FINAL.pdf

26. Brooks Barnes, "Podcast Start-Up Tests Paid Subscriptions," *New York Times*, March 4, 2019, B1.

27. "The Wisdom of Jay Thunderbolt," Third Coast International Audio Festival, accessed May 26, 2019, https://www.thirdcoastfestival.org/feature/wisdom-of-jay-thunderbolt

28. *Love + Radio*, "Jack and Ellen," produced by Brendan Baker, Mooj Zadie, Nick van der Kolk, February 21, 2013, audio, http://loveandradio.org/2013/02/jack-and-ellen/

29. Siobhan McHugh, "The Affective Power of Sound: Oral History on Radio," *Oral History Review* 39.2 (2012): 188–89.

30. Brendan Baker, "Using Music: Brendan Baker," *Transom.org*, February 25, 2014, http://transom.org/2014/using-music-brendan-baker/

31. Jonathan Menjivar, "Using Music: Jonathan Menjivar for This American Life," *Transom.org*, November 11, 2015, http://transom.org/2015/using-music-jonathan-menjivar-for-this-american-life/

32. Baker, "Using Music."

33. "Behind the Scenes with Nick Williams, Nick van der Kolk and Brendan Baker," Third Coast International Audio Festival, accessed May 26, 2019, https://www.thirdcoastfestival.org/article/nick-williams-nick-van-der-kolk-and-brendan-baker-bts

34. Kim Cascone, "The Aesthetics of Failure: 'Post-digital' Tendencies in Contemporary Computer Music," *Computer Music Journal* 24.4 (2002), http://subsol.c3.hu/subsol_2/contributors3/casconetext.html

35. Baker, "Using Music."

36. Ibid.

37. This conception of the "participatory mode" comes from documentary film scholar Bill Nichols, *Introduction to Documentary*, 2nd ed. (Bloomington: Indiana University Press, 2010), 179–94.

38. Ibid., 172–94.

39. Nick van der Kolk has suggested that he leaves the interviewer questions in radio pieces even when they are not necessary because he wants the audience to be aware that there is a "guiding hand" involved: Eric McQuade, "Nick van der Kolk, The Art of Podcasting No. 10," *The Timbre*, March 27, 2015, http://thetimbre.com/nick-van-der-kolk-the-art-of-podcasting-no-10/

40. *Love + Radio*, "An Old Lion, or a Lover's Lute," produced by Ana Adlerstein and Nick van der Kolk, December 19, 2014, audio, http://loveandradio.org/2014/12/an-old-lion-or-a-lovers-lute/

41. "Love + Radio," Need Supply Co., par. 4

42. Walter Benjamin, "The Storyteller: Reflections on the Works of Nikolai Leskov," in *Illuminations: Essays and Reflections*, ed. Hannah Arendt, trans. Harry Zohn (New York: Schocken, 1969), 84.

43. Ibid., 86–87.

44. Jonathan Sterne, *The Audible Past: Cultural Origins of Sound Reproduction* (Durham, NC: Duke University Press, 2003), 10–11.

45. Don Ihde, *Listening and Voice: A Phenomenology of Sound* (Athens: Ohio University Press, 1976), 58.

46. Sterne, *The Audible Past*, 15.

47. Fred Bayles, *Field Guide to Covering Local News* (Thousand Oaks, CA: CQ, 2012), 20.

48. Raymond Williams, *Marxism and Literature* (New York: Oxford University Press, 1977), 132.

49. Raymond Williams, *Culture and Society: 1780–1950* (New York: Penguin, 1963), 314.

50. Raymond Williams, *The Long Revolution*, new ed. (Cardigan, UK: Parthian, 2011), 61–94.

51. This is all the more reason why it is essential that media studies scholars, librarians, and archivists develop a cohesive plan for preserving podcasts and other forms of internet-only radio. Jeremy Wade Morris has developed one such project, PodcastRE: http://podcastre.org/

52. Donald A. Ritchie, *Doing Oral History: A Practical Guide*, 2nd ed. (New York: Oxford University Press, 2003), 19.

53. Valerie Raleigh Yow, *Recording Oral History: A Practical Guide for Social Scientists* (Thousand Oaks, CA: Sage, 1994), 4.

54. Paul Thompson, *The Voice of the Past: Oral History*, 3rd ed. (New York: Oxford University Press, 2000), 172.

55. Clifford Geertz, *The Interpretation of Cultures* (New York: Basic, 1973), 3–5.

56. Robert Perks and Alistair Thomson, "Critical Developments: Introduction," in *The Oral History Reader*, ed. Robert Perks and Alistair Thomson, 2nd ed. (New York: Routledge, 2006), 1–3.

57. Yow, *Recording Oral History*, 1–31.

58. Nancy Lamb, *The Art and Craft of Storytelling: A Comprehensive Guide to Class Writing Techniques* (Cincinnati, OH: Writer's Digest, 2008), 2–3.

59. Amy E. Spaulding, *The Art of Storytelling: Telling Truths through Telling Stories* (Lanham, MD: Scarecrow, 2011), 4.

60. Benjamin, "The Storyteller," 87.

61. Spaulding, *The Art of Storytelling*, 9.

62. "About Us," This American Life, accessed March 16, 2016, http://www.thisamericanlife.org/about/about-our-radio-show

63. Jay Allison, "Intro to Ira Glass's Manifesto," *Transom.org*, June 1, 2004, http://transom.org/2004/ira-glass/; William H. Siemering, "National Public Radio Purposes, 1970," *Current*, May 17, 2012, http://current.org/2012/05/national-public-radio-purposes/

64. Jason Loviglio, "Public Radio, This American Life, and the Neoliberal Turn," in *A Moment of Danger: Critical Studies in the History of U.S. Communication Since World War II*, ed. Janice Peck and Inger L. Stole (Milwaukee: Marquette University Press, 2011), 296.

65. Ira Glass, "Mo' Better Radio," *Current*, May 25, 1998.

66. Jack Mitchell, *Listener Supported: The Culture and History of Public Radio* (Westport, CT: Praeger, 2005), 184; Joe Richman and Bill Siemering, "In Conversation: Joe Richman & Bill Siemering," *Transom.org*, September 15, 2015, http://transom.org/2015/in-conversation-joe-richman-bill-siemering/; *The Pub*, "'The Pub' #49: Bill Siemering, Author of NPR's 1970 Mission Statement," produced by Adam Gagusea, December 17, 2015, audio, http://current.org/2015/12/the-pub-49-bill-siemering-author-of-nprs-1970-mission-statement/

67. Chris Higgins, "What is the Story Structure for a Typical Episode of This American Life," *Quora.com*, August 19, 2014, https://www.quora.com/What-is-the-story-structure-for-a-typical-episode-of-This-American-Life

68. Robert Lamb, "An Interview with Radiolab's Jad Abumrad," *Stuff to Blow Your Mind*, March 4, 2011, http://www.stufftoblowyourmind.com/blog/interview-radiolab-jad-abumrad/

69. Jason Mittell, "Previously On: Prime Time Serials and the Mechanics of Memory," in *Intermediality and Storytelling*, ed. Marina Grishakova and Marie-Laure Ryan (New York: De Gruyter, 2011), 92.

70. Edison Research, *The Podcast Consumer*, marketing report, May 2015, http://www.edisonresearch.com/wp-content/uploads/2015/06/The-Podcast-Consumer-2015-Final.pdf; "Podcast Listeners Demographics Report," Midroll, updated March 2019, http://awesome.midroll.com/

71. *Love + Radio*, "Choir Boy," produced by Brendan Baker, Katie Mingle, and Nick van der Kolk, August 25, 2014, audio, http://loveandradio.org/2014/08/choir-boy/

72. Loviglio, "Public Radio," 296.

73. Michael McCauley, *NPR: The Trials and Triumphs of National Public Radio* (New York: Columbia University Press, 2012).

74. Loviglio, "Public Radio," 294.

75. *Love + Radio*, "The Silver Dollar," produced by Brendan Baker and Nick van der Kolk, February 27, 2014, audio, http://loveandradio.org/2014/02/the-silver-dollar/

76. Loviglio, "Public Radio," 302.

77. Nick Couldry, *Why Voice Matters: Culture and Politics after Neoliberalism* (Thousand Oaks, CA: Sage, 2010), 1–13.

78. Ibid., 109–10.

79. Ibid., 114.

80. Ibid., 130.

81. Chenjerai Kumanyika, "Chenjerai Kumanyika: Manifesto (Vocal Color in Public Radio)," *Transom.org*, January 22, 2015, http://transom.org/2015/chenjerai-kumanyika/; Julie Shapiro, "Women Hosted Podcasts," *Transom.org*, February 26, 2013, http://transom.org/2013/women-hosted-podcasts/

82. Loviglio, "Public Radio," 301.

83. Olga R. Rodriguez, "'Ear Hustle' Host Is Freed, but Prison Podcast Will Go On," *AP News*, January 2, 2019, https://www.apnews.com/cb78634a65524494a74e10714660c2c8

84. *Talking Shop*, "Radiotopia at Kickstarter: A Discussion of the New Golden Age of Radio," moderated by Jake Shapiro, November 11, 2014, video, https://www.kickstarter.com/blog/radiotopia-at-kickstarter-a-discussion-on-the-new-golden-age-of

85. Bolter and Grusin, *Remediation*, 21–44.

86. Nicholas Carr, *The Shallows: What the Internet is Doing to Our Brains* (New York: W.W. Norton, 2010); Michael Z. Newman, "New Media, Young Audiences, and Discourses of Attention: From *Sesame Street* to 'Snack Culture,'" *Media Culture & Society* 32.4 (2010): 581–96; S. Elizabeth Bird, "Tabloidization: What Is It, and Does it Really Matter?," in *The Changing Faces of Journalism: Tabloidization, Technology, and Truthiness*, ed. Barbie Zelizer (New York: Routledge, 2009), 40–50; Anne Dunn, "Telling the Story: Narrative and Radio News," *Radio Journal* 1.2 (2003): 114–17.

87. Naomi Sharp, "The Future of Longform," *Columbia Journalism Review*, December 9, 2013, http://www.cjr.org/behind_the_news/longform_conference.php; Michael Blanding, "The Value of Slow Journalism in the Age of Instant Information," *NiemanReports*, August 19, 2015, http://niemanreports.org/articles/the-value-of-slow-journalism-in-the-age-of-instant-information/

88. danah boyd, "Participating in the Always-On Lifestyle," in *The Social Media Reader*, ed. Michael Mandiberg (New York: NYU Press, 2012): 71–76; David Dowling, "Escaping the Shallows: Deep Reading's Revival in the Digital Age," *Digital Humanities Quarterly* 8.2 (2014), http://www.digitalhumanities.org/dhq/vol/8/2/000180/000180.html

89. Nayomi Reghay, "The Power of Slow: 'Serial' and the Future of Podcasts," *The Kernel*, September 13, 2015, http://kernelmag.dailydot.com/issue-sections/staff-editorials/14287/serial-future-of-podcasts/

90. Alison Griswold, "MailKimp, SchmailChimp: Podcast Ads Are Rambling and Unpredictable, So Why Do Sponsors Love Them?," *Slate*, December 14, 2014, http://www.slate.com/articles/business/ten_years_in_your_ears/2014/12/podcast_advertising_how_serial_and_other_shows_benefit_from_their_rambling.html

91. Joe Lambert, *Digital Storytelling: Capturing Lives, Creating Community*, 4th ed. (New York: Routledge, 2013), 4.

92. Kate Lacey, *Listening Publics: The Politics and Experience of Listening in the Media Age* (Malden, MA: Polity, 2013), 7–8.

Conclusion

1. Bertolt Brecht, "The Radio as a Communications Apparatus," in *Brecht on Film and Radio*, trans. and ed. Marc Silberman (London: Methuen, 2000), 42–43.

2. Ibid., 41–42.

3. Ibid., 45, 43.

4. John Durham Peters, *Speaking Into the Air: A History of the Idea of Communication* (Chicago: University of Chicago Press, 1999), 19.

5. Brecht, "Communications Apparatus," 44.

6. Martin Spinelli, "Democratic Rhetoric and Emergent Media: The Marketing of Participatory Community on Radio and the Internet," *International Journal of Cultural Studies* 3.2 (2000): 274.

7. There are many qualifiers to these statements. There are various barriers that impede internet radio production, especially when considered historically in 1990s and early 2000s. Even today, internet radio production and distribution are more expensive than many realize, at least if one wishes to create high production quality content that is distributed to an audience of considerable size. Music radio on the internet is prohibitively expensive (see chapter 5).

8. Big John Trimble, *Golden Voice: The Autobiography of Halls of Fame Radio Renegade and Legendary Personality Big John Trimble* (Bloomington, IN: AuthorHouse, 2009), 191.

9. Bill Kirkpatrick, "Localism in American Media, 1920–1934" (PhD diss., University of Wisconsin–Madison, 2006).

10. Henry Cassirer, "Radio as the People's Medium," *Journal of Communication* 27.2 (1977): 154–57.

11. "The Infinite Dial" is the name of Edison Research and Triton Digital's annual "internet audio" marketing report. See https://www.edisonresearch.com/infinite-dial-2019/

12. Sites like TuneIn or Apple's Podcasts app act as gatekeepers and can promote some producers' content while diminishing or outright excluding others'. It is still easier to find major broadcast network content online than independent or amateur productions—they may share the same platform but that does not mean they are equally discoverable.

13. Michele Hilmes, "Foreword: Transnational Radio in the Global Age," *Journal of Radio Studies* 11.1 (2004): vi.

14. Hilmes points out how ironic it is that radio technology and production

have historically been so firmly and exclusively controlled by nation-states in most parts of the world.

15. Michele Hilmes, "Rethinking Radio," in *Radio Reader: Essays in the Cultural History of Radio*, ed. Michele Hilmes and Jason Loviglio (New York: Routledge, 2002), 2; Michele Hilmes, "The New Materiality of Radio: Sound on Screens," in *Radio's New Wave*, ed. Jason Loviglio and Michele Hilmes (New York: Routledge, 2013), 45.

16. Jonathan Sterne, *The Audible Past: Cultural Origins of Sound Reproduction* (Durham, NC: Duke University Press, 2003), 15.

17. Kate Lacey, "Ten Years of Radio Studies: The Very Idea," *Radio Journal* 6.1 (2008): 21–32; Norma Coates, "Sound Studies: Missing the (Popular) Music for the Screens?," *Cinema Journal* 48.1 (2008): 123–30.

18. WNYC, "WNYC—Podcasting Pioneer and On-Demand Audio Leader—Introduces 'WNYC Studios,'" press release, October 13, 2015, http://www.wnyc.org/press/wnyc-studios/101315/

19. Goli Sheikholeslami, "Changes at This American Life," *ThisAmericanLife.org*, July 9, 2015, http://www.thisamericanlife.org/blog/2015/07/changes-at-this-american-life

20. Nicholas Quah, "What Spotify's $230 Million Gimlet Deal Means for the Podcast Industry," *Vulture*, February 6, 2019, https://www.vulture.com/2019/02/spotify-gimlet-media-podcast-deal.html

21. "Public Radio Exchange," Atlantic Public Media, accessed March 26, 2016, http://atlantic.org/web-projects/public-radio-exchange

22. Public Radio Exchange, "PRX Offers 'Pubcatcher' Podcast Tool," press release, July 14, 2005, http://blog.prx.org/2005/07/prx-offers-pubcatcher-podcast-tool/

23. Public Radio Exchange, "MacArthur Foundation Announces Support for The Moth Radio Hour," press release, March 31, 2010, https://blog.prx.org/2010/03/macarthur-foundation-announces-support-for-the-the-moth-radio-hour-from-prx/

24. Justin Ellis, "Can't Stop, Won't Stop: PRX Introduces an App for Unending Audio Storytelling," *NiemanLab*, July 9, 2013, http://www.niemanlab.org/2013/07/cant-stop-wont-stop-prx-introduces-an-app-for-unending-audio-storytelling/

25. MacArthur Foundation, "Public Radio Exchange Receives $350,000 Grant," press release, January 27, 2004, https://www.macfound.org/press/press-releases/macarthur-foundation-provides-350000-for-public-radio-exchange-an-online-service-to-link-producers-to-public-radio-stations-and-listeners-throughout-the-country-january-27-2004/

26. Knight Foundation, "PRX Launches Radiotopia, New Podcast Network of Story-Driven Public Radio Shows by Industry's Best Emerging and Established Talent," press release, February 4, 2014, http://knightfoundation.org/press-room/press-release/prx-launches-radiotopia-groundbreaking-podcast-net/

27. "About," Radiotopia, accessed June 19, 2018, https://www.radiotopia.fm/about/

28. Justin Ellis, "Welcome to Radiotopia, a Podcast Network with the Aesthetics

of Story-Driven Public Radio," *NiemanLab*, February 4, 2014, http://www.nieman-lab.org/2014/02/welcome-to-radiotopia-a-podcast-network-with-the-aesthetics-of-story-driven-public-radio/

29. Sam Levin, "Building Public Radio 2.0," *East Bay Express*, November 12, 2014, http://www.eastbayexpress.com/oakland/building-public-radio-20/Content?oid=4121291

30. Ibid.; Ellis, "Welcome to Radiotopia."

31. Vanessa Quirk, "Interview: Jake Shapiro & Kerri Hoffman, PRX," *Columbia Journalism School: Tow Center for Digital Journalism*, September 30, 2015, http://tow-center.org/interview-jake-shapiro-kerri-hoffman-prx/; Eric Blattberg, "Enter the Podcast Network: A Year of Radiotopia," *Digiday*, January 22, 2015, http://digiday.com/publishers/enter-podcast-network-year-radiotopia/

32. Joseph Lichterman, "How PRX and Radiotopia Are Rethinking the Public Radio Pledge Drive for the Podcast Era," *NiemanLab*, October 27, 2015, http://www.niemanlab.org/2015/10/how-prx-and-radiotopia-are-rethinking-the-public-radio-pledge-drive-for-the-podcast-era/

33. Matthew Lasar, "How to Avoid Breaking the FCC's Ridiculously Lenient Rules against Non-commercial Radio Advertising," *Radio Survivor* (2009, September 28), http://www.radiosurvivor.com/2009/09/28/how-to-avoid-breaking-the-fccs-ridiculously-lenient-rules-against-non-commercial-radio-advertising/

34. Andrew Laupin, "Radiotopia Looks to Make Business of Creative Online Audio," *Current*, February 12, 2014, http://current.org/2014/02/roman-mars-and-prx-launch-radiotopia-a-new-network-of-independent-pubradio-podcasts/; Andy Sturdevant, "An Interview with Roman Mars," *Pollen*, June 2014, http://bepollen.squarespace.com/pollen/an-interview-with-roman-mars-by-andy-sturdevant.html; Quirk.

35. Vignesh Ramachandran, "Radiotopia Sets Record for Publishing and Radio Funding on Kickstarter," *KnightBlog*, November 14, 2014, http://www.knightfoun-dation.org/blogs/knightblog/2014/11/14/radiotopia-sets-record-for-publishing-and-radio-funding-on-kickstarter/

36. Kara Bloomgarden-Smoke, "Can Former Planet Money Host Build the 'HBO of Podcasts'?," *New York Observer*, February 27, 2015, http://observer.com/2015/02/gimlet-media/

37. Kristen Clark, "5 Tips for Making Your Podcast Heard," *MediaShift*, May 5, 2015, http://mediashift.org/2015/05/5-tips-for-making-your-podcast-heard/

38. Jake Shapiro, "A Golden Age: How to Join the Podcast Revolution," *PRX*, September 2, 2014, https://blog.prx.org/2014/09/golden-age-join-podcast-revolu-tion/

39. "This Is the Third Wave of Podcasting: Jake Shapiro," *AsiaRadioToday.com*, September 30, 2015, http://www.asiaradiotoday.com/news/third-wave-podcasting-jake-shapiro

40. Kris M. Markman, "Doing Radio, Making Friends, and Having Fun: Exploring the Motivations of Independent Audio Podcasters," *New Media & Society* 14.4 (2011): 555–57.

41. Tim Wu, *The Master Switch: The Rise and Fall of Information Empires*, 1st Vintage Books ed. (New York: Vintage, 2011).

42. Janko Rottgers, "Luminary Podcast Subscription Service to Launch," *Variety*, March 4, 2019, https://variety.com/2019/digital/news/luminary-podcast-subscription-service-1203154250/

43. Andrew Lapin, "Radiotopia Looks to Make Business of Creative Online Audio," *Current*, February 12, 2014, https://current.org/2014/02/roman-mars-and-prx-launch-radiotopia-a-new-network-of-independent-pubradio-podcasts/

44. Eric Nuzum, "What is alt.NPR?," accessed May 24, 2019, http://www.paulingles.com/Alt-NPR.html

45. Joe Richman, "Manifesto," *Transom*, April 15, 2014, https://transom.org/2014/joe-richman/

46. Tony Perez, "The Science of Story-Telling: A Conversation with Robert Krulwich and Jad Abumrad of Radiolab," *Tin House* 13.3 (2012): 143.

47. Stephanie Reagan, "Conversations: Adam Ragusea," *Material Studies*, January 16, 2016, http://www.material-studies.com/conversations/2016/1/7/adam-ragusea

48. For its part, PRX partnered with Google in 2018 to create the Google Podcasts Creator Program to "remove barriers to podcasting, increase the diversity of voices in the industry, and to make sure content is available for all audiences"—an initiative impacting a half-dozen "teams" per year: https://googlecp.prx.org/

49. Ben Sisario and Michael J. de la Merced, "With Deal for Podcaster, Spotify Transcends Tunes," *New York Times*, February 7, 2019, B1.

50. John L. Sullivan, "The Platforms of Podcasting: Past and Present," *Social Media + Society* (2019), DOI:10.33767/osf.io/4fcgu.

Appendix

1. James Hay and Nick Couldry, "Rethinking Convergence/Culture: An Introduction," *Cultural Studies* 25.4–5 (2011): 473.

2. Beyond those cited elsewhere, notable works in US radio history include David Goodman, *Radio's Civic Ambition: American Broadcasting and Democracy in the 1930s* (New York: Oxford University Press, 2011); Cynthia B. Meyers, *A Word from Our Sponsor: Admen, Advertising, and the Golden Age of Radio* (New York: Oxford University Press, 2013); Elena Razlogova, *The Listener's Voice: Early Radio and the American Public* (Philadelphia: University of Pennsylvania Press, 2011); Alexander Russo, *Points on the Dial: Golden Age Radio beyond the Networks* (Durham, NC: Duke University Press, 2010); Josh Shepperd, "Electric Education: How the Media Reform Movement Built Public Broadcasting in the United States, 1934–1952," PhD diss., University of Wisconsin–Madison, 2013; Michael Stamm, *Sound Business: Newspapers, Radio, and the Politics of New Media* (Philadelphia: University of Pennsylvania Press, 2011); Shawn VanCour, *Making Radio: Early Radio Production and the Rise of Modern Sound Culture* (New York: Oxford University Press, 2018). Other notable works in sound recording and popular music history include Kyle Barnett, *Record Cultures: The Transformation of the U.S. Recording Industry* (Ann Arbor: University of Michigan Press, 2020); Allison McCracken, *Real Men Don't Sing: Crooning in American Culture* (Durham, NC: Duke University Press, 2015); Jennifer Lynn Stoever, *The Sonic Color Line: Race and the Cultural Politics of Listening* (New

York: NYU Press, 2016); David Suisman, *Selling Sounds: The Commercial Revolution in American Music* (Cambridge, MA: Harvard University Press, 2012).

3. Michele Hilmes, *Radio Voices: American Broadcasting, 1922–1952* (Minneapolis: University of Minnesota Press, 1997), xiii.

4. Michele Hilmes, *Network Nations: A Transnational History of British and American Broadcasting* (New York: Routledge, 2012).

5. Jonathan Sterne, *The Audible Past: Cultural Origins of Sound Reproduction* (Durham, NC: Duke University Press, 2003); Jonathan Sterne, *MP3: The Meaning of a Format* (Durham, NC: Duke University Press, 2012).

6. Julie D'Acci, "Cultural Studies, Television Studies, and the Crisis in the Humanities," in *Television after TV: Essays on a Medium in Transition*, ed. Lynn Spigel and Jan Olsson (Durham, NC: Duke University Press, 2004), 418–45.

7. Ibid., 433.

8. Susan J. Douglas, *Inventing American Broadcasting, 1899–1922* (Baltimore: Johns Hopkins University Press, 1987); Brian Winston, *Media Technology and Society: A History from the Telegraph to the Internet* (New York: Routledge, 1998).

9. Paul E. Ceruzzi, *A History of Modern Computing* (Cambridge, MA: MIT Press, 1998); Janet Abbate, *Inventing the Internet* (Cambridge, MA: MIT Press, 1999).

10. James W. Carey, *Communication as Culture: Essays on Media and Society* (New York: Routledge, 1992); John Durham Peters, *Speaking into the Air: A History of the Idea of Communication* (Chicago: University of Chicago Press, 1999).

11. Benjamin Peters, "And Lead Us Not into Thinking the New Is New: A Bibliographic Case for New Media History," *New Media & Society* 11.1–2 (2009): 13–30.

12. Lisa Gitelman, *Always Already New: Media, History, and the Data of Culture* (Cambridge, MA: MIT Press, 2006); Thomas Streeter, *The Net Effect: Romanticism, Capitalism, and the Internet* (New York: NYU Press, 2011); Jay David Bolter and Richard Grusin, *Remediation: Understanding New Media* (Cambridge, MA: MIT Press, 1998).

13. Carolyn Marvin, *When Old Technologies Were New: Thinking about Electronic Communication in the Late Nineteenth Century* (New York: Oxford University Press, 1988), 5.

14. Bolter and Grusin, *Remediation*, 15.

15. Lisa Gitelman and Geoffrey Pingree, "What's New about New Media?," in *New Media, 1740–1915*, ed. Lisa Gitelman and Geoffrey Pingree (Cambridge, MA: MIT Press, 2003), xvi.

Selected Bibliography

Abbate, Janet. *Inventing the Internet*. Cambridge, MA: MIT Press, 1999.

Abel, Jessica. *Out on the Wire: The Storytelling Secrets of the New Masters of Radio*. New York: Broadway, 2015.

Anderson, Tim. *Popular Music in a Digital Music Economy: Problems and Practices for an Emerging Service Industry*. New York: Routledge, 2013.

Arceneaux, Noah. "CB Radio: Mobile Social Networking in the 1970s." In *The Mobile Media Reader*, edited by Noah Arceneaux and Anandam Kavoori, 55–68. New York: Peter Lang, 2012.

Arnheim, Rudolf. "In Praise of Blindness; Emancipation from the Body." In *Radio*, translated by Margaret Ludwig and Herbert Read, 133–203. London: Faber & Faber, 1936.

Barlow, Aaron. *The Rise of the Blogosphere*. Westport, CT: Praeger, 2007.

Beck, Alan. *The Death of Radio?: An Essay in Radio Philosophy for the Digital Age*. Kent, UK: Sound Journal, 2002. http://www.savoyhill.co.uk/deathofradio/

Bell, David. *Science, Technology and Culture*. New York: Open University Press, 2006.

Benjamin, Walter. "The Storyteller: Reflections on the Works of Nikolai Leskov." In *Illuminations: Essays and Reflections*, edited by Hannah Arendt and translated by Harry Zohn, 83–110. New York: Schocken, 1969.

Benkler, Yochai. *The Wealth of Networks: How Social Production Transforms Markets and Freedom*. New Haven, CT: Yale University Press, 2006.

Berry, Richard. "Will the iPod Kill the Radio Star?: Profiling Podcasting as Radio." *Convergence: The International Journal of Research into New Media Technologies* 12.2 (2006): 143–62.

Biewen, John, and Alexa Dilworth, eds. *Reality Radio: Telling True Stories in Sound*. Chapel Hill: University of North Carolina Press, 2010.

Bilton, Nick. *Hatching Twitter: A True Story of Money, Power, Friendship, and Betrayal*. New York: Penguin, 2013.

Bird, S. Elizabeth. "Tabloidization: What Is It, and Does It Really Matter?" In *The Changing Faces of Journalism: Tabloidization, Technology, and Truthiness*, edited by Barbie Zelizer, 40–50. New York: Routledge, 2009.

Bogost, Ian. *Alien Phenomenology, or What It's Like to Be a Thing*. Minneapolis: University of Minnesota Press, 2012.

Bolter, Jay David, and Richard Grusin. *Remediation: Understanding New Media*. Cambridge, MA: MIT Press, 1998.

Bonini, Tiziano. "The 'Second Age' of Podcasting: Reframing Podcasting as a New Digital Mass Medium." *Quaderns del CAC* 41 (2015): 21–30.

Bottomley, Andrew J. "The Ballad of Alan and Auntie Beeb: Alan Lomax's Radio Programmes for the BBC, 1943–1960." *Historical Journal of Film, Radio, and Television* 36.4 (2015): 604–26.

Bourdieu, Pierre. *Distinction: A Social Critique of the Judgement of Taste*. Translated by Richard Nice. Cambridge, MA: Harvard University Press, 1984.

boyd, danah. "A Blogger's Blog: Exploring the Definition of a Medium." *Reconstruction* 6.4 (2006). http://reconstruction.eserver.org/Issues/064/boyd.shtml

Brecht, Bertolt. "The Radio as a Communications Apparatus." In *Brecht on Film and Radio*, translated and edited by Marc Silberman, 41–48. London: Methuen, 2000.

Brewster, Bill, and Frank Broughton. *Last Night a DJ Saved My Life: The History of the Disc Jockey*. New York: Grove, 2000.

Bruns, Axel. "Towards Produsage: Futures of User-Led Content Production." In *Cultural Attitudes towards Communication and Technology 2006*, edited by Fay Sudweeks, Herbert Hrachovec, and Charles Ess, 275–84. Perth, Australia: Murdoch University Press, 2006.

Burkart, Patrick, and Tom McCourt. "Infrastructure for the Celestial Jukebox." *Popular Music* 23.3 (2004): 349–62.

Caplan, Robyn, and danah boyd. *Who Controls the Public Sphere in an Era of Algorithms? Mediation, Automation, Power*. Data & Society Research Institute, May 13, 2016. http://www.datasociety.net/pubs/ap/MediationAutomation-Power_2016.pdf

Carey, James W. *Communication as Culture: Essays on Media and Society*. New York: Routledge, 1992.

Cascone, Kim. "The Aesthetics of Failure: 'Post-digital' Tendencies in Contemporary Computer Music." *Computer Music Journal* 24.4 (2002). http://subsol.c3.hu/subsol_2/contributors3/casconetext.html

Cassirer, Henry. "Radio as the People's Medium." *Journal of Communication* 27.2 (1977): 154–57.

Castells, Manuel. *Networks of Outrage and Hope: Social Movements in the Internet Age*. Malden, MA: Polity, 2012.

Ceruzzi, Paul E. *A History of Modern Computing*. Cambridge, MA: MIT Press, 1998.

Christian, Aymar Jean. *Open TV: Innovation beyond Hollywood and the Rise of Web Television*. New York: NYU Press, 2018.

Coates, Norma. "Sound Studies: Missing the (Popular) Music for the Screens?" *Cinema Journal* 48.1 (2008): 123–30.

Couldry, Nick. *Listening beyond the Echoes*. Boulder, CO: Paradigm, 2006.

Couldry, Nick. "Liveness, 'Reality,' and the Mediated Habitus from Television to the Mobile Phone." *Communication Review* 7.4 (2004): 353–61.

Couldry, Nick. *Why Voice Matters: Culture and Politics after Neoliberalism*. Thousand Oaks, CA: Sage, 2010.

Coward, Rosalind. *Speaking Personally: The Rise of Subjective and Confessional Journal-ism*. Houndmills, UK: Palgrave Macmillan, 2013.

Crary, Jonathan. *Suspensions of Perception: Attention, Spectacle, and Modern Culture*. Cambridge, MA: MIT Press, 1999.

Crisell, Andrew. *Understanding Radio*. New York: Methuen, 1986.

Crofts, Sheri, Jon Dilley, Mark Fox, Andrew Retsema, and Bob Williams. "Podcast-ing: A New Technology in Search of Viable Business Models." *First Monday* 10.9 (2005). http://firstmonday.org/article/view/1273/1193

D'Acci, Julie. "Cultural Studies, Television Studies, and the Crisis in the Humani-ties." In *Television after TV: Essays on a Medium in Transition*, edited by Lynn Spigel and Jan Olsson, 418–45. Durham, NC: Duke University Press, 2004.

Dayan, Daniel, and Elihu Katz. *Media Events: The Live Broadcasting of History*. Cam-bridge, MA: Harvard University Press, 1992.

Dean, Jodi. *Blog Theory: Feedback and Capture in the Circuits of Drive*. Malden, MA: Polity, 2010.

Deuze, Mark. "Media Industries, Work and Life." *European Journal of Communica-tion* 24.4 (2009): 467–80.

Doane, Mary Ann. "Information, Crisis, Catastrophe." In *Logics of Television: Essays in Cultural Criticism*, edited by Patricia Mellencamp, 222–39. Bloomington: In-diana University Press, 1990.

Douglas, Susan J. *Inventing American Broadcasting, 1899–1922*. Baltimore: Johns Hopkins University Press, 1987.

Douglas, Susan J. "Letting the Boys Be Boys: Talk Radio, Male Hysteria, and Po-litical Discourse in the 1980s." In *Radio Reader: Essays in the Cultural History of Radio*, edited by Michele Hilmes and Jason Loviglio, 485–504. New York: Routledge, 2002.

Duncombe, Stephen. *Notes from Underground: Zines and the Politics of Alternative Culture*. New York: Verso, 1997.

Dunn, Anne. "Telling the Story: Narrative and Radio News." *The Radio Journal: International Studies in Broadcast & Audio Media* 1.2 (2003): 113–27.

Ellis, John. *Visible Fictions: Cinema, Television, Video*. New York: Routledge, 1992.

Engelman, Ralph. *Public Radio and Television in American: A Political History*. Thou-sand Oaks, CA: Sage, 1996.

Feuer, Jane. "The Concept of Live Television: Ontology as Ideology." In *Regarding Television: Critical Approaches—an Anthology*, edited by E. Ann Kaplan, 12–22. Frederick, MD: University Publications of America, 1983.

Fink, Howard. "The Sponsor's v. The Nation's Choice: North American Radio Drama." In *Radio Drama*, edited by Peter Lewis, 185–243. London: Longman, 1981.

Florida, Richard. *The Rise of the Creative Class*. New York: Basic, 2002.

Freeman, Bradley Carl, Julia Klapczynski, and Elliott Wood. "Radio and Facebook: The Relationship between Broadcast and Social Media Software in the U.S., Germany, and Singapore." *First Monday* 17.4 (2012). http://firstmonday.org/ojs/index.php/fm/article/view/3768

Fuchs, Christian. "Social Media and the Public Sphere." *tripleC: Communication, Capitalism & Critique* 12.1 (2014): 57–101.

Garnham, Nicholas. *Emancipation, the Media, and Modernity: Arguments about the Media and Social Theory*. New York: Oxford University Press, 2000.

Geertz, Clifford. *The Interpretation of Cultures.* New York: Basic, 1973.

Gillespie, Tarleton. "The Relevance of Algorithms." In *Media Technologies: Essays on Communication, Materiality, and Society,* edited by Tarleton Gillespie, Pablo J. Boczkowski, and Kirsten A. Foot, 167–93. Cambridge, MA: MIT Press, 2014.

Gillmor, Dan. *We the Media: Grassroots Journalism by the People, for the People.* North Sebastopol, CA: O'Reilly Media, 2006.

Gitelman, Lisa. *Always Already New: Media, History, and the Data of Culture.* Cambridge, MA: MIT Press, 2006.

Gitelman, Lisa, and Geoffrey Pingree. "What's New about New Media?" In *New Media, 1740–1915,* edited by Lisa Gitelman and Geoffrey Pingree, xi–xxii. Cambridge, MA: MIT Press, 2003.

Gitlin, Todd. *Inside Prime Time.* Paperback ed. New York: Pantheon, 1985.

Goffman, Erving. *Forms of Talk.* Philadelphia: University of Pennsylvania Press, 1981.

Greengard, Samuel. *The Internet of Things.* Cambridge, MA: MIT Press, 2015.

Guralnick, Elissa S. "Radio Drama: The Stage of the Mind." *Virginia Quarterly Review* 61.1 (1985): 71–94.

Hamilton, James F. "Historical Forms of User Production." *Media Culture & Society* 36.4 (2014): 491–507.

Hamula, Scott R., and Wenmouth Williams Jr. "The Internet as a Small-Market Radio Station Promotional Tool." *Journal of Radio Studies* 10.2 (2003): 262–69.

Hand, Richard J., and Mary Traynor. *The Radio Drama Handbook: Audio Drama in Practice and Context.* New York: Continuum, 2011.

Haring, Kristen. *Ham Radio's Technical Culture.* Cambridge, MA: MIT Press, 2006.

Hart, Jack. *Storycraft: The Complete Guide to Writing Narrative Nonfiction.* Chicago: University of Chicago Press, 2011.

Hartley, John. *Digital Futures for Cultural and Media Studies.* Malden, MA: Wiley-Blackwell, 2012.

Hartley, John, and Kelly McWilliam. "Computational Power Meets Human Contact." In *Story Circle: Digital Storytelling around the World,* edited by John Hartley and Kelly McWilliam, 3–15. Malden, MA: Wiley-Blackwell, 2009.

Hay, James, and Nick Couldry. "Rethinking Convergence/Culture: An Introduction." *Cultural Studies* 25.4–5 (2011): 473–86.

Heise, Nele. "On the Shoulders of Giants?: How Audio Podcasters Adopt, Transform and Re-invent Radio Storytelling." Chapter prepared for Transnational Radio Stories MOOC course, September 2014. https://hamburgergarnele.files.wordpress.com/2014/09/podcasts_heise_public.pdf

Hendy, David. *Radio in the Global Age.* Malden, MA: Polity, 2000.

Hilmes, Michele. "Foreword: Transnational Radio in the Global Age." *Journal of Radio Studies* 11.1 (2004): iii–vi.

Hilmes, Michele. *Network Nations: A Transnational History of British and American Broadcasting.* New York: Routledge, 2012.

Hilmes, Michele. "The New Materiality of Radio: Sound on Screens." In *Radio's New Wave,* edited by Jason Loviglio and Michele Hilmes, 43–61. New York: Routledge, 2013.

Hilmes, Michele. *Only Connect: A Cultural History of Broadcasting in the United States.* 4th ed. Boston: Wadsworth, 2014.

Hilmes, Michele. *Radio Voices: American Broadcasting, 1922–1952*. Minneapolis: University of Minnesota Press, 1997.

Hilmes, Michele. "Rethinking Radio." In *Radio Reader: Essays in the Cultural History of Radio*, edited by Michele Hilmes and Jason Loviglio, 1–20. New York: Routledge, 2002.

Hutchby, Ian. *Media Talk: Conversation Analysis and the Study of Broadcasting*. New York: Open University Press, 2006.

Ihde, Don. *Listening and Voice: A Phenomenology of Sound*. Athens: Ohio University Press, 1976.

Jackson-Pitts, Mary, and Ross Harms. "Radio Websites as a Promotional Tool." *Journal of Radio Studies* 10.2 (2003): 270–82.

Jenkins, Henry. *Convergence Culture: Where Old and New Media Collide*. New York: NYU Press, 2006.

Jenkins, Henry. "Convergence? I Diverge." *MIT Technology Review*, June 1, 2001.

Jenkins, Henry, Sam Ford, and Joshua Green. *Spreadable Media: Creating Value and Meaning in a Networked Culture*. New York: NYU Press, 2013.

Jordan, Tim. *Hacking: Digital Media and Technological Determinism*. Malden, MA: Polity, 2008.

Kahn, Douglas. "Introduction: Histories of Sound Once Removed." In *Wireless Imagination: Sound, Radio, and the Avant-Garde*, edited by Douglas Kahn and Gregory Whitehead, 1–29. Cambridge, MA: MIT Press, 1992.

Kait, Casey, and Stephen Weiss. *Digital Hustlers: Living Large and Falling Hard in Silicon Alley*. New York: Regan, 2001.

Keith, Michael C., and Christopher H. Sterling. "Disc Jockeys (DJs or Deejays)." In *Encyclopedia of Radio*, vol. 1, edited by Christopher H. Sterling, 102–5. New York: Routledge, 2004.

Kelly, Kevin. *New Rules for the New Economy*. New York: Viking Penguin, 1998.

Kern, Jonathan. *Sound Reporting: The NPR Guide to Audio Journalism and Production*. Chicago: University of Chicago Press, 2008.

Kirby, Jason. "More on Pandora: Genres, Genomes, and Musical Taste." *Scholars' Lab*, February 15, 2010. http://scholarslab.org/digital-humanities/more-on-pandora-genres/

Kirkpatrick, Bill. "Localism in American Media, 1920–1934." PhD diss., University of Wisconsin–Madison, 2006.

Kompare, Derek. "Flow." In *Keywords for Media Studies*, edited by Laurie Ouellette and Jonathan Gray, 72–74. New York: NYU Press, 2017.

Kozloff, Sarah. *Invisible Storytellers: Voice-Over Narration in American Fiction Film*. Berkeley: University of California Press, 1989.

Kroon Lundell, Asa. "The Design and Scripting of 'Unscripted' Talk: Liveness versus Control in a TV Broadcast Interview." *Media, Culture & Society* 31.2 (2009): 271–88.

Kumanyika, Chenjerai. "Chenjerai Kumanyika: Manifesto (Vocal Color in Public Radio)." *Transom.org*, January 22, 2015. http://transom.org/2015/chenjerai-kumanyika/

Lacey, Kate. *Listening Publics: The Politics and Experience of Listening in the Media Age*. Malden, MA: Polity, 2013.

Lacey, Kate. "Ten Years of Radio Studies: The Very Idea." *The Radio Journal: International Studies in Radio & Broadcast Media* 6.1 (2008): 21–32.

Lamb, Nancy. *The Art and Craft of Storytelling: A Comprehensive Guide to Class Writing Techniques*. Cincinnati, OH: Writer's Digest, 2008.

Lange, Patricia G. "Videos of Affinity on YouTube." In *The YouTube Reader*, edited by Pelle Snickars and Patrick Vonderau, 70–88. Stockholm: National Library of Sweden, 2009.

Lind, Rebecca Ann, and Norman J. Medoff. "Radio Stations and the World Wide Web." *Journal of Radio Studies* 6.2 (1999): 203–21.

Lochhead, Judy. "Introduction." In *Postmodern Music/Postmodern Thought*, edited by Judy Lochhead and Joseph Auner, 1–12. New York: Routledge, 2002.

Loviglio, Jason. "Public Radio in Crisis." In *Radio's New Wave: Global Sound in the Digital Era*, edited by Jason Loviglio and Michele Hilmes, 24–42. New York: Routledge, 2013.

Loviglio, Jason. "Public Radio, This American Life, and the Neoliberal Turn." In *A Moment of Danger: Critical Studies in the History of U.S. Communication since World War II*, edited by Janice Peck and Inger L. Stole, 283–306. Milwaukee: Marquette University Press, 2011.

Loviglio, Jason. *Radio's Intimate Public: Network Broadcasting and Mass-Mediated Democracy*. Minneapolis: University of Minnesota Press, 2005.

Loviglio, Jason. "Sound Effects: Gender, Voice, and the Cultural Work of NPR." *The Radio Journal: International Studies in Broadcast & Audio Media* 5.2–3 (2008): 67–81.

Lovink, Geert. *Social Media Abyss: Critical Internet Cultures and the Force of Negation*. Malden, MA: Polity, 2016.

Lüders, Marika. "Conceptualizing Personal Media." *New Media & Society* 10.5 (2008): 683–702.

Lum, Casey Man Kong. "Notes toward an Intellectual History of Media Ecology." In *Perspectives on Culture, Technology, and Communication: The Media Ecology Tradition*, edited by Casey Man Kong Lum, 1–60. New York: Hampton, 2006.

Madison, Ed, and Ben DeJarnette. *Reimagining Journalism in a Post-truth World*. Santa Barbara, CA: Praeger, 2018.

Malamud, Carl. *A World's Fair for the Global Village*. Cambridge, MA: MIT Press, 1997.

Manovich, Lev. "The Algorithms of Our Lives." *Chronicle Review*, December 16, 2013. http://chronicle.com/article/the-algorithms-of-our-lives-/143557

Markman, Kris M. "Doing Radio, Making Friends, and Having Fun: Exploring the Motivations of Independent Audio Podcasters." *New Media & Society* 14.4 (2011): 547–65.

Markman, Kris M., and Caroline E. Sawyer. "Why Pod?: Further Explorations of the Motivations for Independent Podcasting." *Journal of Radio & Audio Media* 21.1 (2014): 20–35.

Marvin, Carolyn. *When Old Technologies Were New: Thinking about Electronic Communication in the Late Nineteenth Century*. New York: Oxford University Press, 1988.

Marwick, Alice E. *Status Update: Celebrity, Publicity, and Branding in the Social Media Age*. New Haven, CT: Yale University Press, 2013.

McCauley, Michael. *NPR: The Trials and Triumphs of National Public Radio*. New York: Columbia University Press, 2012.

McClung, Steven. "College Radio Station Web Sites: Perceptions of Value and Use." *Journalism & Mass Communication Educator* 56.1 (2001): 62–73.

McCourt, Tom, and Eric W. Rothenbuhler. "Burnishing the Brand: Todd Storz and the Total Station Sound." *The Radio Journal: International Studies in Broadcast & Audio Media* 2.1 (2004): 3–14.

McHugh, Siobhan. "The Affective Power of Sound: Oral History on Radio." *The Oral History Review* 39.2 (2012): 187–206.

McHugh, Siobhan. "Editorial: Podcasting as the New Space for Crafted Audio." *RadioDoc Review* 3.1 (2017).

McHugh, Siobhan. "How Podcasting Is Changing the Audio Storytelling Genre." *The Radio Journal: International Studies in Broadcast & Audio Media* 14.1 (2016): 65–82.

McLuhan, Marshall, and Quentin Fiore. *The Medium Is the Massage: An Inventory of Effects*. New York: Bantam, 1967.

Merton, Robert K. *The Sociology of Science: Theoretical and Empirical Investigations*. Chicago: University of Chicago Press, 1973.

Middleton, Richard. *Studying Popular Music*. Bristol, PA: Open University Press, 1990.

Milner, Greg. *Perfecting Sound Forever: An Aural History of Recorded Music*. New York: Faber & Faber, 2009.

Mitchell, Jack. *Listener Supported: The Culture and History of Public Radio*. Westport, CT: Praeger, 2005.

Mittell, Jason. "Previously On: Prime Time Serials and the Mechanics of Memory." In *Intermediality and Storytelling*, edited by Marina Grishakova and Marie-Laure Ryan, 78–98. New York: De Gruyter, 2011.

Morris, Jeremy Wade. "Curation by Code: Infomediaries and the Data Mining of Taste." *European Journal of Cultural Studies* 18.4–5 (2015): 446–63.

Morris, Jeremy Wade. *Selling Digital Music, Formatting Culture*. Berkeley, CA: University of California Press, 2015.

Morris, Jeremy Wade, and Devon Powers. "Control, Curation and Musical Experience in Streaming Music Services." *Creative Industries Journal* 8.2 (2015): 106–22.

Moscote Freire, Ariana. "Remediating Radio: Audio Streaming, Music Recommendation and the Discourse of Radioness." *The Radio Journal: International Studies in Broadcast & Audio Media* 5.2–3 (2007): 97–112.

Murthy, Dhiraj. *Twitter: Social Communication in the Twitter Age*. Malden, MA: Polity, 2013.

Myers, Kellen. "The RIAA, the DMCA, and the Forgotten Few Webcasters: A Call for Change in Digital Copyright Royalties." *Federal Communications Law Journal* 61.2 (2009): 431–56.

Neumark, Norie. "Different Spaces, Different Times: Exploring Possibilities for Cross-Platform 'Radio.'" *Convergence: The International Journal of Research into New Media Technologies* 12.2 (2006): 213–24.

Newman, Michael Z. "New Media, Young Audiences, and Discourses of Attention: From *Sesame Street* to 'Snack Culture.'" *Media Culture & Society* 32.4 (2010): 581–96.

Nichols, Bill. *Introduction to Documentary*. 2nd ed. Bloomington: Indiana University Press, 2010.

O'Donnell, Penny. "Journalism, Change, and Listening Practices." *Continuum: Journal of Media & Cultural Studies* 23.4 (2009): 503–17.

O'Reilly, Tim. "What Is Web 2.0? Design Patterns and Business Models for the Next Generation of Software." In *The Social Media Reader*, edited by Michael Mandiberg, 32–52. New York: NYU Press, 2012.

Ong, Walter J. *Orality and Literacy: The Technologizing of the Word*. London: Methuen, 1982.

Ostertag, Bob. *People's Movement, People's Press: The Journalism of Social Justice Movements*. Boston: Beacon, 2006.

Papacharissi, Zizi. *Affective Publics: Sentiment, Technology, and Politics*. New York: Oxford University Press, 2015.

Perks, Robert, and Alistair Thomson. "Critical Developments: Introduction." In *The Oral History Reader*, edited by Robert Perks and Alistair Thomson, 1–8. 2nd ed. New York: Routledge, 2006.

Peters, John Durham. *The Marvelous Clouds: Toward a Philosophy of Elemental Media*. Chicago: University of Chicago Press, 2015.

Peters, John Durham. *Speaking into the Air: A History of the Idea of Communication*. Chicago: University of Chicago Press, 1999.

Peters, John Durham. "Witnessing." *Media, Culture & Society* 23.6 (2001): 707–23.

Priestman, Chris. "Narrowcasting and the Dream of Radio's Great Global Conversation." *The Radio Journal: International Studies in Broadcast & Audio Media* 2.2 (2004): 77–88.

Purdy, Michael. "What Is Listening?" In *Listening in Everyday Life: A Personal and Professional Approach*, edited by Michael Purdy and Deborah Borisoff, 1–20. Lanham, MD: University Press of America, 1991.

Raymond, Eric S. "The Cathedral and the Bazaar." *First Monday* 3.3 (1998). http://firstmonday.org/article/view/578/499

Razlogova, Elena. "The Past and Future of Music Listening: Between Freeform DJs and Recommendation Algorithms." In *Radio's New Wave: Global Sound in the Digital Era*, edited by Jason Loviglio and Michele Hilmes, 62–76. New York: Routledge, 2013.

Reid, Robert H. *Architects of the Web: 1,000 Days That Built the Future of Business*. New York: Wiley, 1999.

Rheingold, Howard. *The Virtual Community: Homesteading on the Electronic Frontier*. Revised ed. Cambridge, MA: MIT Press, 2000.

Ritchie, Donald A. *Doing Oral History: A Practical Guide*. 2nd ed. New York: Oxford University Press, 2003.

Rosenberg, Scott. *Say Everything: How Blogging Began, What It's Becoming, and Why It Matters*. New York: Crown, 2009.

Ross, Andrew. "Earth to Gore, Earth to Gore." In *Technoscience and Cyberculture*, edited by Stanley Aronowitz, Barbara Martinsons, and Michael Menser, 111–22. New York: Routledge, 1996.

Ross, Andrew. *No-Collar: The Humane Workplace and Its Hidden Costs*. Philadelphia: Temple University Press, 2004.

Rothenbuhler, Eric, and Tom McCourt. "Radio Redefines Itself, 1947–1962." In *Radio Reader: Essays in the Cultural History of Radio*, edited by Michele Hilmes and Jason Loviglio, 367–88. New York: Routledge, 2002.

Sandvig, Christian. "Seeing the Sort: The Aesthetic and Industrial Defense of 'The Algorithm.'" *NMC Media-N: Journal of the New Media Caucus* 10.3 (2014). http://median.newmediacaucus.org/art-infrastructures-information/seeing-the-sort-the-aesthetic-and-industrial-defense-of-the-algorithm/

Sant, Toni. *Franklin Furnace and the Spirit of the Avant-Garde: A History of the Future*. Chicago: Intellect, 2011.

Sapnar Ankerson, Megan. *Dot-Com Design: The Rise of a Usable, Social, Commercial Web*. New York: NYU Press, 2018.

Scannell, Paddy, ed. *Broadcast Talk*. Newbury Park, CA: Sage, 1991.

Scannell, Paddy. "Editorial." *Media, Culture & Society* 23.6 (2001): 699–705.

Scannell, Paddy. *Radio, Television, and Modern Life: A Phenomenological Approach*. Cambridge, MA: Blackwell, 1996.

Scannell, Paddy. "Review Essay: The Liveness of Broadcast Talk." *Journal of Communication* 59.4 (2009): E1–E6.

Scannell, Paddy. "What Is Radio For?" *The Radio Journal: International Studies in Radio and Broadcast Media* 7.1 (2009): 89–95.

Scannell, Paddy. *Television and the Meaning of Live*. Malden, MA: Polity, 2014.

Schiffer, Michael B. *The Portable Radio in American Life*. Tucson: University of Arizona Press, 1991.

Sconce, Jeffrey. *Haunted Media: Electronic Presence from Telegraphy to Television*. Durham, NC: Duke University Press, 2000.

Shingler, Martin, and Cindy Wieringa. *On Air: Methods and Meanings of Radio*. New York: Oxford University Press, 1998.

Sienkiewicz, Matt. "Start Making Sense: A Three-Tier Approach to Citizen Journalism." *Media Culture & Society* 36.5 (2014): 691–701.

Sieveking, Lance. *The Stuff of Radio*. London: Cassell, 1934.

Silverstone, Roger. *Television and Everyday Life*. New York: Routledge, 1994.

Smith, Andrew. *Totally Wired: The Wild Rise and Crazy Fall of the First Dotcom Dream*. London: Simon & Schuster UK, 2012.

Smith Maguire, Jennifer, and Julian Matthews. "Are We All Cultural Intermediaries Now?: An Introduction to Cultural Intermediaries in Context." *European Journal of Cultural Studies* 15.5 (2012): 551–62.

Spaulding, Amy E. *The Art of Storytelling: Telling Truths through Telling Stories*. Lanham, MD: Scarecrow, 2011.

Spinelli, Martin. "Democratic Rhetoric and Emergent Media: The Marketing of Participatory Community on Radio and the Internet." *International Journal of Cultural Studies* 3.2 (2000): 268–78.

Staiger, Janet, and Sabine Hake. "Preface." In *Convergence Media History*, edited by Janet Staiger and Sabine Hake, ix–xi. New York: Routledge, 2009.

Sterling, Christopher H., and Michael C. Keith. *Sounds of Change: A History of FM Broadcasting in America*. Chapel Hill: University of North Carolina Press, 2008.

Sterne, Jonathan. *The Audible Past: Cultural Origins of Sound Reproduction*. Durham, NC: Duke University Press, 2003.

Sterne, Jonathan. *MP3: The Meaning of a Format*. Durham, NC: Duke University Press, 2012.

Sterne, Jonathan, Jeremy Morris, Michael Brendan Baker, and Ariana Moscote Freire. "The Politics of Podcasting." *Fibreculture Journal* 13 (2008). http://thirteen.fibreculturejournal.org/fcj-087-the-politics-of-podcasting/

Streeter, Thomas. *The Net Effect: Romanticism, Capitalism, and the Internet*. New York: NYU Press, 2011.

Striphas, Ted. "Algorithmic Culture." *European Journal of Cultural Studies* 18.4–5 (2015): 395–412.

Sullivan, John L. "The Platforms of Podcasting: Past and Present." *Social Media + Society* (2019). DOI:10.33767/osf.io/4fcgu.

Taylor, Steve. "'I Am What I Play': The Radio DJ as Cultural Arbiter and Negotiator." In *Cultural Work: Understanding the Cultural Industries*, edited by Andrew Beck, 73–100. New York: Routledge, 2003.

Tebbutt, John. "Imaginative Demographics: The Emergence of a Radio Talkback Audience in Australia." *Media, Culture & Society* 28.6 (2006): 857–82.

Thompson, Paul. *The Voice of the Past: Oral History*. 3rd ed. New York: Oxford University Press, 2000.

Turner, Graeme. "'Liveness' and 'Sharedness' outside the Box." *Flow: A Critical Forum on Television and Media Culture* 13.11 (2011). http://flowtv.org/2011/04/liveness-and-sharedness-outside-the-box/

van Dijck, Jose, and Thomas Poell. "Understanding Social Media." *Media and Communication* 1.1 (2013): 2–14.

Verma, Neil. "The Arts of Amnesia: The Case for Audio Drama, Part One." *RadioDoc Review* 3.1 (2017).

Verma, Neil. *Theater of the Mind: Imagination, Aesthetics, and American Radio Drama*. Chicago: University of Chicago Press, 2012.

Walker Rettberg, Jill. *Blogging*. 2nd ed. Malden, MA: Polity, 2014.

Wilkinson, Jeffrey S., August E. Grant, and Douglas J. Fisher. *Principles of Convergent Journalism*. 2nd ed. New York: Oxford University Press, 2012.

Williams, Raymond. *Culture and Society: 1780–1950*. New York: Penguin, 1963.

Williams, Raymond. *The Long Revolution*. New ed. Cardigan, UK: Parthian, 2011.

Williams, Raymond. *Marxism and Literature*. New York: Oxford University Press, 1977.

Williams, Raymond. *Television: Technology and Cultural Form*. Routledge Classics ed. New York: Routledge, 2003.

Wilson, Michael. *Storytelling and Theatre: Contemporary Storytellers and Their Art*. New York: Palgrave Macmillan, 2006.

Winston, Brian. *Media Technology and Society: A History from the Telegraph to the Internet*. New York: Routledge, 1998.

Wu, Tim. *The Master Switch: The Rise and Fall of Information Empires*. 1st Vintage Books ed. New York: Vintage, 2011.

Index